Multidisciplinary Perspectives on Diversity and Equity in a Virtual World

Emily K. Reeves
Midwestern State University, USA

Christina Janise McIntyre
Midwestern State University, USA

A volume in the Advances in Human and Social Aspects of Technology (AHSAT) Book Series

IGI Global
PUBLISHER of TIMELY KNOWLEDGE

Published in the United States of America by
IGI Global
Information Science Reference (an imprint of IGI Global)
701 E. Chocolate Avenue
Hershey PA, USA 17033
Tel: 717-533-8845
Fax: 717-533-8661
E-mail: cust@igi-global.com
Web site: http://www.igi-global.com

Library of Congress Cataloging-in-Publication Data

Names: Reeves, Emily Kate, editor. | McIntyre, Christina Janise, editor.
Title: Multidisciplinary perspectives on diversity and equity in a virtual world / Emily Kate Reeves, and Christina Janise McIntyre, editors.
Description: Hershey, PA : Information Science Reference, 2022. | Includes bibliographical references and index. | Summary: "This edited book will help readers understand the impact of diversity and equity in the virtual world from a professional perspective which becomes increasingly important as some professions will continue to use virtual platforms in the upcoming years"-- Provided by publisher.
Identifiers: LCCN 2021024372 (print) | LCCN 2021024373 (ebook) | ISBN 9781799880288 (hardcover) | ISBN 9781799880295 (paperback) | ISBN 9781799880301 (ebook)
Subjects: LCSH: Intersectionality (Sociology) | Belonging (Social psychology) | Virtual reality.
Classification: LCC HM488.5 .M86 2022 (print) | LCC HM488.5 (ebook) | DDC 305--dc23
LC record available at https://lccn.loc.gov/2021024372
LC ebook record available at https://lccn.loc.gov/2021024373

This book is published in the IGI Global book series Advances in Human and Social Aspects of Technology (AHSAT) (ISSN: 2328-1316; eISSN: 2328-1324)

British Cataloguing in Publication Data
A Cataloguing in Publication record for this book is available from the British Library.

For electronic access to this publication, please contact: eresources@igi-global.com.

Advances in Human and Social Aspects of Technology (AHSAT) Book Series

ISSN:2328-1316
EISSN:2328-1324

Editor-in-Chief: Mehdi Khosrow-Pour, D.B.A., Information Resources Management Association, USA

MISSION

In recent years, the societal impact of technology has been noted as we become increasingly more connected and are presented with more digital tools and devices. With the popularity of digital devices such as cell phones and tablets, it is crucial to consider the implications of our digital dependence and the presence of technology in our everyday lives.

The **Advances in Human and Social Aspects of Technology (AHSAT) Book Series** seeks to explore the ways in which society and human beings have been affected by technology and how the technological revolution has changed the way we conduct our lives as well as our behavior. The AHSAT book series aims to publish the most cutting-edge research on human behavior and interaction with technology and the ways in which the digital age is changing society.

COVERAGE

- ICTs and human empowerment
- Cyber Behavior
- End-User Computing
- Gender and Technology
- Human Development and Technology
- Information ethics
- Technology Dependence
- Cyber Bullying
- Cultural Influence of ICTs
- Computer-Mediated Communication

IGI Global is currently accepting manuscripts for publication within this series. To submit a proposal for a volume in this series, please contact our Acquisition Editors at Acquisitions@igi-global.com or visit: http://www.igi-global.com/publish/.

The Advances in Human and Social Aspects of Technology (AHSAT) Book Series (ISSN 2328-1316) is published by IGI Global, 701 E. Chocolate Avenue, Hershey, PA 17033-1240, USA, www.igi-global.com. This series is composed of titles available for purchase individually; each title is edited to be contextually exclusive from any other title within the series. For pricing and ordering information please visit http://www.igi-global.com/book-series/advances-human-social-aspects-technology/37145. Postmaster: Send all address changes to above address.

Titles in this Series

For a list of additional titles in this series, please visit: http://www.igi-global.com/book-series

Technological Influences on Creativity and User Experience
Joshua Fairchild (Creighton University, USA)
Information Science Reference • © 2022 • 305pp • H/C (ISBN: 9781799843542) • US $195.00

Human Factors Issues and the Impact of Technology on Society
Heather Lum (Embry-Riddle Aeronautical University, USA)
Information Science Reference • © 2021 • 333pp • H/C (ISBN: 9781799864530) • US $195.00

Ubiquitous Technologies for Human Development and Knowledge Management
Hakikur Rahman (Institute of Computer Management and Science, Bangladesh)
Information Science Reference • © 2021 • 396pp • H/C (ISBN: 9781799878445) • US $195.00

Technological Breakthroughs and Future Business Opportunities in Education, Health, and Outer Space
Angus Hooke (Australian Institute of Higher Education, Australia)
Business Science Reference • © 2021 • 353pp • H/C (ISBN: 9781799867722) • US $225.00

Machine Law, Ethics, and Morality in the Age of Artificial Intelligence
Steven John Thompson (University of California, Davis, USA & University of Maryland Global Campus, USA)
Engineering Science Reference • © 2021 • 266pp • H/C (ISBN: 9781799848943) • US $295.00

Examining the Socio-Technical Impact of Smart Cities
Fenio Annansingh (York College, City University of New York, USA)
Information Science Reference • © 2021 • 231pp • H/C (ISBN: 9781799853268) • US $190.00

IGI Global
PUBLISHER of TIMELY KNOWLEDGE

701 East Chocolate Avenue, Hershey, PA 17033, USA
Tel: 717-533-8845 x100 • Fax: 717-533-8661
E-Mail: cust@igi-global.com • www.igi-global.com

Table of Contents

Section 2
Authenticity: Creating and Maintaining Equitable Personal Connections

Section 3
Creative Solutions: Reimagining the Future

Detailed Table of Contents

The purpose of this chapter is to reflect on and respond to the work done by the authors of the subsequent chapters. The experiences and work of the contributors of this book highlight the general skills necessary for diversity, equity, and inclusion (DEI) and affirm the premises of DEI work. The chapters also include innovative ideas that can increase engagement with the DEI agenda. Finally, the chapters provide some exciting possibilities on how the digital world can further the current DEI discourse.

This chapter investigates the hypothesis that the COVID-19 pandemic was the perfect storm due to the misalignment of competing elements of the US healthcare system, the economic commoditization of disease, the economic commoditization of healthcare delivery, and inadequate data to inform medical decision making on a mass scale. The culmination of a decades-long devolution away from patient care to healthcare or more appropriately sick-care created a system that was unable to quickly find the common ground needed to deal with the pandemic known as COVID-19.

The COVID-19 pandemic created a historic disruption to contemporary society including how, where, and when we work. Given the ubiquity of human capital, most if not every society was crippled by the displacement of the workforce with historic impacts on productivity; GDP in the UK will be at its lowest in 300 years, requiring the largest peacetime debt accumulation in history. Stimulus packages occurred in many countries as a result of the inability to access the workplace, particularly school, restaurant, or travel. Airline travel in the US fell by a precipitous 93% at its nadir, the cruise industry collapsed, and trans-national crossing all but ceased to exist. Along with the freeze in people movement, supply chains were disrupted including components necessary for both treatment and vaccination. The shrinkage of the world we had grown up with became the catalyst for the first pandemic in a century.

Restrictions from the COVID-19 pandemic forced many teachers across the United States to teach their students remotely using online teaching strategies. Analyzing teachers' perceptions and expectations of online teaching before and during the pandemic help stakeholders understand how to better handle the challenges of online learning. The literature review examines the traditional differences between teaching online and in-person courses, challenges faced when teaching online courses, and teacher perceptions of online learning. The survey method was used to collect data on the experiences of online teaching before and during the pandemic from ninth through twelfth grade teachers in a North Texas school district. The study results show that factors such as a teacher age or years of experience had less influence determining if they were prepared for online teaching. Rather, factors like a teacher school campus, prior experience, and access to support structures correlated more to teachers having a high level of preparedness for online teaching.

Due to the COVID-19 crisis, the mental health profession has shifted to online service provision, or telehealth. The aim of this chapter is to describe the COVID-19 crisis and subsequent changes that occurred to mental health service provision; the benefits and drawbacks to telehealth from practical, ethical, and cultural perspectives; and

Preface

Almost every citizen of the world has been impacted as a result of the COVID-19 pandemic. In many cases, this included a shift from face-to-face interactions to a virtual platform. Understanding the impact of diversity and equity in the virtual world from a professional perspective is nascent and should be closely studied as professions continue to use virtual platforms in the upcoming years.

Multidisciplinary Perspectives on Diversity and Equity in a Virtual World takes a close look at equity and diversity in virtual settings across professions from multiple perspectives to better understand the impact moving online has on diverse populations. This text provides insight and context in a timely way by creating a knowledge base to work from while decision makers continue to work towards equity and diversity in the workplace—be it online or face-to-face. Covering topics such as diverse healthcare, remote teaching, and culture of work, this book was designed to serve as the ideal resource for human diversity scholars, university faculty, instructional designers, software developers, students, academicians, researchers, and decision. The text is organized into three sections: Disequilibrium: A Time of Transitions, Authenticity: Creating and Maintaining Equitable Personal Connections, and Creative Solutions: Reimagining the Future. Contributors from different professions and fields with varying perspectives have been included to provide a broad range of ideas about equity and diversity in a virtual world.

Chapter 1. *Response to Diversity in a Digital World*

Chapter 1 is a response chapter that situates the pieces of the book in the new landscape of Covid-era scholarship on diversity and equity. This chapter is a synthesis not only of the pieces in the book but of the ways in which they inform and challenge existing scholarship on diversity and equity.

DISEQUILIBRIUM: A TIME OF TRANSITIONS

Chapter 2. *Virtual Medicine, Virtual Disease, Real Consequences: Was COVID-19 the Perfect Storm of Dysfunction in American Healthcare?*

The COVID-19 pandemic accelerated the evolution of medical practice with further loosening of the traditional doctor-patient relationship through an increasingly nebulous system of providers to a now virtual world where is difficult to discern if the provider is human or machine. The consequences of this evolution to a virtual practice whereby skill, intelligence and experience are replaced by guidelines, checklists and computer administered algorithms are explored. Humanistic bias has been replaced by human influenced computer administered care with less clearly defined but more harshly delivered biases in both diagnosis and care. This complicates existing biases in access to care as providers seek to survive the downturn in revenue which led to the closure of a significant number of practices. Access to care changed from the subjective thoughtful assessment of an experienced professional to a remote analysis of tests leading to procedures in patients who in essence directed their own care without a face-to-face encounter.

Chapter 3. *How COVID-19 Changed the Culture of Work*

Immediate COVID work related issues include the rapid dispersion of a previously collective workforce to remote locations, specifically home, either because of known exposure and/or infection or merely the risk of infection, even though months later those risks are barely quantifiable. The movement of the workforce from complex instant social interactions to virtual interactions, while aided by technology, exposed both technological limitations, uneven distribution of technology and the modes by which those virtual interactions took place. It also required a re-evaluation of the term inclusion in the workplace; perhaps for the better. What are the consequences of social isolation at work when combined with the additional pressures of being at home on a constant basis, sometimes for weeks? Return to work guidelines for those required to return to conventional workplace environments, particularly those returning to healthcare will be discussed.

Chapter 4. *From In-Person to Online Teaching: Experiences of Secondary Educators During the COVID-19 Pandemi*c

As schools are continuing to navigate between offering in-person, online, and hybrid courses, it is important to examine teacher experiences to inform how changes to the learning environment currently affects teachers' experiences in the classroom. Examining the experience of teachers transitioning from in-person to online teaching can provide educators and administrators with research-based data leading to the development of new training courses designed for online teaching. The purpose of this study will be to understand the experiences of traditional in-person secondary educators transitioning to teaching online classes in North Texas school districts through collecting their responses to open-ended questions.

Chapter 5. *Covid and Counseling: A Shift to Telehealth During a Global Crisis*

The COVID-19 global pandemic that emerged in the early months of 2020 is dramatically impacting the lives of citizens around the globe. The virus has taken the lives of millions and has impacted economies worldwide. The crisis has led to a fast-paced shift in service provision for mental health counseling in the United States. What once was almost exclusively a face-to face profession, is now primarily provided through online video conferencing platforms. For client and clinician safety to stop the spread of the coronavirus, many HIPAA compliant software systems have emerged to meet the growing demand for video conferencing approaches to psychotherapy. This rapid shift has led to major changes in the ways in which the counseling relationship manifests in practice.

AUTHENTICITY: CREATING AND MAINTAINING EQUITABLE PERSONAL CONNECTIONS

Chapter 6. *Digital Access to Culturally Relevant Curricula: The Impact on the Native and Indigenous Student*

Native and Indigenous students in low SES urban, rural, and reservation areas have the lowest retention-to-graduation rate and college completion rate of any demographic in the United States. The primary contributing factor is lack of culturally relevant curricula in a culturally sustaining pedagogy to present their culture in a positive representation. The second contributing factor is a lack of access to computer and internet resources to foster and bolster online, remote, and distance learning.

Chapter 7. *Is Telehealth an Effective Way to Increase Accessibility to Counseling Services?*

Mental illness significantly affects one in 10 persons globally and accounts for 30% of the global non-fatal disease burden. Mental health issues impact both urban and rural areas. However, residents living in rural areas tend to have higher rates of mental and medical health concerns than their urban counterparts. Researchers have identified barriers to providing mental healthcare services to individuals living in rural areas including increased consumer vulnerability, higher visibility, more substantial stigma, dual relationships, lack of accessibility, and geographic isolation from services. Researchers concluded that in the United States, public mental health systems are overwhelmed by primary mental healthcare. An overwhelmed mental healthcare system creates a significant burden on mental health providers serving underserved communities. Because of many rural individuals' opposing belief systems, finding an unconventional service delivery for clinical mental health counseling services may assist in improving overall healthcare outcomes. Barriers experienced by individuals living in rural communities are complex and seemingly insurmountable. As the number of Americans with Internet access continues to increase it creates an opening for utilizing telehealth technologies. Telehealth, supported by federal policy, can be a practical solution to the lack of mental health providers. Despite the significant barriers to providers and clients utilizing telehealth services, telehealth technology remains an innovative option for clinical mental healthcare services. Telehealth services continue to grow and become more widely accessible by millions.

Chapter 8. *The Equity of Remote Teaching During COVID-19*

The diversity of the students in the public education classroom is extreme, therefore the diversity of the students that are remote is the same. The support services that some students are required to receive, such as Special Education and Tier 2 interventions are a challenge for public educators to provide when a lack of parental guidance at home is preventing that. Students need structure and guidance when they are at home learning on a virtual platform. It is necessary and the only fair thing to do for students is to provide them that structure.

Chapter 9. *Analyzing Diversity and Equity in Virtual Classrooms*

Across the globe education has been affected the most by this new shift to all virtual everything. Our traditional way of teaching is not currently an option for all students, due to the global pandemic. In a Georgia school district, there about 32,000 students all of which were given the option to sign up for virtual learning for the Fall 2020 school year. About 65% of the districted opted to do virtual learning while the other 35% returned to face-to-face instruction. The district was faced with the challenge of supplying technology to the vast number of students that opted for virtual learning. They solved this deficit by distributing laptops, mobile hotspots, along with positioning buses in neighborhoods with mobile Wi-Fi to meet the needs of all students and families. Meeting the needs of technology for the students was just the beginning, the true task was soliciting certified educators who would be willing to transition from face-to-face teaching to virtual teaching.

Chapter 10. *Diversity and Equity in Healthcare: Are We Doing Enough?*

The significance of this chapter is to bring awareness to healthcare disparities that transpire because of the lack of equity and diversity. This awareness will occur through the identification and exploration of diversity and equity in healthcare facilities. It will look at the impact of the presence of diversity and equity on both consumers and hospital personnel. It will uncover factors that contribute to negative outcomes at the virtual level. Lastly this chapter will uncover ways to reduce inefficiencies and negative outcomes caused by the lack of diversity and equity in healthcare.

CREATIVE SOLUTIONS: REIMAGINING THE FUTURE

Chapter 11. *Closed Captioned: Designed Constraints Impact on Millennial Preservice Teachers' Creativity*

This chapter examines constraints impact on millennial preservice teachers' (PST) creativity when captioning a New Yorker Magazine cartoon. Research suggested Millennials manifest a reduction in the number of unique ideas, frustration tolerance, synthesis, articulation, and open-mindedness. Educator Preparation Programs (EPP) failed to educate millennial PST with instructional practices that promote creativity. The study utilized the concepts of constraint-based model of novelty (C-BMN; creativity problem, constraints, variability, and problem spaces) as the framework to examine the topic, along with the NEW rubric to assess data relating to constraints impact on creativity among 90 millennial PST. The C-BMN posits that design constraints positively influence an individual's creative problem solving. Kruskal-Wallis H and ANOVA inferential statistics compared differences in creativity as measured by the NEW Rubric scores for novelty, effectiveness, wholeness, and total creativity. Results indicated a significant positive difference between the output-constraint group and control group caption creativity scores, but not between the input-constraint group and the control group. The study confirmed earlier C-BMN research and concluded that constraints promote creativity. This study has implications for researchers, EPP administrators, and educational stakeholders to integrate constraint-based teaching practices to promote creativity.

Chapter 12. *Promoting Equity in Elementary School Mathematics: Vignettes From Virtual Learning Environments*

The COVID-19 pandemic of 2020 forced elementary teachers into having to teach mathematics in virtual learning environments (VLEs). These VLEs included opportunities for teachers to interact in real-time with students as well as opportunities to integrate various technology-rich tools. This chapter provides a synthesis of research related to teaching elementary school mathematics in a VLE as well as three vignettes which describe the tasks, approach to teaching, and resources used in each situation. These vignettes shed light on multiple issues related to philosophies and beliefs of mathematics teaching, equitable (or lack their of) approaches to teaching mathematics, as well as meeting students' individual needs.

Chapter 13. *Challenges and Victories of Virtual Environments With Students With Disabilities*

The COVID19 pandemic that began in 2020 altered the educational environment for everyone. Moving from an in-person environment to a virtual one can be highly problematic for individuals (specifically, students with disabilities) who's teachers most often use direct instruction strategies that require face-to-face interaction. This chapter will help the reader understand the variations of the characteristics of the different disability categories served under the Individuals with Disabilities Education Act (IDEA). This chapter will provide the readers with instructional strategies that effectively mitigate difficulties with instruction in a virtual environment. Additionally, this chapter will examine the challenges observed in students with disabilities that make accessing the virtual educational setting problematic. These areas include race, culture, disability, and low socioeconomic factors. Finally, this chapter will provide the reader with tips from teachers and students as they have struggled to navigate this new educational environment.

Chapter 14. *Envisioning the Rural Virtual College Classroom: Some Proposals for Equity and Inclusivity*

One demographic who has been most profoundly affected by the global pandemic is rural undergraduate students. Along with the widespread difficulties faced by undergraduates in general, rural students are often place-bound and attend small, open-enrollment, non-residential campuses. Such students are often first generation, working class, at risk, or some combination of these factors. Such students also face additional issues of accessibility, given that the pandemic has rigidly enforced the need for high-speed internet, a utility often taken for granted at metropolitan, residential, or large main campuses. What particular kinds of educational challenges do these rural students face? Moreover, what personal difficulties must these students encounter while learning from home? How is education affected when students must balance their academic and personal lives? How can professors rise to meet these challenges? In what ways can the classroom serve to ensure these students are not further marginalized? And how can higher education evolve to meet these students' needs? This chapter explores these questions in order to make some proposals for rural higher education in a time of crisis. Specifically, the chapter focuses on synchronous (real-time) learning in rural higher education, but the proposals have wider significance for educators, students, and decision makers. Ultimately, the chapter argues that pedagogical creativity, professional flexibility, and empathy can counteract the often impersonal landscape of online learning. A creative, flexible, and empathetic virtual landscape benefits rural students immensely, increasing the likelihood of their retention and their academic success.

Chapter 1
Response to Diversity in a Digital World

Angela Malone Cartwright
Alderson Broaddus University, USA

Geoff Paul
University of Sunderland in London, UK

ABSTRACT

The purpose of this chapter is to reflect on and respond to the work done by the authors of the subsequent chapters. The experiences and work of the contributors of this book highlight the general skills necessary for diversity, equity, and inclusion (DEI) and affirm the premises of DEI work. The chapters also include innovative ideas that can increase engagement with the DEI agenda. Finally, the chapters provide some exciting possibilities on how the digital world can further the current DEI discourse.

INTRODUCTION

The diverse perspectives represented in this book speak to the growing field of interest in the topic of diversity in a digital world. Though interest in diversity in a digital world has witnessed recent growth, the advent of the Covid-19 global pandemic has brought the issue to fore as professionals across many disciplines were, with little warning, thrust into a digital work environment. Currently, the pandemic continues to rage in some parts of the world, while in other countries due to national vaccine roll-out programmes have taken tentative steps toward a new normal. However, despite the development of vaccines there remains many

DOI: 10.4018/978-1-7998-8028-8.ch001

uncertainties and consequently, the chapters in this book raise more questions than they answer, due to the unprecedented challenges of the pandemic; the questions they raise add necessary context and dimension to both the pre-pandemic discussions and possible post-pandemic challenges surrounding diversity in a digital world.

The experiences and work of the contributors of this book highlight the general skills that we know are necessary for diversity, equity, and inclusion (DEI) work, including reflection, comfort with ambiguity, and flexibility in thought and practice, and creative problem-solving. They also affirm what are considered the premises of DEI work, including the importance of communication, providing appropriate support during the work, the necessity of focusing on systemic issues, and the limitations of government policy alone to address DEI issues.

The insight in the chapters also provides the reader with innovative ideas that can increase engagement with the DEI agenda. Using pandemic experiences to illustrate complex theories like intersectionality, which refers to the ways in which the various aspects of our identity interact to create a unique experience with a specific focus oppression, as well as the role of structural inequities in disenfranchisement and the internal complexity of DEI work, may increase both understanding and engagement. Finally, the chapters provide some exciting possibilities on how the digital world can further the current DEI discourse, including mechanisms to undermine structural barriers and address existing needs in novel ways.

Much of this reflective chapter is written for readers who benefit from unearned privilege based on societal norms and expectations. While the burden of DEI agenda has rested on the shoulders of colleagues who suffered under the structural discrimination that supports systemic inequities; this burden must now be shifted to those associates that currently enjoy the benefits of structural discrimination (Eddo-Lodge, 2018).

DIVERSITY IN A DIGITAL WORLD UTILIZES GENERAL DEI SKILLS

Due to the Cycle of Social Distance (Cartwright & Reeves, 2018) significant personal work must be undertaken to disrupt said cycle through meaningful engagement with DEI work. The Cycle of Social Distance (Cartwright & Reeves, 2018) adapted from Bucher's (2014) work on positionality, is the common set of circumstances that leads to the development of identities that keep us "insulated" from those we, for any number of reasons, see as "outsiders"; this is analogous to Moral Panic Theory (Cohen, 1972) and the identification of 'folk devils'. As Bucher (2104) explains, the process is continuous and nonlinear, though it is discussed linearly for the purposes of this discussion. The effect of the Cycle of Social Distance is succinctly explained by DiAngelo (2018): "if I am not aware of the barriers you face, then I won't see them, much less be motivated to remove them. Nor will I be motivated to remove the barriers if they provide an advantage to which I feel entitled" (p. xiii).

Figure 1.
Adapted from Bucher, 2014

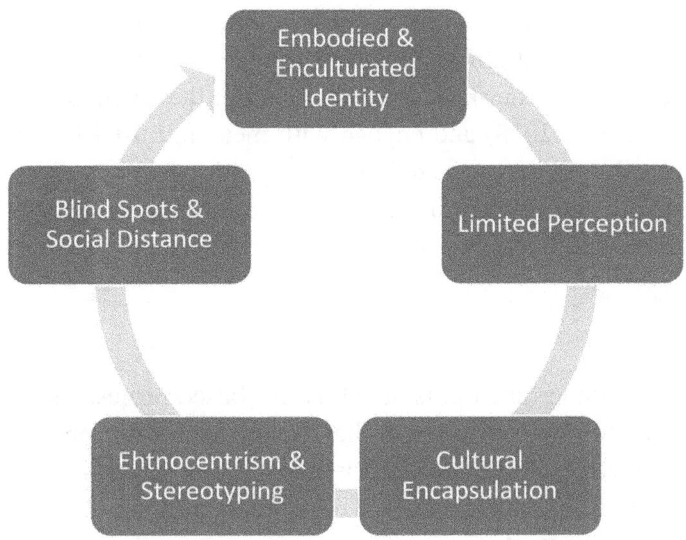

Each person is born in a particular time, place, and body; these embodied and cultural experiences create our positionality, or universal perspective. As with all perspectives, they are naturally bounded by a range of intrinsic and extrinsic elements. We can only see what we can see; we can only know what we can know.

Embodied and enculturated experiences limit the ability to perceive that which does not impact directly, as well as the interpretation of perception possibilities. Due to cognitive filtering; we cannot see everything, so we consciously or unconsciously choose the point of focus. Filtering is both a cause and effect of confirmation bias, where experiences and information that reinforce personal or key opinions are sought. Confirmation bias reifies identities because interpretations are sought that reinforce emotional uplift and avoid cognitive dissonance (Cartwright & Reeves, 2018).

Limited perception leads to cultural encapsulation, and consequently the engagement in homogenous socialisation and therefore a reduction in any cross-cultural contact that contains meaningful impact. Homogenous socialisation processes lead to a skewed worldview because of an inability to challenge the innate culture which itself acts as a reference framework to contrast other cultural standards. This innate reference framework underpins an ethnocentrism philosophy whereby our perspective is "*natural*," and thus superior; and consequently "*different*" becomes equated to "*less than.*" Ethnocentrism forms the basis for stereotyping, the formation of judgement criteria and the creation of cultural blind spots and social distance, where avoidance to challenges of personal embodied and enculturated identities effectively reinforces the cycle of social distance.

To engage successfully in meaningful DEI work, personal Cycles of Social Distance (Cartwright & Reeves, 2018) must be interrupted. Because the cycle is continuous and nonlinear, it can be disrupted at any time. As the chapters in this book have illustrated, many of the skills needed to disrupt the Cycle of Social Distance (Cartwright & Reeves, 2018) and engage with meaningful DEI work in a digital world are identical to those cultivated to engage with DEI work. We must learn to think both reflectively and introspectively, sit comfortably in spaces of ambiguity and contradiction, and think creatively to solve problems.

Reflection

Much DEI work is undertaken in isolated, issues-based professional development sessions; however, continuous DEI work that is embedded within professional culture is much more effective. Successful DEI work is a mindset approach rather than a base of knowledge from which we could draw upon, although knowledge is never wasted! Intercultural competence is rooted in personal growth, in the ability to reflect on individual thoughts and assumptions, to identify root-causes and evaluate its legitimacy in societal contexts. Practitioners in the field of DEI will have a lifetime of enculturation to examine and implement as considered relevant.

Reflection requires evaluation on contemporary practices, including ideologies and assumptions and personal and societal legitimacy. A key premise is to focus on self-evaluation, as opposed to external factors, which usually remains outside of personal control, and expands the sense of responsibility to include other members of society. June (chapter 14) encourages a reflective approach that prioritizes the needs of others in her chapter, *Envisioning the Rural Virtual College Classroom: Some Proposals for Equity and Inclusivity*. There are unmistakable parallels between her discussion of educators' experiences with moving instruction online and the experiences of others in engaging in reflective DEI work.

Therefore, without fears of compromising their rigor, professors can safely revisit their teaching philosophies and practices to consider how best to approach a student body in crisis. Instructors can make the path smoother for both the students and themselves by exploring a greater degree of creativity, flexibility, and empathy (June, this volume).

Several elements of June's statement reflect the realities of engaging in DEI work. She proceeds to discuss how hesitation within the realms and resistance to DEI work is based in *fear*. The above quote alludes to the premise that fear can be based in challenges to our identities, challenges to tradition, and the general challenge of change itself. These fears, while not explicitly unjust, support the status quo and could be argued as oppressive. Reflection requires confrontation of personal fears, including their genesis and their impact on others, and to accept responsibility for personal fear-based attitudes and actions.

Further analysis of June's (chapter 14) work suggests the reality of doing DEI work is reflected within students, colleagues, clients, friends, and family members who struggle under the burden of systemic inequities and structural oppression and that they are *in crisis*. While a detailed discussion of the crises that face the members of every group is outside the scope of this chapter, a relevant example is the experience of the Asian American community in the wake of the Covid-19 pandemic. The Center for the Study of Hate and Extremism (2021), out of California State University San Bernardino, reported in March 2021 that anti-Asian hate crimes increased by 145% over the previous year. In their civil rights lawsuit against former US President Trump, the Chinese American Civil Rights Coalition highlight the critical role of the Trump administration in anti-Asian attitudes and behaviours, following repeated use of the terms: China Plague, Chinese Virus, China Virus, Wuhan Virus, and Kung Flu (Court Listener, 2021). Faced with skyrocketing levels of violence, harassment, and discrimination, the crisis experienced by Asian American communities was exacerbated by the complicity of their nation's highest office (Court Listener, 2021). Reflection requires acknowledgement that experiences outside of personal space are real, valid, and connected to individual and collective behaviours.

June (chapter 14) proposes that educators have the power to facilitate "smoother" experiences for both their students *and themselves*; this speaks to the mutually beneficial nature of DEI work on a personal level, which will be discussed below. In addition, the organizational and institutional benefits of DEI, including increased innovation and productivity, broader range of skills and customer understanding, expanded talent pool, and employee satisfaction, are well documented and accepted in the business world, as they are noted in the relatively conservative Forbes business publication (Cole, 2020).

Comfort With Ambiguity

Whilst undertaking informed self-reflection on owned perspectives and experiences, most people will be challenged with some acknowledgement of privilege, whether it be based in race, gender, social class, country of origin or residence, health and ability, or any number of identify characteristics that imbue them with unearned privilege. For those who have not felt privileged, or who have not considered systemic inequities, this experience can be quite challenging; for those who have lived with systemic inequities, this lack of awareness on the part of the privileged can be quite challenging. Engaging in DEI work is challenging; indeed, DiAngelo (2018) reminds us that discomfort is necessary for the work to occur. The same societal and internal barriers that contribute to the experiences in the Cycle of Social Distance make it difficult to perceive, acknowledge, and support others in their process of engagement. Because we cannot experience what others experience, we must take them at their word; our experiences may contain contradictions, which requires comfort with ambiguity.

DEI work seeks to move marginalized voices toward the center; and that the goal is not to recenter the narrative around new voices, but to decenter the narrative. Decentering the narrative necessarily means that the discourse will have intrinsic conflict, as the experiences that voice the narrative have frequently been in dissension. DEI work will be particularly difficult until we learn to hold multiple, sometimes conflicting, truths simultaneously. Bonnie King (chapter 5, this volume) and Alistair Fyfe (chapter 2, this volume) illustrate this tension in their chapters, C*ovid and Counseling: A Shift to Telehealth During a Global Crisis* and *Covid-19: Virtual Medicine, Virtual Disease, Real Consequences*, respectively. In their discussions of both the US healthcare system and virtual/telehealth care, they discuss the tension between the benefits and challenges of virtual/telehealth care. Their chapters provide an excellent example of two conflicting truths: virtual/telehealth will revolutionize healthcare for some Americans, and yet simultaneously, limit the efficacy of healthcare for other Americans.

In C*ovid and Counseling: A Shift to Telehealth During a Global Crisis,* King (chapter 5, this volume) discusses ways in which telehealth benefit rural and minority communities. She explains that the acceptance of telehealth prompted by pandemic social distancing will allow those in rural areas unprecedented access to mental health services; telehealth services also create a space in which minority populations have access to mental health providers with shared languages and cultural backgrounds. King (chapter 5, this volume) addresses potential challenges in telehealth care, including connectivity issues, confidentiality, and third-party payer restrictions.

In *Covid-19: Virtual Medicine, Virtual Disease, Real Consequences*, Fyfe (chapter 2, this volume) also discusses the impact of telehealth on the patient experience, but instead focuses on the challenges of providing quality patient experience in a telehealth environment. He highlights the potentially harmful impact of inconsistency in healthcare provision, lack of provider knowledge of history and context, including timelines, and challenges patients may experience regarding expressing their experiences, feelings, and sensations, due to both lack of nonverbal communication and inability to articulate emotional sentiments. Many of Fyfe's (chapter 2, this volume) concerns relate to the relationship that develops between a patient and his/her healthcare provider, as well as the relationships that develop between providers.

A similar tension can be observed in the field of education. In *From In-Person to Online Teaching*, Fenderson (chapter 4) found that teachers identify lack of opportunities to collaborate with professionals as a hinderance in their feelings of efficacy. Similarly, healthcare providers rely on the expertise, perspective, and support of colleagues to provide quality patient experience, in a comparable manner that educators rely on professional relationships to provide the best possible educational experiences for students. When these collaborative relationships are disrupted by the lack of communal office space and time, the ability of healthcare providers and educators to meet their own standards of care can be compromised. Wilson (chapter 8, this volume) further discusses the potential limitations of the virtual environment in meeting social and emotional needs in her chapter, *Equity of Remote Teaching*. It is important to note that she discussed the social and emotional wellbeing of both students and teachers, highlighting the relational nature of quality education, which again is also evident in quality healthcare experiences.

Another example of truths in tension lies between access versus differentiation. As June (chapter 14, this volume) discusses in *Envisioning the Rural Virtual College Classroom: Some Proposals for Equity and Inclusivity*, virtual educational opportunities dramatically increase access to students living in rural areas. At the same time, several of the contributors raised concerns about providing adequate differentiation in learning and assessment for students in virtual learning environments. Polly, Colonnese, Bacak, Holshouser, & Lewis (chapter 12, this volume) raises concerns regarding differentiation for gifted students in *Virtual Mathematics Instruction: Vignettes Related to Tasks, Instruction, and Equity*, while Cavitt and Kovatch (chapter 13, this volume) raise concerns for students with High, Low, and Very Low Incidence disabilities in *Challenges and Victories of Virtual Environments with Students with Disabilities*. The impact of teacher feelings of efficacy around differentiation in online environments is discussed in Fenderson's (chapter 4, this volume) *From In-Person to Online Teaching*.

Examples from both the medical and educational fields illustrate, conflicting experiences, even realities, are not mutually exclusive. They result from different embodied and encultured experiences, as do many of the experiences highlighted by DEI work. Awareness of and comfort with conflicting truths, in addition to the ambiguity they necessitate is an important part of the personal development necessary for meaningful engagement in DEI work.

Flexibility in Thought and Practice: Envisioning the Rural Virtual College Classroom

In situations where ambiguity is pervasive and change is in process, flexibility is essential. Correspondently, society has had to remain flexible in both thought and practice, when engaging in DEI work whilst navigating the widespread COVID-induced changes within the ambiguity of unclear and evolving guidelines. As June (chapter 14, this volume) detailed in *Envisioning the Rural Virtual College Classroom: Some Proposals for Equity and Inclusivity*, willingness to change past practices is necessary to meet the needs of the present. She encourages the need to incorporate flexibility in expectations, systems and structures, and communication styles and platforms; consequently, feedback from stakeholders should inform the formats of flexibility, as should a clear, collaboratively developed understanding of priorities.

In a point that parallels with DEI work, June (chapter 4, this volume) presents the argument that greater flexibility is necessary when working with populations that are in crisis. It is worth having a brief discussion here on the difference between equality and equity; equality, despite its good intentions, can produce a rigid framework that does not meet the needs of the people it purports to serve. Equality is related to a procedural understanding of justice, as opposed to a distributional understanding of justice, which is aligned more with the concept of equity. Procedural justice maintains that the outcome of the system cannot be held as the assessment of its justice (Miller, 1999). Merrett (2004) describes procedural justice as that which provides "equality of opportunity" (p. 94). He describes distributional justice as that which provides "equality of outcomes" and focuses on "relational justice" (Merrett, 2004, p. 94). A procedural approach, a focus on equality, disregards embodied and enculturated experiences that provide the context in which their policies and procedures exist. Conversely, a distributional approach, one focused on equity, is firmly rooted in the embodied and enculturated experiences, attempting to provide bespoke support in pursuit of universal goal attainment. A commitment to equity requires flexibility in thought and action as we seek to meet the variety of needs represented by our diverse stakeholder groups.

A key component of flexibility of thought is the willingness to interrogate the realms of possibility and reasonableness in each situation. In *Is Telehealth an Effective Way to Increase Accessibility to Counseling Services?*, Fox (chapter 7, this volume) calls for variety in the ways mental health needs are identified, as well as increased global recognition of mental health issues. She argues that meeting global mental health targets is possible if properly resourced. As opposed to the prevailing US philosophy of mental health care being a luxury available to those who can afford it, Fox (chapter 7, this volume) questions the diverse ways in which mental health care can be provided, especially with the increased acceptance and structures for telehealth. In a similar argument, though made in the context of workplace expectations, Fyfe (chapter 3, this volume) discusses the changing perspective of reasonable expectations around telecommuting and work-life balance in his *How Covid-19 Changed the Culture of Work*. The rapidity with which the "status quo" around work culture adapted to the pandemic environment speaks to our capacity to respond when motivated.

Creative Problem-Solving

Flexibility is a key component of creative problem-solving. Many of the contributing authors allude to creative problem-solving; however, Hinchman's (chapter 11, this volume) *Closed Captioned: Designed Constraints Impact on Millennial Preservice Teachers' Creativity* provides a thorough argument for the necessity of intentionally developing creative problem-solving skills. He also makes a compelling argument for developing skills related to ambiguity tolerance (as discussed above). The changes experienced due to the pandemic have forced adaption in significantly altered circumstances and could be seen as a global application of the Constraint-Based Model of Novelty, as discussed by Hinchman (Chapter 11, this volume). Working towards a comprehensive DEI approach requires a similar process, presenting challenges that require a solution that works within the situational constraints. Frequently, DEI initiatives face several constraints, some of which will be discussed below, and finding ways in which to meet the DEI needs of stakeholders within these constraints is a perpetual challenge.

DIVERSITY IN A DIGITAL WORLD AFFIRMS WHAT WE ALREADY KNOW ABOUT DEI WORK

The contributed chapters illustrate ways in which general DEI skills are beneficial in pursuit of diversity in a digital world. Likewise, these chapters affirm general approaches to DEI work, including the importance of communication, support, and a focus on systemic issues, as well as acknowledging the limitations of government policies.

Importance of Communication

As is discussed above, some of the most important and challenging work of DEI is learning to acknowledge and interrogate inherent assumptions. Authentic conversations are not possible until personal experiences are viewed through the lived experiences of others. When we can truly hear what is being shared with us, we can respond to it, not react to our own emotions. Intentional communication practices will further support the difficult dialogues that are required during DEI work.

Requires Support: From In-Person to Online Teaching

DEI work is relational. It is through relationships that awareness of external experiences is highlighted; it is through relationships that allies are developed. DEI work is also intensely personal, as it requires substantial reflection for meaningful engagement. Both aspects of the work, the relational and the individual, are challenging, and support is required for successful engagement. Consequently, when engaging in DEI work, support on both an individual and institutional level is important. Examples of both are included in Fenderson's (chapter 4, this volume) *From In-Person to Online Teaching;* he shares teachers' experiences of support in the transition to online learning environments, including personal collaborations and mentoring relationships, as well as institutional professional learning communities.

Focus on Systemic Issues

As previously discussed, one of the reasons DEI work is so challenging is that it requires reflection on our role in systemic inequities. While many who enjoy unearned privileges have taken refuge in individual-level conceptions of oppression, it is now critical to examine complicity in systemic oppression (DiAngelo, 2018). Systemic problems require systemic solutions and approaching DEI issues on a systemic level can support those who benefit from unearned privilege in acknowledging their privilege. As previously discussed, the Cycle of Social Distance (Cartwright & Reeves, 2018) discussed the role of confirmation bias toward the information and experiences that reify positive perceptions of ourselves, just as individuals are attracted to premises that preserves a positive sense of self and conversely resist propositions that challenges a positive self-perception. A systemic approach can support the privileged in overcoming their natural tendency toward resistance. It can, however, be difficult to see systemic privilege and oppression, as they have become normalized.

Swagerty's (chapter 6, this volume) *Digital Access to Culturally Relevant Curricula: The Impact on the Native and Indigenous Student* and Fyfe's (chapter 2, this volume) *Covid-19: Virtual Medicine, Virtual Disease, Real Consequences* provide excellent examples of systemic inequities. Swagerty (chapter 6, this volume) discusses the systemic causes of educational underperformance in Native and Indigenous communities, including the historic use of "education" to perpetrate cultural genocide combined with contemporary lack of fiscal support to develop community infrastructure. Acknowledging the impact of the long-standing explicit policy of erasure, as well as the effects of the current policy of benign neglect, educators cannot hope to make progress in closing the educational opportunity gap for Native and Indigenous students.

Similarly, Fyfe's (chapter 2, this volume) indictment of the US healthcare system details the systemic inequities that exacerbated the health crisis of the recent pandemic.

This perspective investigates the hypothesis that the Covid-19 pandemic was the perfect storm due to the misalignment of three competing elements of the US healthcare system: the economic commoditization of disease, the economic commoditization of healthcare deliver, and the inadequate data to inform medical decision-making on a mass scale. The culmination of these decades-long devolution away from patient care to health care, or more appropriately, sick care, created a system which was overwhelmed and unable to quickly find the common ground needed to deal with the pandemic.

In his chapter, Fyfe (chapter 2) details the process of healthcare commodification where profit becomes the primary goal with the resultant dehumanizing of patients. As with Swagerty's (this volume) *Digital Access to Culturally Relevant Curricula: The Impact on the Native and Indigenous Student*, there is clarification on the hidden backstory that informs some of the most urgent and persistent questions of contemporary society: why do some student groups underperform in our schools, and why is healthcare so difficult for substantial portions of our population to access? Following a detailed examination of systemic inequities, the answer is disturbingly simple: because the system was designed that way.

In discussions about DEI, it is often argued that the system is broken; while that may be a helpful rallying cry, it is untrue. The system is not broken; it is functioning exactly as it was designed. It does not need to be fixed, it must be dismantled and rebuilt with people repositioned as the highest goal, not profit. Commodification of goods and services that are necessary for an adequate standard of living, such as infrastructure and healthcare, leads to dehumanization in pursuit of profit by normalizing the idea that human wellbeing is less important than profit margins. The destructive forces of unchecked capitalism are at the root of most systems of oppression, dehumanizing both the oppressed and the oppressor (Freire, 2000).

In his seminal work, *Pedagogy of the Oppressed*, Freire (2000) discusses the dehumanizing effects of oppression. Unsurprisingly, the oppressed are dehumanized in the eyes of their oppressors and those who are complicit with the oppressors; however, Freire (2000) points out that the oppressors are themselves dehumanized by the oppression of others, as the act of oppression keeps them from being fully human. Dismantling oppression allows us all to be fully human, to enjoy the full complexity and beauty of the human experience.

Limitations of Government Policy to Address DEI Issues

Because systemic oppression is most often supporting and established by governments, it is important to consider the limitations and implications of government policy to address DEI issues. The tension between the two positions is beautifully articulated by two civil rights activists, the Reverend Dr Marin Luther King, Jr and Audre Lorde. On the occasion of the conference of his honorary doctorate degree from the University of Newcastle on Tyne, King (1967) expressed hope in the power of government policy to effect change in systems of oppression:

Well, it may be true that morality cannot be legislated but behaviour can be regulated. It may be true that the law cannot change the heart but it can restrain the heartless. It may be true that the law cannot make a man love me but it can restrain him from lynching me; and I think that is pretty important also. And so, while the law may not change the hearts of men, it does change the habits of men if it is vigorously enforced, and through changes in habits, pretty soon attitudinal changes will take place and even the heart may be changed in the process (p. 2).

Two decades later, Lorde (1984), commenting on the pervasive heteronormativity and homogeneity in academic feminism at the 1979 Second Sex Conference in New York, famously asserted that "the master's tools will never dismantle the master's house" (p. 113). Lorde's commentary on systemic oppression within communities that identify as anti-oppressive indicate that King's optimism, expressed in the years immediately following passage of the US Civil Rights Act, was not entirely justified.

Swagerty (chapter 6) discusses the limitations of government policy in addressing systemic inequities in *Digital Access to Culturally Relevant Curricula: The Impact on the Native and Indigenous Student.* Despite two pieces of legislation, the *Telecommunications Act of 1996* and the *Repack Airwaves Yielding Better Access for Users of Modern Services Act of 2018*, aimed at ensuring all Americans had "reasonably comparable" access to communication infrastructure, many Americans living on Tribal Lands lack sufficient access.

While legislation like the Civil Rights Act of 1964 and the various telecommunication legislation have had a demonstrable positive effect (Swagerty, chapter 6, *Digital Access to Culturally Relevant Curricula: The Impact on the Native and Indigenous Student)*, it is also clear that even if legislation could be perfectly enforced, it would not necessarily lead to positive cultural change; it is possible to obey the letter of DEI policy whilst failing to embrace the inclusive spirit of the policy. Cultural change, whether it be in an institution or larger society, also requires reflective work on the individual level.

NEW INROADS TO INCREASE ENGAGEMENT WITH DEI WORK

The chapters contributed to this book have affirmed that diversity in the digital world is firmly grounded in the sphere of DEI work. They also provide insights that enable the establishment of new inroads in increased engagement with DEI work. This has permitted the shared experiences of the pandemic and resulting lockdowns to develop understanding and empathy for intersectionality, the role of structural inequities in disenfranchisement, and the internal complexity of DEI work.

Intersectionality

Lorde's (1984) commentary on academic feminism, discussed above, illustrates the concept of intersectionality; many white feminists were blind to the ways in which their own privilege silenced their colleagues of color who were experiencing oppression based on both gender *and* race. Understanding intersectionality increases the depth of engagement possible with DEI work, so any opportunity to expand understanding is a valuable occasion to be seized.

The experiences of working from home during the pandemic, as discussed in Fyfe's (chapter 3) *How Covid-19 Changed the Culture of Work*. Fyfe (chapter 3, this volume) outlines the benefits and challenges of the experience, including the impact on parenting. The tension that many experienced in balancing their caring responsibilities, whether they be parenting or elder care responsibilities, with their work responsibilities provides a relatable example of navigating two identities in one space.

Systemic Inequity and Disenfranchisement

Though the importance of addressing systemic inequity has been discussed above in depth, the opportunities to invite engagement with DEI work through a deeper understanding of systemic inequities bears additional attention. Just as an understanding of intersectionality increases meaningful engagement with DEI work, so does understanding of systemic inequities, and the experiences of the pandemic provide new opportunities for personal experience to lead to understanding.

Many of the contributing authors discussed issues related to internet access, particularly as it relates to rural inhabitants. The digital divide, created by infrastructure limitations related to profit margins, as described in King's (chapter 5) *Covid and Counseling* and Swagerty's (chapter 6, this volume) *Digital Access to Culturally Relevant Curricula* provide an example of structural barriers to success for those who have limited experience and understanding of systemic inequities. Introducing the concept of systemic inequities based on an external identity factor such as geography provides a non-threatening way to engage with the concept, creating opportunities to engage with systemic inequities based on internal, or immutable, identify factors.

Complexity Within DEI

The divide between those who acknowledge and understand systemic inequities and those who do not has become quite partisan in both the US and the UK after years of Trumpism and Brexit-focused Toryism, respectively. In these politically charged environments, where critical theory is challenged, DEI is a partisan issue and has become reduced to a considerable extent, on race and immigration status. Awareness of the oppression faced by communities of color and immigrant communities has significantly increased in recent years and their plight deserves urgent attention. The legitimate focus on race and immigration, however, has resulted in a narrow understanding of diversity and a lack of engagement with the needs of other under-served communities.

The contributing authors provide insight into the needs of communities that remain under-served and under-prioritized. In Cavitt and Kovach's (chapter 13, this volume) *Challenges and Victories*, the needs of students with learning differences are brought to the fore; the overview of relevant legislation and judicial action provide a sobering reminder that the struggle for equitable treatment for those with disabilities remains. Swagerty's (chapter 6, this volume) *Digital Access to Culturally Relevant Curricula* and Fox's (chapter 7, this volume) *Is Telehealth an Effective Way to Increase Accessibility to Counselling Services?* raise the needs of the Native and Indigenous community and rural communities, respectively. By expanding the conception of diversity and inviting additional voices and perspectives, it is possible to increase engagement with DEI work by demonstrating the pervasiveness of system inequity and importance of intersectionality.

EMBRACING DIGITAL OPPORTUNITIES IN DEI

The chapters contributed to this book affirm that diversity in a digital world is firmly aligned with existing knowledge in DEI work; they also point to exciting new opportunities for DEI work. The unique environment of the digital world provides new opportunities for undermining structural barriers and meeting existing needs in new ways.

Undermining Structural Barriers

The structural barriers that lead to systemic inequity have been thoroughly discussed in previous sections; however, the potential for undermining these structural barriers that exists in the digital world is a new development that deserves further exploration. Bain's (chapter 10, this volume) *Diversity and Equity in Healthcare* contains a fascinating discussion about the National Medical Association, a predominantly Black physician network, and the American Medical Association, a predominantly White physician network. Racism kept physicians of color from joining the AMA, which limited their access to new developments in medicine, negatively impacting their patients' experiences.

While the high cost of professional associations and journal subscriptions continues to be a barrier to access, the digital world has the potential to shatter barriers to knowledge sharing and increase equity in every academic and professional discipline. To some degree, this has been fuelled by the advent of MOOCS and the relative dispersion and increased access to knowledge. In addition to increased internet access, the rise of virtual communications precipitated by the pandemic lockdown has further eroded barriers to knowledge sharing, as well as geographic and political boundaries to collaboration.

Innovative Approaches to Existing Needs

Just as the digital world expands boundaries and undermines structural barriers, it also provides innovative opportunities to meet existing needs. The potential for telehealth to revolutionize access to healthcare, particularly mental health services, is discussed by several chapter contributors. In addition to connecting patients from under-served groups with providers that share their racial, cultural, linguistic, etc. background, it also has the potential to address another of area of opportunity in addressing healthcare disparities: workforce diversity, as discussed by Bain (chapter 10, this volume).

The necessity of continuing training for healthcare professionals during the pandemic lockdown led to the development of virtual learning pedagogies in a previously overwhelmingly hands-on discipline. Creative problem-solving to meet the needs of various stakeholder groups is a productive effect of our collective pandemic experiences, as is illustrated by several contributing authors, such as Baily and Sims (chapter 9, this volume) *Analysing Diversity and Equity in virtual Classrooms.* The lessons we learned about facilitating learning for healthcare providers can be utilized to continue expanding access to training and education to under-served populations, narrowing the demographics gap between patients and providers.

CONCLUSION

The chapters contributed to this book firmly establish that diversity in a digital world is aligned with established theory and practice in DEI work and provides additional avenues for engagement with DEI work, asking us to confront old challenges in new ways as we also learn to confront new challenges. They highlight the ways in which digital communications and information-sharing provide new opportunities to resolve societal DEI issues while also reminding us that structural inequities must still be addressed for the potential of DEI work to be fully realized. The chapters in this book are an important contribution to the understanding and approaches to DEI work in various contexts and will support professionals across fields in interrogating the ways in which they can work for equity in their fields.

REFERENCES

Bucher, R. D. (2014). *Diversity consciousness: Opening our minds to people, cultures, and opportunities*. Pearson.

Cartwright, A. M., & Reeves, E. K. (2018). *Everyday diversity: Developing cultural competency & information awareness*. Cognella Academic Publishing.

Center for the Study of Hate and Extremism. (2021). *Fact Sheet: Anti-Asian prejudice March 2021 Center for the Study of Hate & Extremism*. Center for the Study of Hate and Extremism. https://www.csusb.edu/sites/default/files/FACT%20SHEET-%20 Anti-Asian%20Hate%202020%20rev%203.21.21.pdf

Chinese Americans Civil Rights Coalition, Inc. v. Donald J. Trump. (2021). Cohen, (1972).

Cole, B. M. (2020, September 15). 8 Reasons Why Diversity and Inclusion Are Essential to Business Success. *Forbes*. https://www.forbes.com/sites/biancamillercole/2020/09/15/8-reasons-why-diversity-and-inclusion-are-essential-to-business-success/?sh=2280c1ae1824

Court Listener. (2021). *Chinese Americans Civil Rights Coalition, Inc. v. Donald J. Trump*. Court Listener. https://www.courtlistener.com/?type=r&q=1%3A21-cv-04548&type=r&order_by=score%20desc

DiAngelo, R. (2018). *White fragility: Why it's so hard for White people to talk about racism*. Penguin Books Ltd.

Freire, P. (2000). Pedagogy of the oppressed: 30th anniversary edition. *Continuum*.

King, M. L. (1967, November 13). *Speech on receipt of Honorary Doctorate in Civil Law* [Address]. University of Newcastle upon Tyne, UK. https://www.ncl.ac.uk/media/wwwnclacuk/congregations/files/Transcript%20of%20Dr%20Martin%20Luther%20King%20Jr%20speech%2013th%20November%201967.pdf

Lorde, A. (1984). *Sister outsider: Essays and speeches*. Ten Speed Press.

Merrett, C. D. (2004). Social justice: What is it? Why teach it? *The Journal of Geography*, *103*(3), 93–101. doi:10.1080/00221340408978584

Miller, D. (1999). *Principles of Social Justice*. Harvard University Press.

Section 1

Disequilibrium: A Time of Transitions

Chapter 2
Virtual Medicine, Virtual Disease, Real Consequences:
Was COVID-19 the Perfect Storm of Dysfunction in American Healthcare

Alistair Fyfe
https://orcid.org/0000-0001-6810-4090
New York Life, Dallas, USA

ABSTRACT

This chapter investigates the hypothesis that the COVID-19 pandemic was the perfect storm due to the misalignment of competing elements of the US healthcare system, the economic commoditization of disease, the economic commoditization of healthcare delivery, and inadequate data to inform medical decision making on a mass scale. The culmination of a decades-long devolution away from patient care to healthcare or more appropriately sick-care created a system that was unable to quickly find the common ground needed to deal with the pandemic known as COVID-19.

INTRODUCTION

The COVID-19 pandemic is currently associated with the loss of more than 500,000 lives in the US in the past year with the largest number of both cases and fatalities in the developed world (Covid-19 May, n.d.) in the context of a medical system which spends twice as much per capita as comparable countries. That this occurred in the context of record numbers of healthcare organizations being ranked by organizations

DOI: 10.4018/978-1-7998-8028-8.ch002

such as JD Power as superior quality institutions (presumably based on care delivery) is an apparent further contradiction.

Individual patient exposure to SARS-CoV2 would seem to be random and while risks for the development of mild or severe infection have been published; what accounts for our inability to prevent those infected from dying and is that inability due to structural deficits within the US healthcare system that can be defined and modified. Problematic is the racial and socio-economic distribution of both infection and death were not evenly distributed and represented both risk factor distribution increasing infection rates and inequality in access to healthcare (Keckley, 2020). The severity of disease onset and widely published symptoms would not favor cultural factors as contributing to recognition of infection but may have impacted early arrival to healthcare. The pandemic highlighted the urban-rural healthcare divide in unexpected ways: the early high death rate in New York City, associated with enhanced transmission in a crowded landscape, together with lack of familiarity with infection, testing, and treatments contrasts with the winter 2020/2021 outbreak in rural areas associated perhaps with complacency and lack of access to tertiary care, but benefiting from rapid advances in treatment and disease modification therapies.

This perspective does not address the factors which led to infection, but to the inability of the healthcare system to respond which may have contributed to such a large number of fatalities. It is not appropriate to assess fatality rate since testing leading to discovery evolved rapidly over time and at differing rates around the world so the actual number of infections may never be fully known. The defining difference with this pandemic, compared to prior pandemics or other disasters is the speed and intensity with which it consumed medical resources. It disrupted healthcare on an international scale, unlike local disasters, or even national disaster when other entities, including international entities like the World Health Organization (WHO) or Red Cross/Crescent can relocate supplies across borders in a timely manner. COVID-19 overwhelmed all healthcare everywhere and within weeks. Supply chains were disrupted by national embargoes, "just-in-time" economic principles suddenly not only irrelevant but disastrous, and every person, hospital and every state were left to fend for themselves.

This perspective investigates the hypothesis that the COVID-19 pandemic was the perfect storm due to the misalignment of three competing elements of the US healthcare system; the economic commoditization of disease, the economic commoditization of healthcare delivery, overwhelmed in the face of inadequate data to inform medical decision making on a mass scale. The culmination of these decades long devolution away from patient care to health care, or more appropriately sick-care, created a system which was unable to quickly find the common ground needed to deal with the pandemic known as COVID-19.

These perspectives are a result of my 40 years in academic practice, group practice, solo practice and as an employed physician. More recently it relates to my activities as both a health insurance medical director (facilitating healthcare decisions) and in disability claims insurance (paying for the complications of those decisions). It brings perspectives from time practicing in socialized healthcare (Canada), private healthcare (US) and a hybrid system (Australia). It is not a critical comparison as all healthcare systems were stressed by the pandemic in different ways. I have supported my perspectives with examples which are by no means all encompassing.

ECONOMICS AND THE AMERICAN PROVIDER

Bryan in his historical overview (Bryan, 2011) documents the evolution of medicine from the historical doctor patient relationship largely devoid of outside influence through the development of scientific underpinnings to medicine as a migration from the art form to a measured entity, then as a commodity. He opines "my sole concern – revolves around the business and financial dimensions of medical practice" The physician-patient interaction became codified, its value assigned and then discounted as economic necessity dictated. The same objectification occurred with tens of thousands of procedures assigned CPT coding; each with its complexity and legal implications codified, negotiated, and valued. Bryan's concerns have materialized as one follows the evolution of healthcare delivery in economic terms.

Historically the relationship between physician and patient was sacrosanct and information was exchanged for value within this relationship (Figure 1)

With the advent of The Great Society in 1960s, recognizing that some employer funded plans such as Kaiser Permanente existed, medical insurance and Medicare became the patients surrogate for payment but with no direct relationship between provider and payor. Figure 2.

The arrival of Hillarycare in 1992 sought to consolidate healthcare decision making with payors as a way of controlling costs (Managed Care Organizations). This arrangement solidified the payor-provider relationship such that direct payments to the provider were associated with the ability to intervene in care through specific contracts which specified reimbursement rates, but with incremental conditions attached related primarily to documentation, but then to oversight and recoupment. An example includes the attestation on checks "this payment is made with Federal Funds. Fraud in procuring…this check is punishable under the US Criminal Code. By endorsing this check, the payee is subject to the terms of the participating agreement and regulation".

Figure 1. The historical physician-patient relationship exchanged medical care for value (including barter)

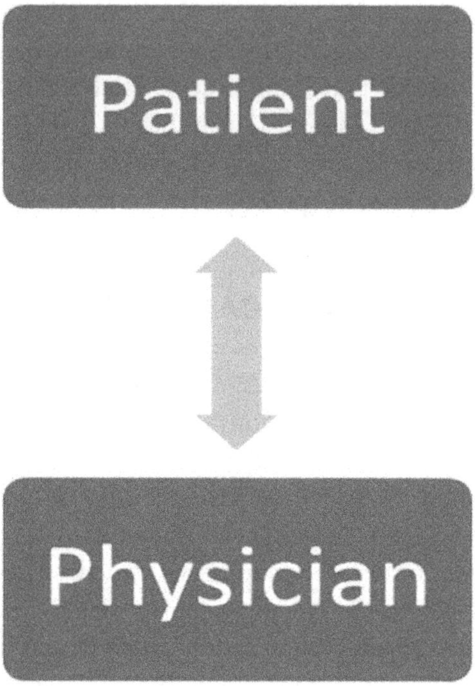

Figure 2. Payor organizations in the form of Medicare and other insurance carriers re-imbursed patients for all or part of their medical expenses

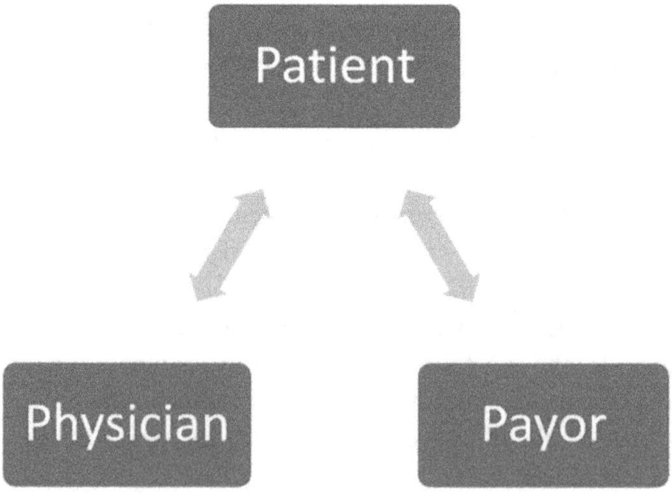

Figure 3. The formation of the direct relationship between physician and payor has roots in Hillarycare circa 1990

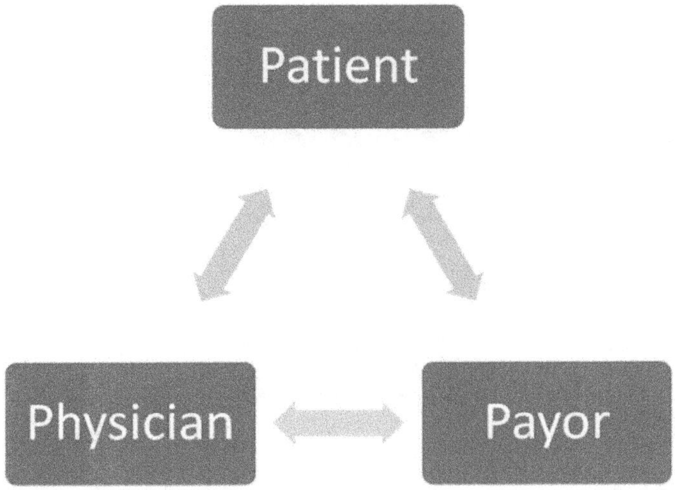

A series of legal determinations in favor of patients denied care by the Managed Care Organizations led to payors developing in-house medical expertise as justification for controlling the care requested by the treating physician. Since MCOs received fixed payment per-member-per month premiums; the so-called capitated care, they assumed financial risk for illness. The treating physician at this point remained autonomous but effectively decided, by diagnosing sickness (or wellness) whether the hospital was enriched (remember hospitals account for 80% of healthcare expenditure) or the MCO became enriched by retaining the fixed block grant type payments from employers.

Obamacare in 2009 switched the accountability and financial risk to providers, primarily large hospital entities as the medical home for accountable care. Hospitals purchased physicians and their practices frantically to consolidate referrals and amass "covered lives" for which "per-member-per-month" fees were now directed. By 2015 up to 72% of physicians (depending on specialty) were owned by a hospital entity, and increasingly since by private equity investors. The consolidating of the spending power to include the decision makers into shareholder funded entities required a reciprocal increase in the number and specialization of medical "directors" at shareholder funded health insurance organizations for both oversight and cost containment. The net consequence of these changes is an adversarial economic relationship between large entrepreneurial entities controlled by non-medical interests. The migration of traditional Medicare (which does not have medical directors) to the Medicare-Advantage type HMO plans is a further transfer of financial risk to private entities. Figure 5.

Figure 4. As hospitals became a major controlling interest in healthcare, the independent physician remained the arbiter of illness

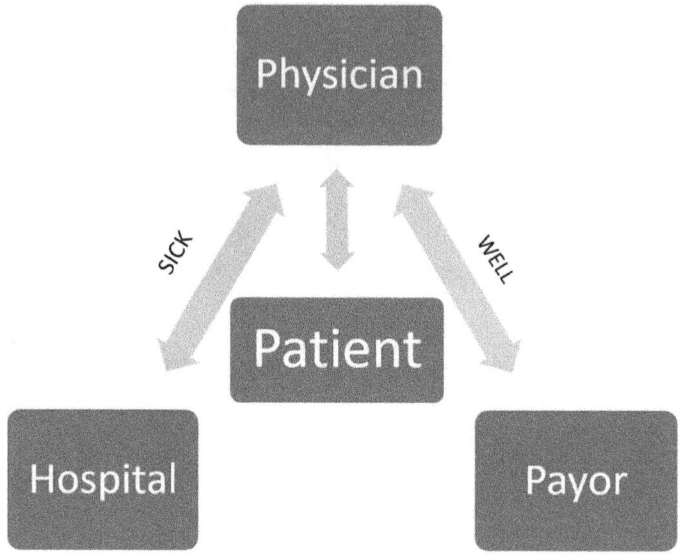

Figure 5. The medical relationship between the patient and physician has been minimized by the scale of the economics in "accountable care"

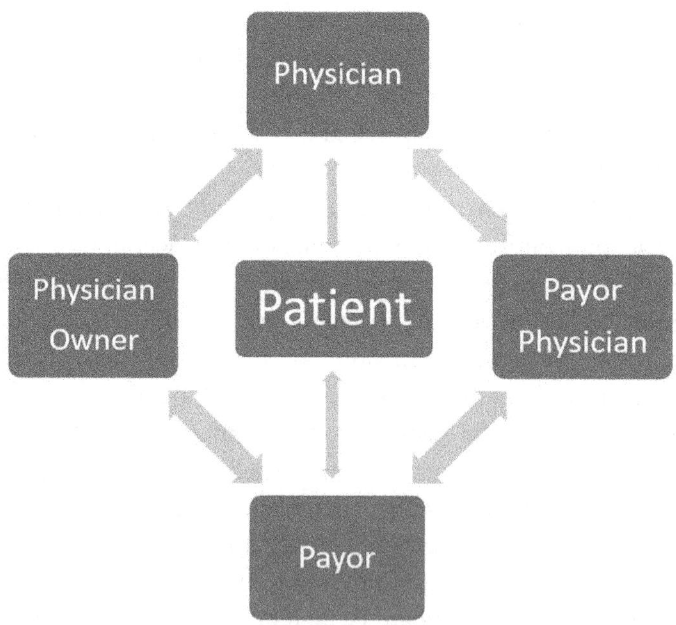

What has this to do with "medical professionalism" which is contemporaneously defined as the "primacy of patient welfare, patient autonomy and social justice". It is because these altruistic professional goals cannot easily co-exist in a capital driven system and I suspect are an uncomfortable topic at corporate meetings with respect to the new mantra; "we are tasked with the financial stewardship" of shareholder investments. The pursuit of volume in response to changes in re-imbursement, often skirt the pay-to-play laws known as Stark and anti-kickback (HHBP).

VIRTUAL PROFESSIONALISM

Bryan re-interprets the doctor patient relationship from the modern perspective (Generation X) and quotes Gourevitch in predicting that physicians will be replaced by healthcare technicians (Gourevitch, 1999). Physician "professionalism" and putting the patient first is both costly and unpredictable regarding shareholder returns. It is widely known that each practicing physician in the US has at least 10 independent individuals tasked with their personal oversight and regulation. While the true corporate ownership of physicians is banned in many states, it occurs through shell corporations. The employment of "mid-levels" such as physician assistants or nurse practitioners (with the same prescribing power) is not constrained by historical anachronisms; neither physician assistants or nurse practitioners have defined or published articles of professionalism to this date. In a personal conversation with the developers of Google glass, the fundamental question arose, "with so much information available so easily on the internet, how much knowledge does a practitioner need to safely practice" (Chicago 2013). The issue becomes not so much the generation of data, but the ability to determine its veracity and applicability. The suspension of the belief and trust in science generated by the climate change controversy was carried into COVID-19 science, with every provider entitled to their own personal interpretation of the data and absent guidelines, their own personal application to their patients. It seemed initially that 50 years of evidence-based medicine was discarded, precisely because there was no evidence.

The efficient use of human capital in medicine is to put the creation of value in the hands of the least expensive "provider" supported by corporate guidelines, flowcharts, and checklists. An Op-ed in the prestigious New England Journal of Medicine during the prequel to Obamacare stated specifically that legal exposure resulted from not following the determined checklist whereas following an incorrect checklist was defensible. As we speak payors are closing this "loophole" by insisting that nurse practitioners bill under their own National Provider Identifier (at their assigned lower re-imbursement rates) rather than continuing to bill under the more lucrative physician NPI number (TMA 5/4/2021). Not only has the delivery of medicine become a commodity, but each diagnosis is also a commodity with its own re-imbursement. Note that checklist medicine is not to produce excellence in any way other than to reduce error. The diversity in healthcare delivery was attributed humanistic bias which became replaced by human influenced computer algorithms which administer care with less clearly defined but more harshly delivered biases in both diagnosis and care options; your procedure or medication is either approved or denied. Thus, the economics of medicine as a consumer (or producer) of 16% of GDP placed medical decision making in the hands of the least experienced defended by guidelines and checklists.

COVID-19 disrupted the system in ways not previously imagined by closing physician practices and any non-urgent hospital visits through government mandate. Some 37 million patients became unemployed and many lost both their employer mandated health insurance and access to care. Many patients were lost to regular follow-up and patients were seen virtually, predominantly by telephone(s) by providers they did not know (United Department of Justice, 2017). The relaxation of the rules for telemedicine unraveled the previously sacrosanct provider-patient relationship even further. A large percentage of telemedicine visits were performed by providers who had no prior knowledge or relationship with the patient (ibd). Virtual medicine unilaterally documents symptoms with no ability to verify the time-intensity relationship which is so critical to any assessment of pathology or functional capacity. The adage that 75% of communication is non-verbal further exacerbated the ability to collect accurate data, fundamental to the commoditization process. This is further exacerbated by the inability of patients to see ongoing providers, their appearance at acute care clinical and ER is further devoid of contextual information that may elucidate that many of the symptoms attributed to COVID-19 infection and more specifically the new so-called "long hauler" status existed for months or years prior to COVID-19 infection.

Those conversations were not recorded by the participating provider but often a designated "scribe" who may not even be in the US. This raises serious issues related to the assignment of liability in such circumstances. If a medical record is written by a scribe in another country paid for by the employer of the physician assistant who is overseen by a provider who is also not on-site, where does accountability lie? In the era of electronic medical records, the ability of computers to generate unsubstantiated medical information ensnared one large provider (Healthcare Provider Solutions, n.d.) the cut-and-paste insertion of incorrect information makes the current medical record as a transmitter of meaningful information dubious. In a recent example, two visits to a hospital emergency room generated 128 pages of medical records. The inaccuracy of computer-generated medical records has led to the imposition of so called "meaningful use" criteria, the purpose of medical records having devolved so far from being either meaningful or useful.

COMMODOTIZATION OF DISEASE

Simultaneously with the commoditization of the healthcare delivery process the patient, their disease, and complexity became increasingly codified and so called "relative value units" assigned. These units became the currency of payment under the diagnosis related groups (DRGs) which form the basis of hospital re-imbursement. With the progressive consolidation of healthcare providers and payors, why is there not a typical supply and demand relationship? Supply and demand rely on price transparency and despite many efforts to encourage and legislate price transparency, this remains and illusion. Ironically the published Medicare re-imbursement rates for more than 10,000 procedures are readily available and indexed by county. While not readily accessible, payor contracts typically treat trigger off this list and pay (a small amount) above Medicare rates and sometimes below. It is generally accepted, and many procedures are condemned to oblivion when Medicare fails to allocate a specific billable code for re-imbursement. In a fashion Medicare (which by law must pay the lowest price for services) sets the lower limit of reimbursement for any procedure. The upper limit has no bounds and whose burden falls primarily on the uninsured who are billed at multiples of the Medicare floor rate. The authority of Medicare to both assess the viability of procedures and assign a CPT code generally determines if the procedure will be reimbursable. Multiple procedures have been discarded in the dustbin of history having failed to meet meaningful outcomes criteria. Private payors add both outcomes and value-based criteria to procedures before determining authorization strategies. The alignment of providers with large shareholder entities who have no vested interest in curbing demand and through the promotion of screening programs seek to increase the number of asymptomatic people accessing healthcare. The decades long controversy surrounding coronary calcification and calcium scoring is testament to this struggle between early detection and the costs of early intervention.

The widespread use of electronic medical and health records with data mining of existing patients to find more valuable diagnoses further dehumanizes the patient under the guise of population-based health or "big data". The use of IBM Watson Big Blue by hospital systems is well known (Health Provider Solutions, n.d.). The spin-off Watson Healthcare now has its own hospital quality awards.

A significant constraint on data mining is the factual basis of the underlying data which is subject to documentation errors. The system rewards the assessment of many diagnoses even if incorrect, rather than a fewer number of correct ones. There are cases where insurers, perhaps inadvertently based on improper documentation, up charged Medicare for complex care (Keckley, 2020). A medical coding company was held liable. Both hospital systems, medical coding companies and insurers have been subject to the legal ramifications of such "upcoding" and "over-utilization. The declaration of a disaster by either local or federal authorities requires an immediate cessation of prior authorization requirements for medical procedures and medications, in the case of COVID-19 at a federal level. Oftentimes this creates an immediate avenue for exploitation so that experimental procedures which had previously been banned or limited can be performed with no questions asked. Payors early in the pandemic petitioned for ongoing control over the "medication supply" under the guise of preventing hoarding of medications such as hydroxychloroquine but maintained control over the most expensive biologics and generated record profits during the pandemic. To their credit, they relaxed payments for diagnosis, care, and treatment for COVID related illness, more than compensated by the abolition of elective procedures for weeks or months. The loss of 37 million employee/employer premiums appears not to have been significant factor.

SCIENCE, STATISTICS AND GUIDELINE BASED MEDICINE

The scientific basis by which technicians become decision makers results from the adoption of "checklist medicine" (Gawande, n.d.) formulated by guideline-based practices, citing evidence-based medicine derived from scientifically conducted clinical trials. The patient decision is therefore at least three steps removed from the original data upon which the decision is made. Clinical trials are graded in guidelines using a Class I (best evidence) to Class IV (studies that are non-comparative, unrepresentative of the population of interest, with major biases or confounding, lacking useful measures of effect, or lacking measures of effect estimate stability. In everyday practice the class of evidence supporting a decision point is often ignored. Most providers can find some guideline somewhere which supports their personal mode of practice, and when all else fails, "guidelines are just that, guidelines". The use of hydroxy-chloroquine in the treatment of COVID-19 was a stark reminder of treatment decisions being made ahead of the science which was subsequently interpreted "as needed".

While rarely acknowledged, clinical trials provide the answer to what is the average outcome of the intervention in the average patient. This is an unacceptable standard in practice when patients and payors demand 95% or greater. No hospital, practitioner or insurance company would be rewarded for the delivery of "average". Therefore, the guidelines upon which check-list medicine is practiced are themselves inimical to the quality demanded, even by governmental authorities such as the Center for Medicare Services. There is further lay confusion when clinical trial data do not translate to reality: for example, COVID vaccine efficacy of 90+% in clinical trials, when associated with 50% vaccine resistance among a large portion of the US population will likely not get to herd immunity. As in other efficacious therapies the total clinical efficacy is diluted by the ability to implement. While clinician scientists report intention to treat, this is not the outcomes measure patients want, they want absolute risk assessments, like the risk of getting COVID with any specific activity; a guideline not published to this day. The recent stand-off between Dr Fauci and Rep Jim Jordan in Congress highlighted this disconnect between science and the day-to-day reality in which decisions must be made. Decisions cannot be made on relative risk; for example, am I more likely to die skydiving or hand-gliding has no relationship to knowing the absolute risk of death in commercial airline travel. In practice extrapolation from clinical trials is rampant and only egregious violations are exposed (Department of Justice, n.d) The retraction of COVID-19 related data from prestigious medical journals early in the pandemic harmed the legitimacy of both the science and the role of peer-review (Others et al., 2020). Since much of the data related to COVID-19 has not been peer-reviewed, a common response is "who has the experience to peer review anything to do with COVID-19" The lack of scientific comparison to prior coronavirus outbreaks SARS and MERS is noteworthy. Many observations are published primarily in the media referencing impending publication on pay-for-publish sites like MedRxiv.org. Patients then have access to vast amounts of conflicting information on the internet with which inform their side of the conversation. Social media has become a major arbiter of information/ fact/truth? in COVID-19, so much so that the CDC guidelines on "long haulers" reference social media studies or sites in 17 of their 30 quoted references (CDC, n.d.). No attention appears to have been made to the fact that more than 500,000 social media websites are fictitious people/patients rented by pharmaceutical companies (Oxford University, 2018).

Checklist based care was developed from the airline industry and the safety record derived from such practices. Checklists are generated by the airplane manufacturer; most physicians don't believe there is a manufacturer for the human body which humorously makes the application of checklists at its best bootstrapping and at its worst ludicrous. The Boeing 737max debacle is evidence that checklists are only as good as the disclosure and understanding of the underlying system. Checklists work safely and efficiently in common conditions for which decades of research have clearly defined the central determinants of both survival and quality of life, and clearly defined (often in case law) the outer boundaries of acceptable medical practice and resource utilization. What happens when there is no data, there is no experience, and specifically in COVID-19 the data being published is incomplete and redacted as quickly. News reports document checklists placed in the hands of 2nd year medical students tasked with delivering care as their first taste of "medical practice". The failure of the check-list guru Atul Gawande to create the revolutionary medical system (HAVEN) backed by Berkshire Hathaway, Chase and Amazon is testament to the fact that check lists do, as their name suggests, check existing systems, they do not create new information or systems.

What about when the evidence is correct, and the patient data is manipulated to fit the guideline to affect an economic outcome? A recent settlement documented that 457 hospitals were found to have implanted defibrillator inappropriately with respect to the published guideline (Morse, n.d).

SUMMARY

This perspective seeks to provide an overview of the tension and dysfunction in the US healthcare system that contributed to the excess deaths noted during the COVID-19 pandemic. Three intersecting factors were discussed; the relentless pursuit of value at the cost of the physician-patient relationship, the economic valuation of disease more than the patient, and inability of science to quickly supplement outcomes-based interventions. While the development of treatment strategies occurred at a historic pace with the development and implementation of effective vaccines, many lives were lost needlessly in the first year. The pandemic is clearly not resolved, the failure to achieve herd immunity will produce ongoing infection over the next few years.

It is a noteworthy to compare the COVID-19 pandemic to the disruption seen with the appearance of HIV in the 1980s; it has taken two decades to make the disease compatible with long term survival, though no cure is yet available. When sick-care is less an altruistic endeavor and merely another, albeit very high-priced commodity, the populace is forced to make difficult choices in how and where they choose to invest their time and resources. The contemporaneous suspension in belief of generic benevolence, the altruistic nature of science and government were additional complications impeding both control of the infection and remediation of the consequences.

REFERENCES

Atul Gawande. (n.d.). *The checklist manifesto*. http://atulgawande.com/book/the-checklist-manifesto/

Bryan, C. (2011). Medical Professionalism Meets Generation X. In Texas Heart Institute Journal (pp. 465–470). Texas Heart Institute.

Centers for Disease Control and Prevention. (n.d.). *Post-covid conditions*. Centers for Disease Control and Prevention. https://www.cdc.gov/coronavirus/2019-ncov/long-term-effects.html

Gourevitch, D. (1999). The history of medical teaching. *Lancet*, *354*(Suppl), SIV33. https://doi.org/10.1016/s0140-6736(99)90376-8

IBM. (n.d.). *Healthcare provider SOLUTIONS: Watson Health*. https://www.ibm.com/watson-health/solutions/healthcare-provider

Johns Hopkins Coronavirus Resource Center. (n.d.). *COVID-19 map*. https://coronavirus.jhu.edu/map.html

Lopez Bernal & Shimabukuro. (2020, June 18). Retraction: Cardiovascular disease, drug therapy, and mortality In Covid-19. *New England Journal of Medicine*. https://www.nejm.org/doi/full/10.1056/nejmc2021225

Morse, S. (n.d.). *457 hospitals settle for $250 million over IMPROPER implantable cardioverter Defibrillator surgeries*. Healthcare Finance News. https://www.healthcarefinancenews.com/news/457-hospitals-settle-250-million-over-improper-implantable-cardioverter-defibrillator-surgeries#:~:text=than%20%24250%20million.-,The%20biggest%20hospital%20chain%20settlements%20are%20HCA%2C%20at%20%2415.8%20million,cardioverter%20defibrillator%20cost%20about%20%2425%2C000.&text=The%20whistleblowers%20have%20received%20more%20than%20%2438%20million%20from%20the%20settlements

Paul Keckley. (2020, May 18). *COVID-19 unintended Consequences: UNFLATTERING attention to the gap between "haves" and "have nots"*. https://www.paulkeckley.com/the-keckley-report/2020/5/18/covid-19-unintended-consequences-unflattering-attention-to-the-gap-between-haves-and-have-nots

Schulte, F. (2021, May 17). *Humana health Plan Overcharged Medicare by nearly $200 Million, federal audit finds*. Kaiser Health News. https://khn.org/news/article/humana-health-plan-overcharged-medicare-by-nearly-200-million-federal-audit-finds/

Texmed. (n.d.). *Texas medicine today*. https://www.texmed.org/tmt/

The United States Department of Justice. (2017, May 31). *Electronic health RECORDS vendor to pay $155 million to Settle false Claims Act Allegations.* https://www.justice.gov/opa/pr/electronic-health-records-vendor-pay-155-million-settle-false-claims-act-allegations

University of Oxford. (n.d.). *Social media manipulation by political actors an industrial scale problem.* https://www.ox.ac.uk/news/2021-01-13-social-media-manipulation-political-actors-industrial-scale-problem-oxford-report

Warner-Lambert to pay $430 million to resolve criminal & civil health care liability relating to off-label promotion. (n.d.). https://www.justice.gov/archive/opa/pr/2004/May/04_civ_322.htm

Chapter 3
How COVID-19 Changed the Culture of Work

Alistair Fyfe
https://orcid.org/0000-0001-6810-4090
New York Life, Dallas, USA

ABSTRACT

The COVID-19 pandemic created a historic disruption to contemporary society including how, where, and when we work. Given the ubiquity of human capital, most if not every society was crippled by the displacement of the workforce with historic impacts on productivity; GDP in the UK will be at its lowest in 300 years, requiring the largest peacetime debt accumulation in history. Stimulus packages occurred in many countries as a result of the inability to access the workplace, particularly school, restaurant, or travel. Airline travel in the US fell by a precipitous 93% at its nadir, the cruise industry collapsed, and trans-national crossing all but ceased to exist. Along with the freeze in people movement, supply chains were disrupted including components necessary for both treatment and vaccination. The shrinkage of the world we had grown up with became the catalyst for the first pandemic in a century.

INTRODUCTION

The COVID-19 pandemic created a historic disruption to contemporary society including how, where and when we work. Given the ubiquitous role of human capital, most if not every society was crippled by the displacement of the workforce resulting in historic impacts on productivity; GDP in the UK will be at its lowest in 300 years, requiring the largest peacetime debt accumulation in history (Johns Hopkins Coronavirus Resource Center, n.d.). The shrinkage of the world we had

DOI: 10.4018/978-1-7998-8028-8.ch003

grown up with became the catalyst for the first pandemic in a century. Stimulus packages occurred in many countries because of the inability to access the workplace, schools, restaurants, and travel. Airline travel in the US fell by a precipitous 93% at its nadir, the cruise industry collapsed and trans-national tourism all but ceased to exist. Along with the suspension of the movement of people, supply chains were disrupted, specifically early for the raw materials and trans-national collaboration necessary for diagnosis, mitigation, treatment, and vaccination. Nations scrambled to restart long dormant industries required for the production of protective equipment and ventilators.

This perspective represents my personal view; garnered through a typical MBA education, supplemented by 30 years in the workforce of multiple institutions, and more recently by 5+ years assessing disability claims and return to work issues, most specifically return to work after COVID and "long-COVID" syndromes.

It is almost impossible to have easy answers, those who reached to science forgot that scientists look outwards, searching for anomalies, anecdotes and outliers which provide sensationalist findings, rather than the inward-looking consensus building middle ground which generally comes after years of both positive and negative data. Governments around the world intervened in archetypical ways, dictators with edicts, socialist economies with "what's best for everyone" and in the land of freedom who made all men equal in their pursuit of happiness, chaos reigned. The pandemic is still in play, mutating, evolving and consuming resources on a historic scale. There is no precedent to inform this current calamity, there is no "helping the needy" this pandemic in a sense it became a great equalizer because we all became needy, if not for socialization rather than material goods. Perhaps this is the modern iteration of the "Germs" popularized in the book of the same name (Keckley, 2020) which transformed the planet in ways still evident. The demonstrated inequity in delivery of healthcare, and the inequality in the distribution of vaccine, the playing field remains tilted; Given that work is a common commodity within the individual purview in most if not all countries, supply and demand will arrive at uneasy equilibria. Uneasy because it will be a new norm, it will require new rules, it has shaken the confidence of the many in the reliability of institutions; science to deliver the "truth", government to "manage", the economy to demonstrate "fairness". The ubiquity of misleading news, sensational science and personal tragedy on a daily basis was overwhelming as evident from data on the use of mental health services and associated medications.

EFFECTS OF COVID-19 ON THE WORKPLACE

Work Loss

Since March 2020, 36.5 million jobless claims have been filed. According to the Federal Reserve, 39% of households with income below $40,000 lost their jobs; but only 13% of those above $100,000 (Keckley 2020). One notable behavior seen during this period related to companies moving employees to short term disability as a result of their illness (furlough) rather than unemployment benefits which must be re-imbursed to the state (in general). This provides a potential 2+ year benefit for employees at no cost (since premiums had already been paid) rather than weeks or months of state-sponsored unemployment. The recent designation of the "long-COVID" syndrome as a disability (Wagner, 2021) requiring assessment for long-term disability will have catastrophic effects on insurers, one company's reported income fell by 90% from December 2019 to December 2020 (Cigna, 2021). The delivery of timely stimulus payments and unemployment benefits, while alleviating immediate cash-flow problems, has created inflationary pressure. Businesses struggle to hire at pre-pandemic wages, price inflation has not replaced the loss of demand, and employees demand higher wages to maintain a pre-pandemic standard of living.

Work and the Workplace

For the first time in a millennium since agrarian societies were disrupted by the industrial revolution, work became separated from the workplace. The workplace, which over that era developed its own identity, its own culture and set of rules which persisted with the movement from primary to secondary and then to tertiary industries. Specifically, over the past 40 years while the technology boom introduced the information age, the workplace remained as the center for productivity; even if that production was "out-sourced" to overseas workplaces with their own different sets of rules, but expected to follow the parent culture, enforced generally through infamous QA and QC standards, worklists and checklists found in binders full of policy documents. Much import over the past decade by companies like Adobe, Google, Microsoft, Apple, and Facebook has been given to office environments that are both environmentally sound, foster teams and cohesion while simultaneously reducing stress through work-out facilities and made to order kitchens staffed by gourmet chefs. Business administration courses studied corporate culture, evaluated serendipity, teamwork, soft and hard power as means for productivity enhancement.

Employers in office settings have spent decades determining the optimum arrangement of employees to improve productivity; cubicle size was carefully reduced by square inches, computer screens, chair height and other ergonomics carefully honed by high priced consultants promising value. Cubicles became open concept, yoga balls for chairs and gyms and kitchens for lunch time creativity periods abounded amongst "new tech" and became a major recruiting point for new employees.

In a single month this was completely undone, employees relegated to their sofa, kitchen table, even a bed table as their new office space. Dress codes and casual Fridays became a thing of the past; the sales of "leisure wear" skyrocketed. The toxicity inherent in certain corporate cultures came to light during movements such as "Black lives Matter" and "Me Too" as COVID came into existence, and contemporary society was forced to examine in public forums the private "rules" which controlled clubby closed industries like "Hollywood USA", Google, Microsoft, and Apple. Crash courses on inclusion and boundaries became synonymous with the close of 2019.

The form and function of the contemporary workplace becomes important as we reassess how and why we return to a workplace, what do we retain and what do we retrain. A simple classic dichotomy would suggest that the extroverts are much more likely to want to return to the workplace than introverts. Recent studies have shown significant resistance to the return to the prior corporate structures, power-sharing arrangements and traditional measures of productivity and compensation (scientific America). How do power-sharing and hierarchy occur in a remote environment, two to three days per week, versus full time office? Is productivity going to be measured separately from the social aspects of career development, characterized as teamwork, character building etc?

Work and Workflow

Best practices, 6-Sigma and other efficiency tools evaluated decades of process to streamline workflow and introduce just-in-time principles not just for manufacturing, but increasingly to human interactions as seen for example in check-list medicine typical of surgical care flows, or claims-management software systems controlling information flow. The ability of these systems to work on remote "desktops" coupled with high-speed internet services has minimized the disruption inherent in complex processes involving multiple brokers when working remotely. When productivity is measured, it now must be more able to separate individual work product from workflow, the end-result becomes a much more complicated metric.

Measuring Work

Historically in industry, the end-result became the normative measurement of productivity. As processes became more complex, the individual contribution became more difficult to measure, but the number of Tesla's leaving the production line remains the reported measure, for example. Likewise in the service industry, including medicine, every service has been allocated a RVU or relative value unit, which integrates the average time taken, the skill-set used, the malpractice exposure for each and every one of more than 10,000 procedures or interactions. Once the system was codified, discounting became the norm; salaries were dependent on the total RVU's generated by a provider.

The separation of work from the workplace has had impact when considering how to measure work, perhaps for the better. Contemporary measures of work related to time-in, time-out of the workplace. The transition to home forced organizations to re-consider measures of productivity. One of the unexpected benefits for many companies, was an increase in productivity of up to 40% from remote workers; with the simultaneous realization that businesses lose more than $500 billion yearly because of workplace distractions. Increased productivity does not appear to have been mitigated by home distractions and remote employees report better mental health (Brenan, 2021) and an average savings of $4000.00 per year compared to in-office workers. Recent claims that home-based call centers were being remotely monitored and time-based alarms inserted into calls at intervals corresponding to pre-determined allotments of time represent a new intrusion by companies into the home which will need to be rationalized. This literally brings the workplace stress into the home, and the long-term consequences of this will need to be addressed.

Workplace Stress

The pandemic by separating work from the workplace has allowed the independent contributions of working and the workplace to perceived stress. Workplace stress clearly affects individuals differently, but is somewhat measurable across and organization, the diversity of situations represented by "remote work" cannot be so easily assessed. Claims of disability caused by work stress are more difficult to assess when working remotely when the stress of a workplace culture is replaced by stressors in the home.

After illness, a return to work becomes a two-part process - a return to the workflow and a return to the workplace. If one works remotely, and the location (generally home) causes stress, can the employee claim workers compensation or disability for that stress. Control of the workplace allows harmful behaviors to be addressed in office situations, such control is lost when workers are remote. How is stress transferred at a distance when 75% of communication is non-verbal?

Remote Work

The immediate COVID work related issues revolved around the rapid dispersion of a previously collective workforce to remote sites, (generally home), either because of exposure and/or infection or risk of infection, even though a year later the absolute risk of infection remains unquantified (Courtney, 2021). Countries, states, and cities went into lockdown disrupting movement of goods, services, and people. The number of people working at home full time went from 17% in 2019 to 66% or two thirds of the US workforce estimated to number 100 million people. Most have not returned as of August 2021, and some 36 million are expected to continue to work remotely for the foreseeable future (Flexjobs, 2019).

The movement of the workforce from complex instant social interactions to virtual interactions, while aided by technology, exposed both technological limitations, uneven distribution of technology and the modes by which those virtual interactions took place. It also required a re-evaluation of the term inclusion in the workplace; perhaps for the better. What are the consequences of social isolation at work when combined with the additional pressures of being at home on a constant basis "locked-down", sometimes for weeks?

Movement of the workforce to home changed, perhaps permanently, the meaning of terms like office, collegiality, serendipity, the watercooler, and the other myriad of social interactions at work which have been studied over the past 50 years, published in many books, and which form the staple of MBA programs worldwide. What has happened to the office power structure, how do anecdotal learning, informal learning and the social interactions which form company culture occur in a virtual world. Moving the workforce to home, also changed the meanings of key terms like work/life balance and parenting in unexpected ways as shown in Figure 1.

Figure 1.

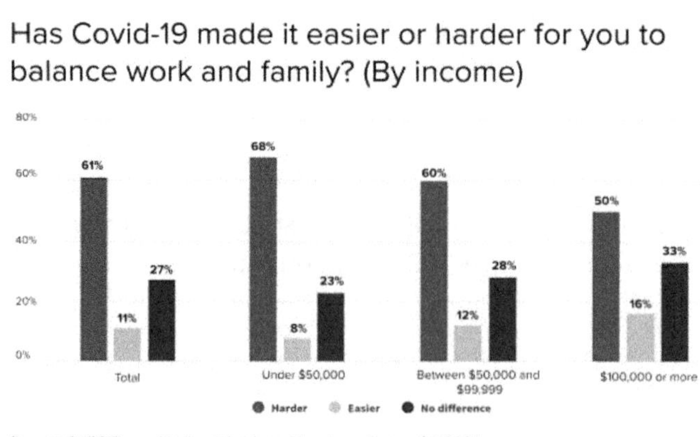

On a global scale, remote working has forever changed the meaning of commuting and overtime. One silver lining of the pandemic is that it necessitated human behavioral changes that have led to slowed deforestation rates, reduced air pollution, and improved water quality all over the world. This positive environmental impact is due, in part, to reducing traffic congestion and air pollution from commuting. When 3.9 million employees work from home at least half time, they reduce greenhouse gas emissions by the equivalent of taking more than 600,000 cars off the road for an entire year. Considering exp that 1 in 4 Americans are expected to work remotely in 2021 (approximately 39 million), that number jumps to 6,000,000 cars (Ioannou, 2021). With an estimated 13 to 27 million people working from home in coming years, remote work could reduce commuting miles by 70 to 140 billion every year (Automotives New Reality, 2020). The alteration of traffic patterns, reduced road use, maintenance costs and need for new construction fundamentally alter the American landscape, and have already cause significant migration patterns from large, more expensive cities to smaller cities with a resurgence of rural America. The proposed infrastructure initiative to provide widespread broadband and/or satellite internet will further incentivize population dispersion. Work no longer becomes the overriding factor on determining geographic location. By making environmentally sound choices—like opting to use less printer paper and monitoring their air conditioning, heating, and lighting—remote workers have the same potential impact on air quality as planting an entire forest of 91 million trees (Flexjobs, 2021).

These changes have significant impact immediately on the need for office space, and change the demand for ancillary services, restaurants and even the bar for after work meetings. Long term this trend will produce meaningful impact on how we view the need for and use of cities in general.

Lastly corporate and individual expenditure has changed significantly, it is unclear when the corresponding taxation rules will change to reflect the change in burden of routine daily expenses to the individual. The current taxation rules do not permit employees from claiming the expenses inherent in working remotely. Some employers are formulating re-imbursement strategies.

There are downsides to remote work, increased isolation, and overwork. Other risks are quoted as visibility anxiety, which may produce career development difficulties. The lack of structure may contribute to burnout. Isolation, including the remote hiring of new employees without face-to-face interviews, may miss subtle clues important in team building. How do non-verbal aspects of team empowerment translate to e-mail or Zoom? The ramifications of what is printed are quite differently codified than casual or business conversation, leading to enhanced use of the words “retraction, walk-back and gas-lighting”.

RETURN TO THE WORKPLACE

Why Return?

A significant number of executives (68%) say employees should be in the office at least three days a week to maintain a distinct company culture. No scientific validation is produced for this position; perhaps it is because 3/5 workdays is the number closest to 68% of executives. It is entirely likely that this represents the best "business as usual" scenario since the previous company culture benefits executives primarily; given the evolution of both the company, and its executives. Does this culture routinely benefit the employees? The pandemic created a window into this question since executives were left with no option but to disperse their employees. Given the issues with organizational culture present in the year leading up to the pandemic exemplified by the necessity for race, gender, sensitivity and inclusivity training, one wonders if that "culture" is something most wish to return to. The improvements in job satisfaction, mental health and productivity amongst remote workers suggest otherwise. The resistance of employees to return to even the most elaborate and fluid work environments such as Google and Apple are also supportive.

Return to Workplace Guidelines

Guidelines for resumption of the workplace remain fluid, subject to both national and state mandates, but mostly the personal interpretation of the CDC guidelines by human resource managers. The fluidity of even the CDC guidelines which are issued on a weekly basis, has made it difficult to define policy much less enforce it. The return to work has become an ideological battleground in terrain devoid of scientific landmarks. The lack of cohesive science disrupted belief systems and prevented a coherent message from being delivered. The internet provided both instantaneous information about conditions in other countries; it also provided conflicting policies from both current and former administrations in the US, and vastly different interpretations of data and policy at the individual level. A classic example is that American airlines in interpreting the CDC national mask policy states gaiter use as non-compliance, even though they were acceptable at every airport in the US. The pre-occupation in the US with "fake news" for the past 4 years was a distinct barrier, not only to infection control, but more recently facilitated vaccination resistance. These controversies set up an adversarial stance with employers who demand vaccination prior to return to work with the legality that no entity can force any human to undergo experimental treatments (since no vaccine has yet received full FDA approval), dating back to Nuremberg.

The constantly evolving nature of RNA, specifically COVID-19, the development of mutant variants (currently 12) which may not be neutralized by the current vaccines, together with population vaccine resistance leads to a lack of herd immunity will likely allow the virus to become endemic with cyclic outbreaks of COVID illness. The management of time off work, and the guidelines regarding return to work after Illness, Infection or Exposure have rationalized in the form of CDC guidelines.

The large remaining issue is how to manage the fear that those vaccinated may still be become infected, they may still transmit virus and how to handle employee refuse of vaccination based on its relative in-effectiveness. The fact that the current vaccines are only approved for emergency use limits mandates which may be enforced.

What Happens if You Cannot Return to Work

Apart from the social and psychological factors associated with a return to work, the persistence of symptoms from COVID infection may physically limit the individual in performing workplace activities. It would be highly unlikely that someone would be required to be physically active working remotely, although there are reports of someone claiming disability because they were unable to assemble a fighter jet in their living room. By definition, work from home is likely to be sedentary.

The so-called long-COVID syndrome is a constellation of symptoms which persist for weeks to months after infection. Many of these non-specific symptoms such as headache, fatigue, "brain-fog" postural tachycardia, myalgias preceded evidence of infection and are also common to other non-specific disorders such as Lyme disease, chronic fatigue syndrome, POTS, toxic mold, and others. Further complicating the assessment is that many of the symptoms relate to psychological disorders such as depression and anxiety, PTSD etc., conditions that were worsened by the socio-economic consequences of the pandemic, including social isolation, and are not specifically related to the infection. The CDC has published a comprehensive set of assessment tests and tools to gain a better understanding of this syndrome. The reported incidence of this syndrome in 5-50% of patients after infection suggests a large selection bias. In a UK study named the Great British Intelligence Test, some 81,000 people responded but only 518 respondents identified as having had COVID infection, a remarkably small number given that the Johns Hopkins website states the incidence of COVID infection in the UK was around 10%. Clearly there is much to learn about this syndrome, the potential overlap with other post-viral disorders, and the apparent positive impact of vaccination on the symptoms. It must be realized that more than just a damaging infection, SARS-CoV2 and specifically its covering spike protein are functionally active molecules that signal processes in the body that are only now becoming clear. The production of pieces of this spike protein from the mRNA vaccinations (Moderna and Pfizer), viral Trojan Horses (Astra and

Johnson and Johnson) or delivered directly (Novovax) may have biologic effects of their own, represented as the symptoms associated with vaccination.

Clearly long-hauler symptoms would have to be severe to preclude the sedentary type activities inherent in remote work. The CDC has presented methods (including on-line tools) as a way to objectify these deficits and measure recovery. Clinics, websites, and support groups abound to provide a forum for assessment and to evaluation potential treatments. No specific treatment has appeared at this point.

Long Term Remote Work

The shift to remote work will challenge worker's compensation laws to cover injuries at home, although as a sedentary employee, this ought to be a rare event. The laws around taxation and claims for compensation/re-imbursement for use of the home will need to be clarified because the costs of operating at home are not deductible for employees (Bawden-Davis, 2020).

FINAL THOUGHTS

The COVID-19 pandemic has challenged long held beliefs, and resultant workplace cultures, that have withstood generations of change, the information age, 9/11, and the 2008 "depression" in ways not previously imagined. The information age both informed the calamity and provided the connectivity which allowed individual and collective productivity to survive. The dispersion of the workforce as a result of the medical response, also caused socio-economic change on a vast scale with dependent effects on mental health, and the work-life balance. Remote work became more than work from home, it became work-at-home with the entire family, sometimes with stranded guests to diversify the cultural experience.

REFERENCES

Atul Gawande. (n.d.). *The checklist manifesto*. http://atulgawande.com/book/the-checklist-manifesto/

Bawden-Davis, J. (2020). How does working from HOME change tax deductions in 2020? In *Business Class: Trends and Insights*. American Express. https://www.americanexpress.com/en-us/business/trends-and-insights/articles/how-does-working-from-home-change-tax-deductions-in-2020/

Brenan, M. (2021, August 16). *COVID-19 and remote work: An update*. https://news.gallup.com/poll/321800/covid-remote-work-update.aspx

Centers for Disease Control and Prevention. (n.d.). *Post-covid conditions*. Centers for Disease Control and Prevention. https://www.cdc.gov/coronavirus/2019-ncov/long-term-effects.html

Cigna. (2021, February 4). *Cigna reports fourth quarter and full year 2020 Results, expects continued revenue and earnings growth in 2021*. https://www.prnewswire.com/news-releases/cigna-reports-fourth-quarter-and-full-year-2020-results-expects-continued-revenue-and-earnings-growth-in-2021-301221754.html

Courtney, E. (2021, April 29). *Remote work Statistics: Navigating the new Normal: FlexJobs*. FlexJobs Job Search Tips and Blog. https://www.flexjobs.com/blog/post/remote-work-statistics/#:~:text=%E2%80%93%20Remote%20Work%20Is%20Here%20to,be%20working%20remotely%20by%202025

Flexjobs. (2019, December 6). *How remote work supports the fight against climate change: Flexjobs*. FlexJobs Job Search Tips and Blog. https://www.flexjobs.com/blog/post/how-remote-work-supports-un-sustainable-development-goal-no-13-of-combating-climate-change/

Flexjobs. (2021, June 10). *The environmental impacts of remote work: Stats and benefits: Flexjobs*. FlexJobs Job Search Tips and Blog. https://www.flexjobs.com/blog/post/telecommuting-sustainability-how-telecommuting-is-a-green-job/

he United States Department of Justice. (2017, May 31). *Electronic health RECORDS vendor to pay $155 million to Settle false Claims Act Allegations*. https://www.justice.gov/opa/pr/electronic-health-records-vendor-pay-155-million-settle-false-claims-act-allegations

IBM. (n.d.). *Healthcare provider solutions: Watson Health*. https://www.ibm.com/watson-health/solutions/healthcare-provider

Ioannou, L. (2021, February 6). *1 in 4 Americans will be working remotely in 2021, Upwork survey reveals*. CNBC. https://www.cnbc.com/2020/12/15/one-in-four-americans-will-be-working-remotely-in-2021-survey.html

Johns Hopkins Coronavirus Resource Center. (n.d.). *COVID-19 map*. https://coronavirus.jhu.edu/map.html

kpmg.com. (2020). *Automotives New Reality*. https://assets.kpmg/content/dam/kpmg/br/pdf/2020/09/automotives-new-reality.pdf.

Lopez Bernal & Shimabukuro. (2020, June 18). Retraction: Cardiovascular disease, drug therapy, and mortality In Covid-19. *New England Journal of Medicine*. https://www.nejm.org/doi/full/10.1056/nejmc2021225

Morse, S. (n.d.). *457 hospitals settle for $250 million over IMPROPER implantable cardioverter Defibrillator surgeries*. Healthcare Finance News. https://www.healthcarefinancenews.com/news/457-hospitals-settle-250-million-over-improper-implantable-cardioverter-defibrillator-surgeries#:~:text=than%20%24250%20million.-,The%20biggest%20hospital%20chain%20settlements%20are%20HCA%2C%20at%20%2415.8%20million,cardioverter%20defibrillator%20cost%20about%20%2425%2C000.&text=The%20whistleblowers%20have%20received%20more%20than%20%2438%20million%20from%20the%20settlements

Paul Keckley. (2020, May 18). *COVID-19 unintended Consequences: Unflattering attention to the gap between "haves" and "have nots"*. Paul Keckley. https://www.paulkeckley.com/the-keckley-report/2020/5/18/covid-19-unintended-consequences-unflattering-attention-to-the-gap-between-haves-and-have-nots

Schulte, F. (2021, May 17). *Humana health PLAN OVERCHARGED Medicare by nearly $200 Million, federal audit finds*. Kaiser Health News. https://khn.org/news/article/humana-health-plan-overcharged-medicare-by-nearly-200-million-federal-audit-finds/

University of Oxford. (n.d.). *Social media manipulation by political actors an industrial scale problem*. https://www.ox.ac.uk/news/2021-01-13-social-media-manipulation-political-actors-industrial-scale-problem-oxford-report

Wagner, J. (2021, July 26). Biden says long-term effects of Covid-19 can be considered a disability under federal civil rights laws. *The Washington Post*. https://www.washingtonpost.com/politics/biden-ada-long-covid-disability/2021/07/26/972f2a04-ee20-11eb-a452-4da5fe48582d_story.html

Chapter 4
From In-Person to Online Teaching:
Experiences of Secondary Educators During the COVID-19 Pandemic

Douglas Graham Fenderson
Midwestern State University, USA

ABSTRACT

Restrictions from the COVID-19 pandemic forced many teachers across the United States to teach their students remotely using online teaching strategies. Analyzing teachers' perceptions and expectations of online teaching before and during the pandemic help stakeholders understand how to better handle the challenges of online learning. The literature review examines the traditional differences between teaching online and in-person courses, challenges faced when teaching online courses, and teacher perceptions of online learning. The survey method was used to collect data on the experiences of online teaching before and during the pandemic from ninth through twelfth grade teachers in a North Texas school district. The study results show that factors such as a teacher age or years of experience had less influence determining if they were prepared for online teaching. Rather, factors like a teacher school campus, prior experience, and access to support structures correlated more to teachers having a high level of preparedness for online teaching.

DOI: 10.4018/978-1-7998-8028-8.ch004

INTRODUCTION

Background of the Problem

The unexpected and rapid development of the COVID-19 pandemic during 2020 has changed many daily routines and norms, including education. Schools were greatly affected by the pandemic, impacting 1.58 billion learners around the world (United Nations, 2020). School districts rapidly shuttered in-person classes and quickly transitioned to offering online-only courses. While DiRienzo and Lilly (2014) concluded that the method of instruction, either in-person or online, does not affect student learning outcomes, the participants in their study were tenured teachers who were adequately prepared for online teaching. Gudmundsdottir and Hathaway (2020) concluded that the majority of teachers during the pandemic were not adequately prepared; many teachers were forced to change their method of instruction without much prior experience or training.

Problem Statement

As schools are continuing to navigate between offering in-person, online, and hybrid courses, it is important to examine teacher experiences to inform how changes to the learning environment currently affects teachers' experiences in the classroom. Examining the experience of teachers transitioning from in-person to online teaching can provide educators and administrators with research-based data leading to the development of new training courses designed for online teaching (De Paepe, Zhu, & Depryck, 2018; Gudmundsdottir & Hathaway, 2020). These insights can also help develop effective training for new educators who are confronting the current educational environment.

Purpose of the Study

The purpose of this study will be to understand the experiences of traditional in-person secondary educators transitioning to teaching online classes in North Texas school districts through collecting their responses to open-ended questions.

LITERATURE REVIEW

This literature review will discuss differences between online and in-person teaching, along with the theory, challenges, and teachers' pre-pandemic perceptions of online learning.

Comparing Online and In-Person Teaching

Traditionally, all formal learning has been conducted in face-to-face environments, where the instructor teaches the learners through a variety of strategies: lectures, group work, peer collaboration, etc. Then, with the introduction of technology and telecommunication in education, both public and charter schools began offering more online options for courses (Anderson, 2008). For example, some schools have created hybrid, or "blended," courses, utilizing a mixture of face-to-face instruction with online learning (Brodersen & Melluzzo, 2017). However due to the pandemic, many schools closed their physical campuses and only offered online learning options; face-to-face and blended classes were not an option for students (United Nations, 2020).

To understand the differences between online and in-person learning, it is critical to clearly define and identify differences between the learning environments. Outdated ideas and generalizations of online learning exist; Brown and Liedholm (2002) asserted that online courses were easier and students lacked motivation in their courses. However, in a more recent study, DiRienzo and Lilly (2014) concluded that the learning environment did not significantly alter student-learning outcomes. Furthermore, they concluded no discernible difference between student motivation and perceived course difficulty in online and in-person courses. According to the authors, there was no strong correlation between student learning outcomes and the learning environment, either in-person or online. More specifically, Robinson, Kilgore, and Warren, (2017) found no strong correlation between student learning outcomes and the learning environments as long as the instructor structured the course well, communicated effectively, and delivered the content in a clear manner. Knowing this, it is important to examine the various characteristics teachers need to implement in online courses to ensure student success.

The Theory of Online Teaching

Since researchers (DiRienzo & Lilly, 2014; Robinson et al., 2017) have shown that the instructor is the greatest factor in student success regardless of the type of learning environment, it hinges upon experienced teachers to create and organize the content of their courses. Garrison, Anderson, and Archer, (1999) created one of the first fundamental structures that outlines how teachers should structure an online learning environment, see Figure A1. Effective educational experiences occurred when teachers selected content with which students engaged, teachers and students effectively communicated together in a safe online learning environment, and students collaborated on learning together. With continued improvement in technology and best practices over the next decade, Anderson (2008) updated the characteristics of an effective online learning environment (see Figure A2). Anderson improved upon the first model by highlighting the responsibilities for each member of the learning community. For example, his model represents how students engage with the content through simulations, which are organized by the teacher. Results from both studies demonstrate how online learning is most effective when teachers and students engage with each other through the content and communication. Recent studies (Robinson et al., 2017; Park & Kim, 2020) continue to expand on these ideas by analyzing how teachers who provided opportunities for students to have meaningful online collaboration significantly improved student experiences in online courses. These frameworks are theoretical models; in practice, instructors face many challenges when creating online courses.

Challenges of Online Learning

In spite of the increase of online programs over time and improvements in the theory of online teaching, several challenges continue to hinder online teachers: time, technology, and communication (Capdeferro & Romero, 2012; Robinson, et al., 2017; Kewal Ramani et al., 2018; De Paepe et al., 2018). Although these factors challenge in-person teachers as well, these challenges are magnified through distance and asynchronous working time. For example, one research participant in Pramling Samuelsson, Wagner, and Ødegaard's (2020) study noted that it took them more time to establish classroom management with their online students than it would take them in an in-person setting. During virtual calls, direct instruction took more time because each student faced their own individual distractions in their homes, in comparison to the teacher being able to mitigate all distractions in one physical classroom. In this one example, the remote learning distance intensified problems around time, technology, and communication.

Time. For instructors teaching online courses, researchers (Robinson et al., 2017; Martin, Budhrani, & Wang, 2019; Pramling Samuelsson et al., 2020) identified one challenge as the amount of time needed to plan the structure of their courses. One research participant specifically noted that online teaching required more prior planning than in-person teaching (Pramling Samuelsson et al., 2020). Even after teachers completed the design and implementation of a course, they needed time to further refine and improve it (Martin et al., 2019). Online courses required even more time to grade assignments, especially those that required student writing. Due to these increased demands in online instruction, teachers must manage their own time well. Although these time demands for online instructors are similar in some ways to in-person teaching, instructors who are not proficient with online tools require more time to complete tasks, such as planning and grading.

Additionally, teachers must continue to invest their time into personal development (De Paepe et al., 2018). Martin, et al., (2019) found that the most successful online courses were the ones in which teachers were familiar with and utilized the latest online tools and teaching trends. Therefore, as new online tools are developed and online teaching practices are refined, it is imperative for teachers to be up to date on the latest advancements in their field.

Technology. Teachers must have the necessary technological skills to be able teach effectively online. Online teachers need a plethora of skills to navigate online courses. A few of these required skills include: using Web 2.0 tools, video recording and editing, troubleshooting technology problems, designing online course in learning management systems (LMS) and giving online feedback (De Paepe et al., 2018; Martin et al., 2019). Teachers who lack these skills faced a challenge in creating an online learning environment that students can easily navigate (Park & Kim, 2020).

Another technological challenge for educators is ensuring all their students have access to the necessary technology. Elizabeth, a participant in the Robinson et al. (2017) study, reflected that "as much as we like to believe they (students) are placed on a level playing field, the bottom line is not everybody is" (p. 40). As recently as 2015, one group of researchers found that 19% of students did not have reliable Internet access at home, due mostly to expensive costs and the choice to not have it (Kewal Ramani et al., 2018). This technological divide during the pandemic has led to students not engaging in their courses and falling behind in their assignments (Pramling Samuelsson et al., 2020). This often results in teachers spending more time with each individual student, ensuring they understand the content and assignments. Therefore, this gap in accessible technology for students increases the amount of work for teachers.

Communication. Online learning creates new challenges for teachers to communicate with their students, including discussing classroom rules and procedures, assignment instructions, and giving feedback (Capdeferro & Romero, 2012; De Paepe et al., 2018; Martin et al., 2019). Although Capdeferro and Romero (2012) focused on student frustrations with online learning, their results pointed to lack of communication from instructors as the source of these student frustrations. The participants reported that if teachers would have provided clearer instructions and feedback, many of their frustrations would have been mitigated. De Paepe et al., (2018) concluded that frequent communication over email, a direct messaging system, and synchronous virtual meetings positively improved teacher interactions with students. These communication tools and strategies are essential because without students physically present in the classroom, educators must communicate in an adequate, timely, and prompt manner (Martin et al., 2019). Teachers who proactively and effectively communicated to their students, reduced the challenge of communication in an online learning environment.

Teacher Perceptions of Online Learning

After understanding the main challenges of online learning for teachers, analyzing teachers' perceptions, how teachers view their own personal experiences in the classroom, is beneficial because their perceptions play "a major role in how faculty approach online teaching goals, tasks, and challenges" (Martin, Budhrani, & Wang, 2019, p. 111). Not only does understanding teacher perceptions help teachers improve their own craft, but teacher perceptions impact instructional design and best practices in the classroom (Pajares, 1992). An awareness of these perceptions allows researchers to better understand educators' misunderstandings in online teaching and inform educators how to better prepare themselves for online teaching (Gudmundsdottir & Hathaway, 2020; De Paepe et al., 2018). Teacher perceptions of readiness and technology skills are two of the most influential factors in determining educator success (Martin et al., 2019; Santelli, Stewart, & Mandernach, 2020).

Readiness

At the beginning of the pandemic, many teachers were not adequately prepared to teach online courses; Gudmundsdottir and Hathaway (2020) reported that 92% of US teachers had no prior online teaching experience. Despite the overall inexperience, teachers demonstrated positive attitudes regarding online learning. Attitude, along with ability, constituted educators' readiness for online teaching (Martin et al., 2019). Instructors with low perceptions of readiness in online teaching were less adept at promptly giving online feedback, communicating with students, and at designing online learning activities in their LMS. Overall, this resulted in a teachers' poor attitudes and low ability for online teaching. Therefore, teachers must be properly trained and equipped with the necessary skills to teach online courses. Personal development and other informal learning are keys to developing educator confidence and ability in teaching online courses (Martin et al., 2019; Gudmundsdottir & Hathaway, 2020).

Technology Tools

Next, Santelli et al., (2020) found that teachers' perception to effectively use technology tools, such as the LMS, directly enhanced teachers' experience with online learning. Tools that were perceived as easier to use by faculty, such as online grading and feedback tools, positively impacted faculty perceptions of online learning because these tools made time consuming tasks more efficient. Additionally, if school administration or technology faculty assisted educators with technology problems, teachers were more likely to confidently use technology tools (De Paepe et al., 2018).

SUMMARY

This literature review analyzes the theory, challenges, and teachers' perceptions of online learning. Most of the discussed research has been conducted in an environment where students and teachers were able to choose online learning environments, unlike the early stages of the COVID-19 pandemic. Despite this difference, researchers and educators can take the existing research regarding online learning to understand these best practices and how to apply them to the classroom. Unfortunately, despite roughly two decades of research regarding online learning, teachers were largely unprepared to transition to online teaching when schools across the nation forced all teachers and students into remote learning during the pandemic (Gudmundsdottir & Hathaway, 2020; Pramling Samuelsson et al., 2020). It is important to further study the extent to which educators were prepared for online teaching and if they implemented these best practices. The aim of this study is to examine teacher experiences teaching online courses during the pandemic. This research into teacher experiences will inform educators and administrators on how to effectively train and equip new teachers for the current educational environment. Therefore, the research question is: what are the experiences of traditional in-person secondary educators transitioning to teaching online classes in North Texas school districts?

METHODOLOGY

The research uses a qualitative approach to understand educators' vast and complex experiences during the COVID-19 pandemic. The variety of educators' professional and personal experiences has led to a myriad of responses to the pandemic and varying degrees of preparedness to teach online. In this research study, a constructivist worldview will build a narrative around participants' complex experiences; educators' social, cultural, and demographic factors shape their interpretations of their profession and the pandemic. The phenomenological research design will make connections between their experiences of online teaching before and during the pandemic through digital survey data. When appropriate, the researcher will interject their own experiences regarding teaching online courses during the pandemic.

Participants

The researcher collected data after receiving an approval from a large North Texas school district's research committee. The district employs about 3,600 teachers across five high schools, five ninth or ninth/tenth grade campuses, fifteen middle schools, forty-one elementary schools, and eleven additional campuses such as career centers. This research only gathered survey data from the teachers who taught courses for ninth and twelfth grade students. Unlike the diverse ethnic distribution of students, 39% white, 31% Hispanic, 15% Asian, and 11% African American, teachers in the district do not reflect the same level of diversity: 78% white, 12% Hispanic, 2% Asian, and 6% African American. Analyzing gender, 80% of teachers in the district are female and 20% are male. Teachers in the district reflect diversity in their years of experience. While less than 3% of teachers were in their first year of teaching, 23% of teachers had completed 1-5 years, 21% of teachers completed 6-10 years, 33% had completed 11-20 years, and 20% of teachers had over twenty years of experience.

Methods

Phenomenological Design

The phenomenological approach investigates "the lived experiences of individuals around a singular phenomenon as described by the participants" (Creswell & Creswell, 2018, p. 13). In this qualitative research study, the phenomenological design explores educators' experiences with teaching online courses before and during the COVID-19 pandemic. A phenomenological approach is the best method of research to study an area where not much research exists (Donalek, 2004). Very few persons living today have ever experienced a global pandemic. Additionally, the pandemic is an ongoing reality for the researcher and participants, meaning our educational environment and perceptions are continuing to shift during this time.

The Researcher's Role

My experience heavily factors into the design of this qualitative research question. Donalek (2004, p. 516) argues that "research is not truly phenomenological unless the researcher's beliefs are incorporated into the data analysis." As a high school teacher, I have been affected by my district's decision to move to a completely online learning environment. I had never taught online courses or received personal development specifically for teaching online courses prior to the pandemic. My personal experiences influenced my curiosity to discover whether or not other educators have had similar experiences. If, in fact, educators' have had different experiences, I am collecting demographic data to determine if there are any common factors that influenced those differing experiences. Was my experience of transitioning to teaching online as a 20-29, white, married, male with no kids different from the experiences of an educator with different demographics? Or was the pandemic a historical moment that impacted all educators similarly, despite demographic differences?

Data Collection and Recording

Data will be collected from secondary teachers within my own school district, a pool of over 1,000 participants, between February and March of 2021. Before data collection can begin, approval must be secured from a university Institutional Review Board (IRB) and the school district's own research committee. A "Request to Conduct Research" application will be submitted to the school district's research committee, which will include the research proposal, the IRB approval letter, and the data collection instrument, Google Forms. Upon approval, the district will send out an email to all applicable participants, over 1,000 secondary educators, to take part in the voluntary research study. Below the district's communication, the researcher will add an introduction letter requesting teachers to participate in the voluntary research study and linking to the Google Forms questionnaire (Appendix B). The responses are anonymous to ensure the confidentiality of participants' identities. The form will not collect personal identifiers such as name, employee ID number, or email address.

The survey is divided into four main sections. In the first section, participants will answer ten demographic questions in order to analyze whether strong trends within key demographic factors shaped their experiences. Then, participants will answer up to eight open-ended and five closed-ended questions. These questions are divided into three sections: teacher experience prior to remote/online learning, teacher experience during spring 2020 when the school district forced all teachers to remote learning, and teacher experiences during the 2020-2021 school year if they taught at least one online course (see Appendix C). The survey might take anywhere from 15 to 45 minutes depending on the complexity of their responses. Creswell and Creswell (2018) identify the advantage of collecting data through digital interview strategy is that it provides an opportunity for the participants to provide background information to the research topic. Therefore, the questionnaire includes three questions asking participants to add any additional information they deem relevant to teaching online courses during the pandemic.

Data Analysis Procedures and Interpretation

To organize and prepare the data for analysis, the Google Form will transfer and store all data into a Google Sheet. To analyzing the data in the Google Sheet, the research will add new columns to code the responses into broad categories and themes. An advantage for using digital data collection is searching for keywords to help code the responses. Some expected categories include educators with positive or negative experiences teaching online, educators who did and did not take advantage of district resources for transitioning to teaching online, and educators with and without prior online teaching experience. By coding the responses in these themes, the data can be filtered based on the demographic information provided by the participants to identify diverse perspectives. This coding will happen simultaneously with data collection.

The interpretation of the data will include a personal view of the findings, as well as comparing teacher perceptions of online teaching from the literature review. Although there may be many novel experiences of online teaching during the pandemic, some pre-existing problems with online teaching might resurface in the participants' responses like challenges around time, technology, and communication. Last, this research study will most likely raise more questions regarding education and the COVID-19 pandemic, as it remains an ongoing crisis in April 2021 with little clarity on how schools will handle virtual learning next semester. Hopefully, this research can shed some light on what educators, administrators, and policymakers can do to better equip and meet the needs of educators transitioning to online teaching.

Validity and Reliability

In order for the findings to be valid, the researcher will clarify their bias, spend prolonged time in the field, and employ peer debriefing (Creswell & Creswell, 2018). The researcher is currently teaching online courses in their district and have not taught a face-to-face course since the pandemic began. He served as his school campus' virtual teacher liaison in the fall of 2020, communicating district expectations with all online teachers at the campus and providing real-time feedback to virtual administrators. Last, a curriculum coordinator within our district has generously agreed to debrief the data and findings ensuring that it makes sense to a wider audience. His input will be valuable, as he is not eligible for the study as an administrator.

To guarantee qualitative reliability, the Google Form will only be accessible to participants with a district email address, though the form will not collect their email. The Google Form questionnaire will also allow consistency and structure to the interview process, ensuring that each participant answers the same questions. During the analysis of data, extensive note keeping regarding the meanings of the codes will ensure there is little to no drift in their definitions.

Reporting the Findings

The outcomes of this research study will be presented in a narrative format, providing detailed descriptions of the experience of the participants (Creswell & Creswell, 2018). The narrative will analyze the different educators' varying experiences of teaching online courses, analyzing possible demographic factors for this difference in experience. First-person and collective pronouns will be necessary for the researcher to embed their own experiences alongside the experiences of their colleagues in the narration.

Research Questions

What are the experiences of traditional in-person secondary educators transitioning to teaching online classes in North Texas school districts?

Subquestions

- What factors prepared educators to teach online courses?
- What factors caused educators to be unprepared for teaching online courses?
- What demographic factors influenced secondary educators' perception of their own preparedness to teach online courses?

DATA COLLECTION AND ANALYSIS

The data was collected through a Google Form (Appendix C) over a seven-day period; a total of thirty-two participants responded to the survey. Due to the district's rejection of the initial research proposal and other delays, I had limited time to collect data. The district approved the research project on Thursday, April 1 and sent the introduction letter (Appendix B) and survey to all secondary teachers later that afternoon. As a teacher within the district, I directly asked peers and friends to complete the survey.

Analysis

At least one teacher from all but one campus in the district completed the survey, with the largest participation coming from campuses "D" and "E" (see Figure D1). Participants were diverse in their years of educational experience, age, and household income. Participants were fairly homogenous in their gender, race/ethnicity, household, and marital status. Although the participants in the survey did not reflect a diverse group by gender or race/ethnicity, the percentages closely reflect the lack of diversity within the district, as discussed in the Methodology section.

Half of participants expressed they felt unprepared to teach online courses prior to remote learning, about thirty percent were neutral, and about twenty percent felt prepared (see Figure E2). On one hand, five teachers noted that prior experience completing online college courses prepared them for online teaching, and four teachers also noted that their experience implementing the district's LMS system, Canvas, in previous school years prepared them for online teaching. Familiarity with Canvas prior to the pandemic even helped six educators who, overall, felt unprepared to teach online courses. On the other hand, twelve teachers expressed the lack of familiarity with technology and LMS applications as the source of their unpreparedness.

Once the school district decided in March of 2020 to finish the spring semester with remote learning for all teachers and students, a third of respondents felt prepared, while a third felt unprepared (see Figure E3). The two main factors in perceptions of teachers' preparedness during the spring 2020 semester were: prior experiences with online learning, and taking advantage of district trainings for the new technology. One prepared teacher directly acknowledged that "we were all figuring it out as we went." However, for nine prepared and unprepared teachers, issues with new technology tools, such as the district's online communication platform, Webex, remained a problem. Teachers reported feeling unprepared to handle a drop in student participation, as well as restricted with new district curriculum policies for remote learning.

For the 2020-2021 school year, the district rolled out a learning pathway, where parents and students could choose between virtual, a mix of virtual and in-person, or all in-person courses. While teachers reported back to their home campus at the beginning of the fall semester, students did not return until September. Even teachers who only taught in-person courses continued to use online learning, especially when students in the courses had to quarantine from home for several days. However, just four teachers who had only in-person courses completed the survey; twenty-eight of the survey respondents taught online courses in the 2020-2021 school year (see Figure E5). In many cases, teachers volunteered to teach these online courses, which explains why sixty-four percent felt prepared and only fourteen percent responded as unprepared. Teachers noted that new technology tools (a new learning platform, Edgenuity), difficulty providing special education accommodations and modifications, lack of collaboration, and expectations that were too high all contributed to feeling unprepared for teaching online courses. Teachers who responded that they had a high level of preparedness did so because they adapted and learned how to teach online courses along the way.

As seen in the Figures E2, E3, and E5, there was an overall increase in teacher preparedness for online teaching. Twenty-one teachers increased their feeling of preparedness, ten teachers felt that their level of preparedness remained the same over time (of those, five denote a high level of preparedness) and only one participant reported a decrease in their feeling of preparedness teaching online courses. One participant said, "we had been tossed into the deep end by this pandemic and I think that I learned a lot…" By the end of the 2020-2021 academic school year, teachers had mostly positive experiences with online learning because they changed their expectations of student participation, collaborated with other teachers on content and technology tools, and took time to improve their own teaching.

Validity

My own teaching experiences sparked my interest in researching teachers' perceptions and experiences regarding online teaching. Prior to March 2020, I had three-and-a-half years of experience teaching in-person courses. Like some of the participants in this survey, my main exposure to online learning had been taking a few online courses during my undergraduate studies. I was also fortunate enough to complete some personal development in 2019 regarding blended learning, which used similar skills as those required for teaching online courses, like creating online learning content. In 2019, I also took a leadership role in helping train my colleagues on how to use our LMS system. Although I had no direct experience teaching online courses when remote learning began at our district during March 2020, my personal experience helped prepare me to navigate early challenges. Much of my content was already created in our LMS system and I helped create online lessons for the rest of my content team. At the beginning of the 2020-2021 school year, I volunteered to teach online courses and became my campus' virtual teacher liaison. That role required me to attend meetings with the virtual administrators and communicate with all the other virtual teachers at my campus.

This research project was driven by my prolonged time as a teacher in the midst of the COVID-19 pandemic and online teaching. From my experience, I encountered teachers who expressed many different experiences regarding online teaching, and I wanted to better understand what factors caused teachers to have those different experiences. By understanding those differences, I hope to highlight how teachers can be better equipped in the future to handle change in their teaching methods.

DISCUSSION

What are the experiences of traditional in-person secondary educators transitioning to teaching online classes in North Texas school districts?

Despite the diverse backgrounds and experiences, from years of teaching, to levels of education, to age and race, the COVID-19 pandemic and rapid transition from in-person to online teaching "hit [all educators] like a brick to the face." The following subquestions explore which factors caused educators to be prepared or unprepared for online teaching, and which factors affected teachers' experiences teaching online classes.

What factors prepared educators to teach online courses?

When analyzing responses from teachers who rated themselves between 3 and 5 on a scale of preparedness for teaching online courses prior to the pandemic (see Figure E2), two main factors emerged. Prior personal experience with online teaching and access to support structures both prepared teachers to quickly learn and implement online teaching strategies.

Prior personal experience. One-and-a-half years prior to the pandemic, the district implemented Canvas, an LMS, and gradually raised expectations on their staff to use more of its functions and features. A third of all participants, and ten out of the sixteen teachers who felt at least somewhat prepared for online teaching prior to the pandemic, noted that their familiarity with Canvas "helped to make the transition [to online teaching] easier...[they] did not have to learn a new platform at the start of the pandemic." Familiarity and experience with creating online content was a major factor for teachers to feel prepared for teaching online courses, even those who had not formally done so beforehand. Six out of sixteen teachers noted that taking online courses as a student while completing undergraduate and graduate work was another contributing factor for their preparedness for teaching online classes. Last, two respondents had experience designing online curriculum prior to the pandemic. Prior experiences with using online technology tools, completing online courses as a student, and designing online curriculum were all factors that helped teachers be prepared for online teaching heading into the COVID-19 pandemic.

Support Structure. During the pandemic, in both the spring 2020 semester and 2020-2021 school year, teachers reported high levels of preparedness due to having access to and using a variety of support structures. One teacher struggling with the new technology said that "...if not for our amazing librarian, most of us would not have been able to function online." Another educator collaborated with other educators at her school who taught the same content in order to create online lessons. Similarly, many other teachers relied on their school's professional learning communities to create online content and learn how to use the technology tools. However, not all teachers were fortunate enough to have such accessible collaboration; one teacher reached out to an online learning community for support in their subject area because no one else in the district taught a similar course.

What factors caused educators to be unprepared for teaching online courses?

In the survey, various teachers reported that challenges with not having enough time to prepare and learning new technology presented the greatest obstacle to fully preparing for teaching online courses. As noted in the literature review, these challenges are similar to what researchers (Capdeferro & Romero, 2012; Robinson, et al., 2017; Kewal Ramani et al., 2018; De Paepe et al., 2018) found when studying teacher experiences of online teaching prior to the COVID-19 pandemic.

Time. The Friday before teachers and students in the district went on Spring Break in 2020, the superintendent sent a communication that teachers should take home everything they would need to teach remotely. The next week, the district officially decided to finish the school year with remote learning and extended spring break one week for teachers to prepare for the last nine weeks of school. In the survey, five teachers specifically mentioned that the limited time to prepare significantly impacted their experience teaching online. Although these teachers were familiar with the technology and had at least six years of teaching experience, they had never taught an online course before and would have benefitted from having more time to prepare. During this initial phase of online learning, the lack of preparation time was a major factor for teachers. However, by the 2020-2021 school year, most teachers reported that their experience from the spring and time over the summer helped prepare them for teaching online in the fall. Only two teachers mentioned needing more time to prepare and their concern revolved around needing more time to learn how to use the district's newest learning platform, Edgenuity, rather than needing time to learn how to structure and teach an online course like in the previous semester.

Technology. While the challenge of time dissipated, technology remained a constant challenge for teachers. As previously mentioned, the district had previously implemented an LMS, but some teachers did not use Canvas beyond the basics required by the district, and others resisted using this platform altogether. Those who did not take advantage of using Canvas prior to the pandemic had to learn how to take their traditional in-person content and transform it for the online platform. On top of learning how to use Canvas and creating all their lessons online, these teachers also had to learn how to use Webex Teams and Meetings, two new telecommunication tools our district implemented for holding live class meetings. In the spring 2020 semester, nine teachers mentioned the challenge of learning how to use Webex. Learning new technology tools continued into the 2020-2021 school year, where virtual teachers had to learn Edgenuity, a new district mandated technology platform where teachers assign video instructions and assignments to students. This time, almost half, 46% of all teachers responded that learning Edgenuity was challenging. Some respondents wanted Edgenuity trainings earlier in the 2020 summer, or more practice with Edgenuity before implementing it for their students. Other teachers felt restricted by Edgenuity's limitations and, as a result, felt that the system "took the individuality of teaching away." Overall, the majority of teachers responded that learning and effectively using new technology with limited time and less-than-ideal trainings contributed to feeling unprepared for online teaching.

What demographic factors influenced secondary educators' perception of their own preparedness to teach online courses?

Some demographic factors, including age and years of teaching experience, did not have a strong correlation with teachers' levels of preparedness for teaching online courses. The data showed that a teacher's specific school campus strongly correlated between different educator experience. Although more data would be needed to form substantial conclusions regarding the other campuses within the district, a clear difference among teachers emerged from campuses "D" and "E." Teachers at "Campus D" rated themselves at a lower level of preparedness and specifically mentioned difficulties encountered at their campus, whereas teachers at "Campus E" rated their preparedness higher and mentioned positive methods implemented by their campus to help prepare them to teach online course (see Figures F4 and F5).

Age. The collected data did not suggest a strong correlation between age and preparedness to teach online. Although the two respondents in the 60+ age group had preparedness scores between 1 and 3 throughout online teaching, their written responses were polar opposite. One responded that despite the learning curve of new technology, they "have enjoyed teaching virtually." The other respondent likened their experience teaching online courses to "being a new teacher again after all these years… If I have to teach online again, I will retire." From the 50 to 59 age group, nine teachers reported a high level of preparedness for teaching virtually. Notably, one respondent in this group directly mentioned their age as a factor to their unpreparedness before the pandemic and said, "I'm old, technology is not my favorite, and prefer/believe strongly in face-to-face learning." Despite this teacher's initial resistance to the change, over the course of the year they began to feel more comfortable with the technology and teaching online. Only one other teacher in the 50 to 59 age group responded that they were not prepared, but he only taught virtually when required by the district in the spring 2020 semester and returned to in-person teaching for the 2020-2021 school year. On the other end, two of the three 20- to 29-year-old respondents stated that they felt very prepared to teach online, but they each had 1-5 years of prior teaching experience. The other 20- to 29-year-old teacher slowly improved over time from unprepared to neutral (2 to 3) as she completed her first year of teaching and gained more teaching experience. Although anecdotal evidence from a handful of respondents noted that age factored into their preparedness for online teaching, the data does not support a direct correlation between age and teacher's preparedness teaching online courses.

Years of Teaching Experience. Again, the data did not demonstrate a strong correlation between a teacher's years of educational experience and their preparedness to teach online courses. Of the eight teachers in the group with 20+ years of teaching experience, only a quarter reported a feeling of preparedness prior to teaching online courses, but half ended their time teaching online courses feeling prepared (see Figure F1). In the group with 11-20 years of teaching experience, the largest group with thirteen respondents, a similar pattern emerges. Only two teachers felt prepared to teach online prior to the COVID-19 pandemic. However, through the 2020-2021 school year, all but one reported they felt prepared to teach online (see Figure F2). In the group with 6-10 years of experience, while the six respondents reported various levels of preparedness prior to teaching online courses, only one ranked themselves as unprepared by the 2020-2021 school year (see Figure F3). Holistically, each grouping of teachers, no matter their years of experience teaching, increased their level of preparedness over time.

School Campus. Only one single demographic factor had a strong correlation that contributed to different teacher experiences: the specific high school campus at which an educator taught. Three teachers at "Campus D" (see Figure D1) accounted for over half of the teachers (a total of five) who reported low levels of preparedness (1 or 2) in all three categories: prior to teaching online courses, remote learning in the spring 2020 semester, and 2020-2021 school year. One reported that they were "the only teacher in my subject teaching online at my school." Another teacher felt pressured by her campus to pass her students, regardless of whether they had earned a passing grade. This group of unprepared teachers at "Campus D" accounted for 33% of total respondents from that campus. In comparison, eight participants from "Campus E" felt supported by their campus administrators through clear communication of expectations and trainings. One teacher specifically mentioned that collaboration within their professional learning communities at the campus helped them transition their in-person lessons for online learning. At the end of the 2020-2021 school year, no teacher at "Campus E" reported feeling unprepared for online teaching (see Figure F5).

Analyzing the responses from the two campuses demonstrates that campus administrators and school culture factored into teachers having different experiences teaching online courses. Beyond that single factor, no other demographic information yielded a strong correlation between differences in teachers experiences for online teaching. The greatest factor for educator experiences transitioning to teaching online courses during the COVID-19 pandemic was teachers who had prior personal experiences with online courses and familiarity with technology.

Future Research

Although researchers (Capdeferro & Romero, 2012; Robinson, et al., 2017; Kewal Ramani et al., 2018; De Paepe et al., 2018) reported the challenges of online teaching prior to the COVID-19 pandemic, these challenges repeated themselves across an entire North Texas school district when all educators taught online courses in the spring of 2020, and again when the district offered virtual learning to students in the 2020-2021 school year. Future researchers should examine how to design professional development and courses for teacher certification that work to mitigate some of the biggest challenges of online teaching: the lack of experience (time) and working with new technology tools. Future researchers could also examine perceptions and experiences from Kindergarten through eighth grade teachers.

REFERENCES

Anderson, T. (2008). Towards a theory of online learning. In T. Anderson & F. Elloumi (Eds.), *The theory and practice of online learning* (pp. 45–74). Athabasca University Press.

Brodersen, M. R., & Melluzzo, D. (2017). *Summary of research on online and blended learning programs that offer differentiated learning options. REL 2017-228*. Regional Educational Laboratory Central.

Brown, B., & Liedholm, C. (2002). Can web courses replace the classroom in principles of microeconomics? *The American Economic Review*, *92*(2), 444–448. doi:10.1257/000282802320191778

Capdeferro, N., & Romero, M. (2012). Are online learners frustrated with collaborative learning experiences? *The International Review of Research in Open and Distributed Learning*, *13*(2), 26–44. doi:10.19173/irrodl.v13i2.1127

Creswell, J. W., & Creswell, J. D. (2018). *Research design: qualitative, quantitative, and mixed methods approaches*. SAGE Publications, Inc.

De Paepe, L., Zhu, C., & Depryck, K. (2018). Online language teaching: Teacher perceptions of effective communication tools, required skills and challenges of online teaching. *Journal of Interactive Learning Research*, *29*(1), 129–142.

DiRienzo, C., & Lilly, G. (2014). Online versus face-to-face: Does delivery method matter for undergraduate business school learning? *Business Education & Accreditation*, *6*(1), 1–11.

Donalek, J. G. (2004). Phenomenology as a qualitative research method. *Urologic Nursing*, *24*(6), 516–517. PMID:15658739

Garrison, D. R., Anderson, T., & Archer, W. (1999). Critical inquiry in a text-based environment: Computer conferencing in higher education. *The Internet and Higher Education*, *2*(2-3), 87–105. doi:10.1016/S1096-7516(00)00016-6

Gudmundsdottir, G. B., & Hathaway, D. M. (2020). "We always make it work": Teachers' agency in the time of crisis. *Journal of Technology and Teacher Education*, *28*(2), 239–250.

Kewal Ramani, A., Zhang, J., Wang, X., Rathbun, A., Corcoran, L., Diliberti, M., & Zhang, J. (2018). *Student access to digital learning outside of the classroom. NCES 2017-098*. National Center for Education Statistics.

Martin, F., Budhrani, K., & Wang, C. (2019). Examining faculty perception of their readiness to teach online. *Online Learning*, *23*(3), 97–119. doi:10.24059/olj.v23i3.1555

Pajares, F. (1992). Teachers' beliefs and educational research: Clearing up a messy construct. *Review of Educational Research*, *62*(2), 307–332. doi:10.3102/00346543062003307

Park, C., & Kim, D. (2020). Perception of instructor presence and its effects on learning experience in online classes. *Journal of Information Technology Education*, *19*, 475–488. doi:10.28945/4611

Pramling Samuelsson, I., Wagner, J. T., & Eriksen Ødegaard, E. (2020). The coronavirus pandemic and lessons learned in preschools in Norway, Sweden and the United States: OMEP Policy Forum. *International Journal of Early Childhood*, *52*(2), 129–144. doi:10.100713158-020-00267-3 PMID:32836368

Robinson, H. A., Kilgore, W., & Warren, S. J. (2017). Care, communication, learner support: Designing meaningful online collaborative learning. *Online Learning*, *21*(4), 29–51. doi:10.24059/olj.v21i4.1240

Santelli, B., Stewart, K., & Mandernach, J. (2020). Supporting high quality teaching in online programs. *Journal of Educators Online*, *17*(1), 1–15.

United Nations. (2020, August). *Policy Brief: Education during COVID-19 and beyond*. Author.

APPENDIX A

Figure 1. Garrison et al., (1999) Framework for Online Teaching

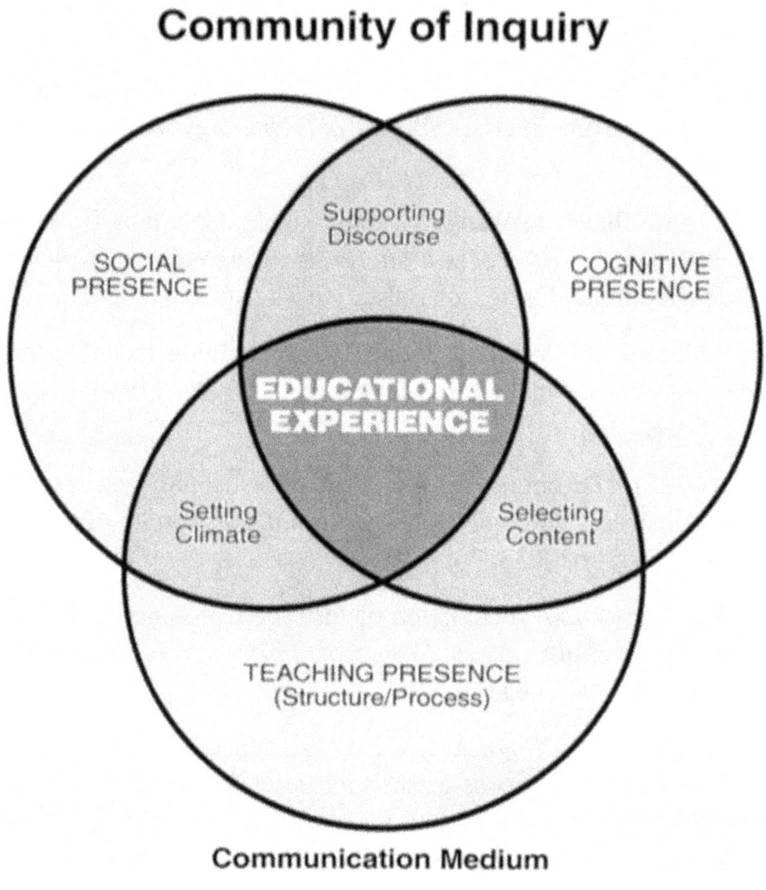

Figure 2. Anderson's (2008) Key Characteristics of Online Learning

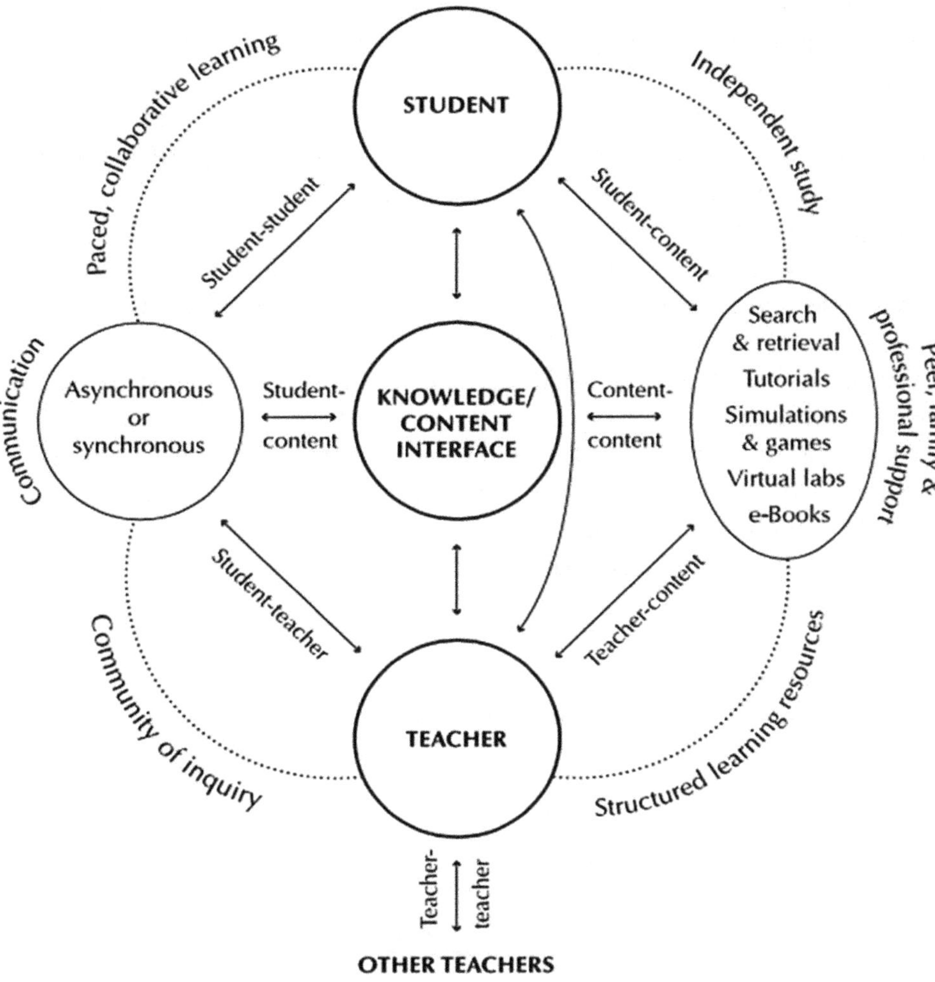

APPENDIX B

The Researcher's Introduction Letter to Potential Participants

Dear Potential Participant,

My name is Douglas Fenderson. I am a graduate of Marcus High School ('12) and am currently working as a history teacher at The Colony High School. As a master's candidate at Midwestern State University in the College of Education, I am requesting your participation in a research study that I am conducting titled, "From In-Person to Online Teaching: Experiences of Secondary Educators During the

COVID-19 Pandemic." The purpose of this study is to understand the experiences of traditional in-person secondary educators transitioning to teaching online classes. Click the "Teacher Experiences of Online Teaching Survey" to complete the survey.

The study involves completing a survey that may last between ten to thirty minutes, depending on the depth and detail of your answers. The interview consists of ten optional demographic questions, between nine and thirteen open-/closed-ended questions based on your teaching experience within the last year and a half. You can complete it at your own pace.

Participation is voluntary and the survey is completely anonymous. The survey will not collect personal identifiers such as name, employee ID number, or email address.

You can also reach out to me at dfenderson0224@d2lmail.msutexas.edu if you have any questions.

Your participation in the research will be of great importance to understand the perceptions and experiences of educators teaching online during this pandemic. The results of the study will be made available to you if desired.

Thank you for your time and participation.

Sincerely,
Douglas Fenderson
Master's Candidate, Midwestern State University

Figure 3.

APPENDIX C

Survey Questions

Section 1 - Consent

1. I am willing to proceed with the survey and consent that my anonymous responses are used in the research project.
   a. Yes
   b. No

Section 2 – Demographic Information

At which campus are you currently teaching?

Years of educational experience

- First year teacher
- 1-5
- 6-10
- 11-20
- 20+

Highest degree held:

- No degree held
- Bachelors
- Masters
- Doctoral

Age

- 20-29
- 30-39
- 40-49
- 50-59
- 60+

Race/Ethnicity (select all that apply)

- White (Eg: German, Irish, English, Italian, Polish, French, etc)
- Hispanic, Latino or Spanish origin (Eg: Mexican or Mexican American, Puerto Rican, Cuban, Salvadoran, Dominican, Colombian, etc)
- Black or African American (Eg: African American, Jamaican, Haitian, Nigerian, Ethiopian, Somalian, etc)
- Asian (Eg: Chinese, Filipino, Asian Indian, Vietnamese, Korean, Japanese, etc)
- American Indian or Alaska Native(Eg: Navajo nation, Blackfeet tribe, Mayan, Aztec, Native Village or Barrow Inupiat Traditional Government, Nome Eskimo Community, etc)
- Middle Eastern or North African (Eg: Lebanese, Iranian, Egyptian, Syrian, Moroccan, Algerian, etc)
- Native Hawaiian or Other Pacific Islander (Eg: Native Hawaiian, Samoan, Chamorro, Tongan, Fijian, etc)
- Prefer not to say
- Other…

Gender

- Female
- Male
- Non-binary
- Prefer not to say
- Other…

Current marital status

- Single, never married
- Married or domestic partnership
- Widowed
- Divorced
- Separated
- Prefer not to say
- Other…

Who currently lives in your primary place of residence? (select all that apply)

- Only yourself
- Spouse/partner
- Kids
- Parents/in-laws
- Extended family
- Non-relative roommates
- Other…

How many people currently live in your primary place of residence (including yourself)?

- 1
- 2
- 3
- 4
- 5
- 6
- 7
- 8
- 9
- 10
- 11+

What is your total household income? (estimate)

- Less than $10,000
- $10,000 to $19,999
- $20,000 to $49,999
- $50,000 to $74,999
- $75,000 to $99,999
- $100,000 to $149,999
- $150,000 or more
- Prefer not to say

Section 3 – Experience Prior to Remote/Online Learning

Have you taught entirely online courses (remote, virtual) before the COVID-19 pandemic?

- Yes, often
- Yes, seldom
- Never

Overall, you felt prepared (equipped, trained to teach online courses before the COVID-19 pandemic?

- 1. Strongly Disagree
- 2. Disagree
- 3. Neutral
- 4. Agree
- 5. Strongly Agree

Please elaborate in what ways you were/were not prepared to teach online courses prior to March 2020:

Is there anything else you would want me to know about your experience and/or perception of online teaching prior to the COVID-19 pandemic?

Section 4 – Experience During Spring 2020 Remote Learning

Overall, you felt prepared (equipped, trained) to teach online courses during remote learning from March 2020 through May 2020.Online courses (remote, virtual, etc.)

- 1. Strongly Disagree
- 2. Disagree
- 3. Neutral
- 4. Agree
- 5. Strongly Agree

Please elaborate in what ways you were/were not prepared to teach online courses after March 2020 when Lewisville ISD transitioned to remote/online learning:

Was there anything that could have been done differently (by yourself, your administration, or your district) to have better prepared you to teach online courses? Please be specific.

Is there anything else you would want me to know about your experience and/or perception of teaching remote/online courses during the COVID-19 pandemic?

Section 5 – 2020-2021 School Year

In this academic year (August 2020 through present), did you teach at least one online course?

- Yes, I taught at least one online course (remote, virtual, hybrid, etc)
- No, I taught all in-person courses (face-to-face)

Section 6 – Experience During the 2020-2021 School Year (online)

Overall, you felt prepared (equipped, trained) to teach online courses during the 2020-2021 school year.

- 1. Strongly Disagree
- 2. Disagree
- 3. Neutral
- 4. Agree
- 5. Strongly Agree

Please elaborate in what ways you were/were not better prepared to teach online courses during the 2020-2021 school year after your initial experiences from the 2020 spring semester:

Was there anything that could have been done differently (by yourself, your administration, or your district) to have better prepared you to teach online (virtual/hybrid) courses in the 2020-2021 school year? Please be specific.

Is there anything else you would want me to know about your experience and/or perception of teaching online courses during the 2020-2021 school year?

Section 7 – Thank you

Thank you for your participation! If you have any questions please reach out to me at dfenderson0224@d2lmail.msutexas.edu.

APPENDIX D

Survey Responses

Demographic Information

Figure 4.

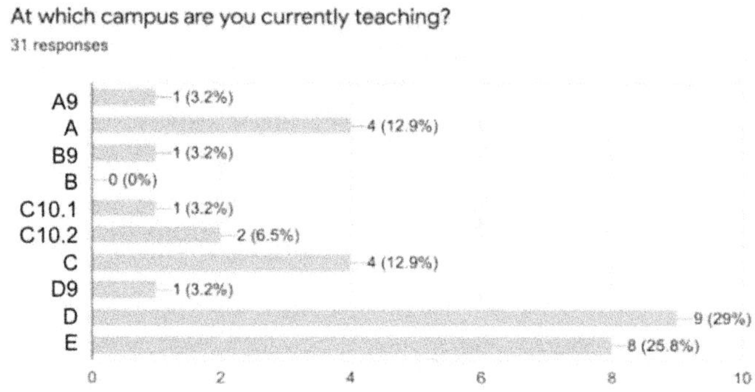

Figure 5.

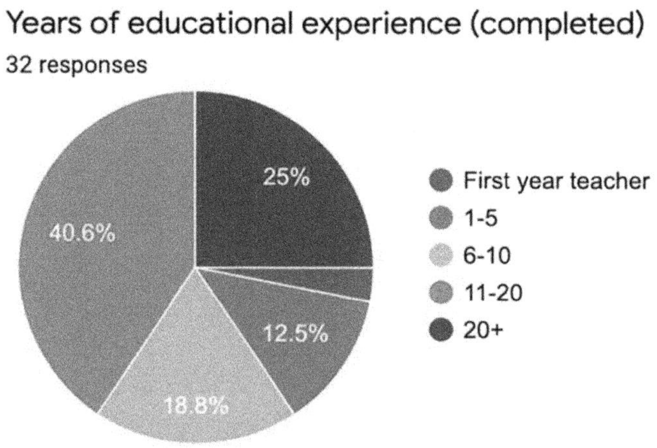

Figure 6.

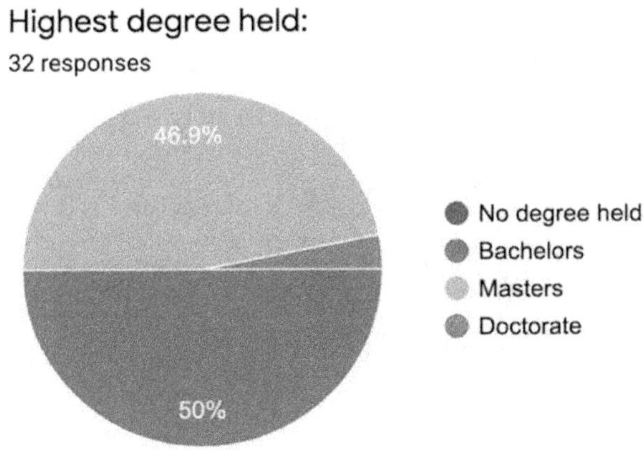

Figure 7.

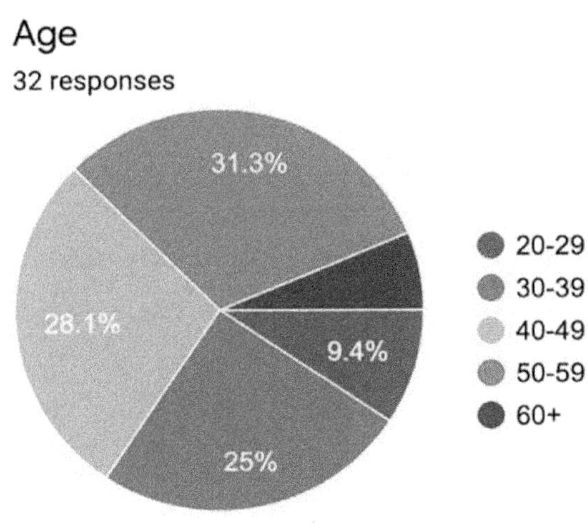

Figure 8.

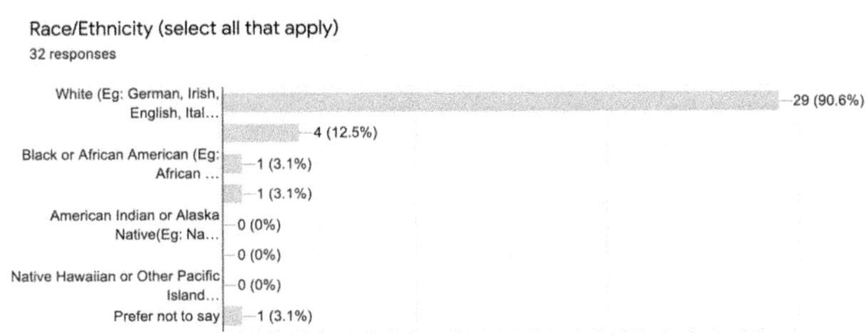

Figure 9.

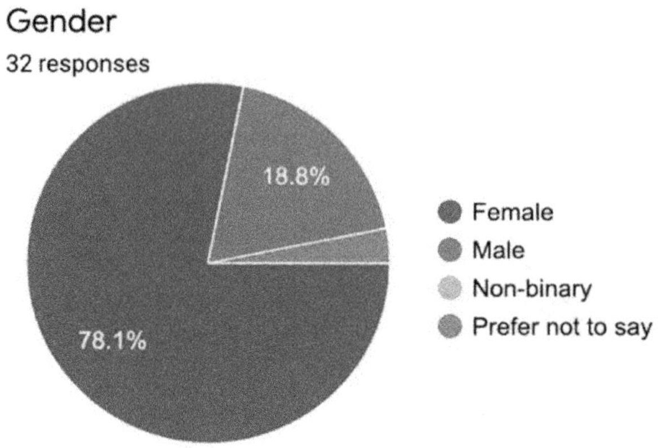

Figure 10.

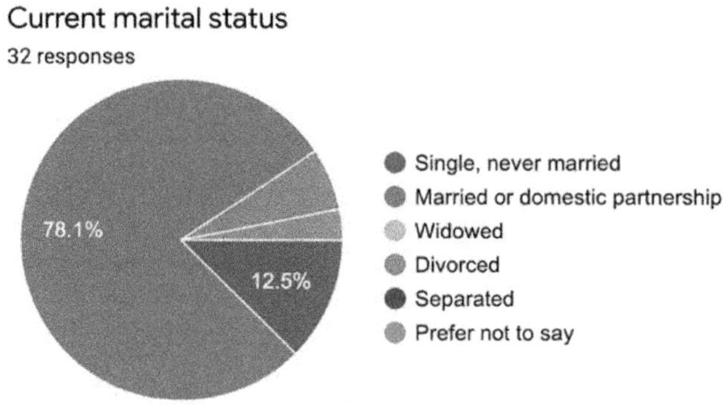

Figure 11.

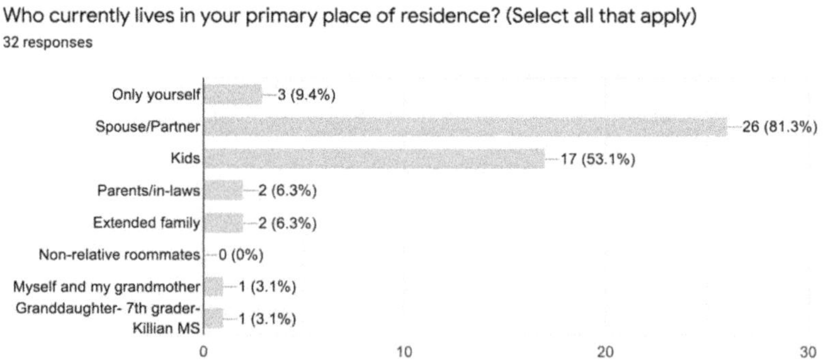

Figure 12.

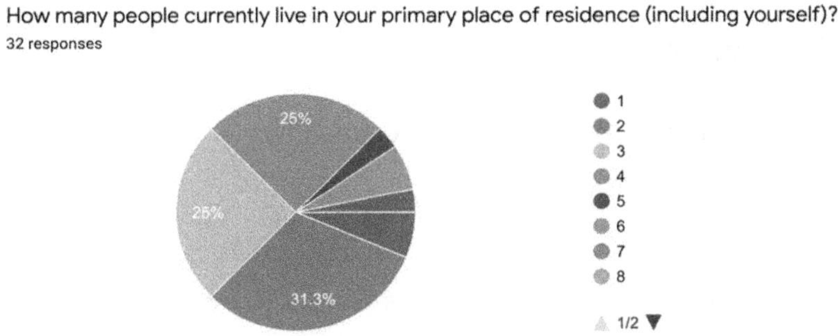

Figure 13.

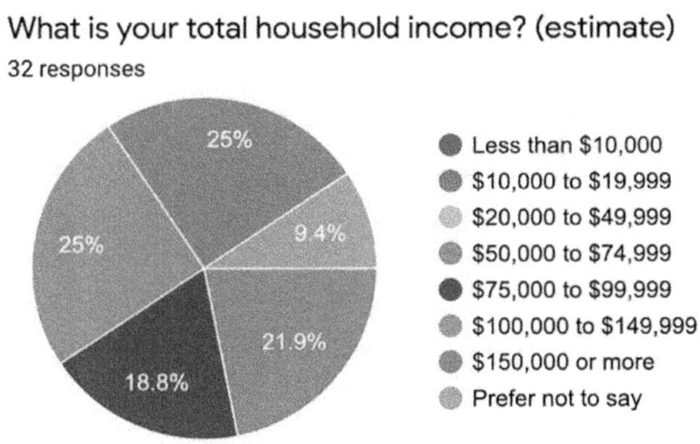

APPENDIX E

Survey Responses

Experience Prior to Remote/Online Learning

Figure 14.

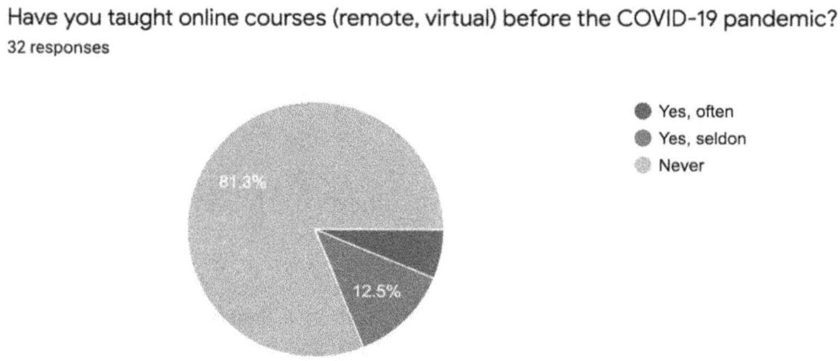

Figure 15.

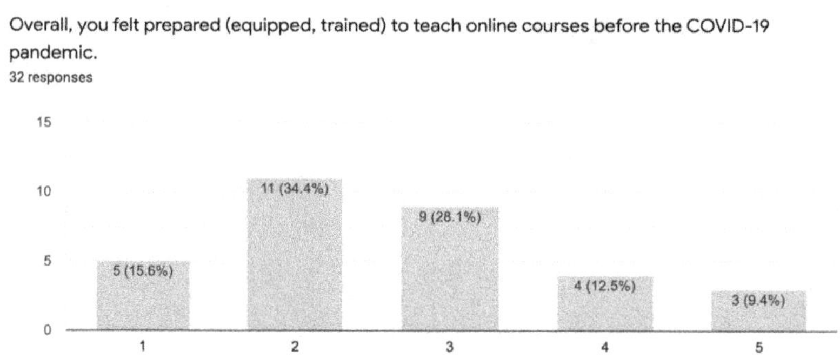

Experience During Spring 2020 Remote Learning

Figure 16.

Overall, you felt prepared (equipped, trained) to teach online courses during remote learning from March 2020 through May 2020.
32 responses

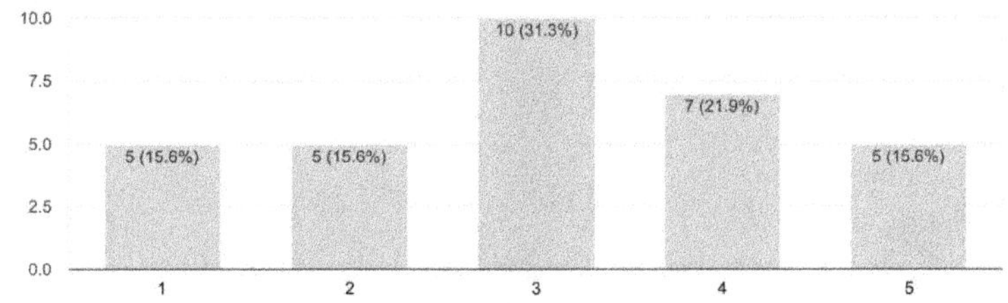

2020-2021 School Year

Figure 17.

In this academic year (August 2020 through present), did you teach at least one online course?
32 responses

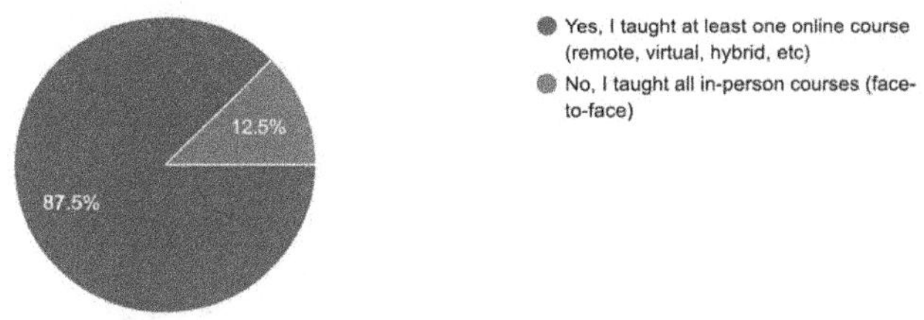

Experience During the 2020-2021 School Year (online)

Figure 18.

Overall, you felt prepared (equipped, trained) to teach online courses during the 2020-2021 school year.
28 responses

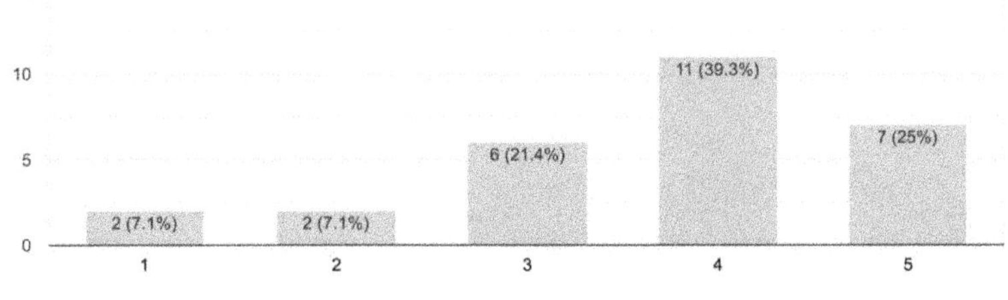

APPENDIX F

Years of Experience and Levels of Preparedness

Figure 19.

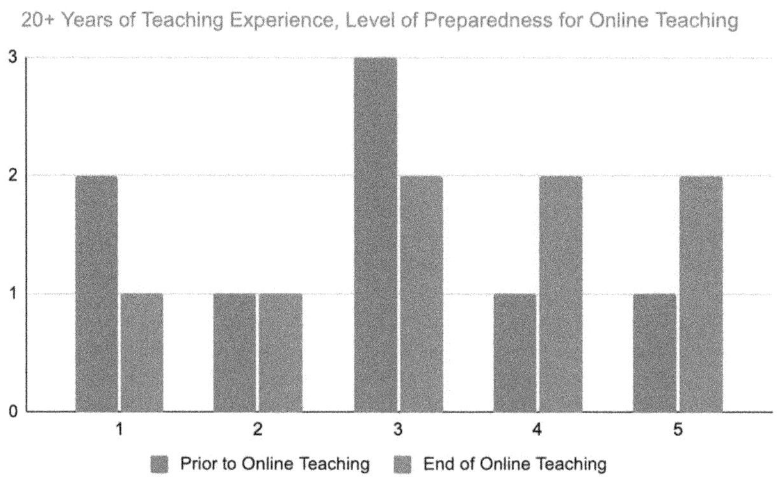

Figure 20.

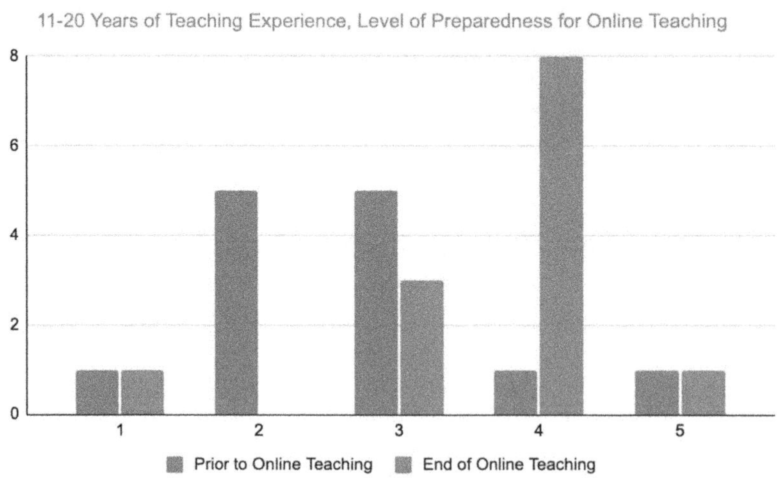

Figure 21.

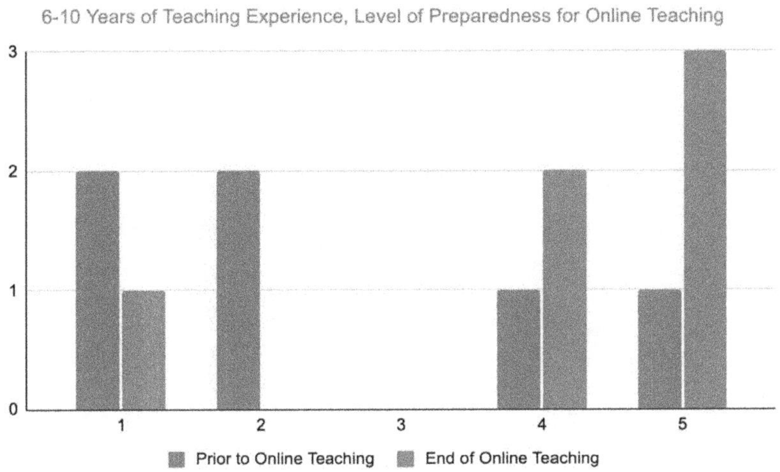

Differences Between Teacher Perceptions at Various School Campuses

Figure 22.

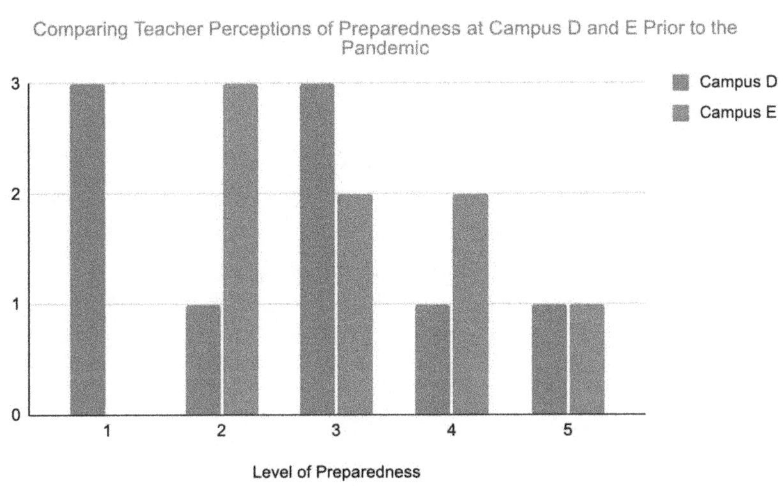

Figure 23.

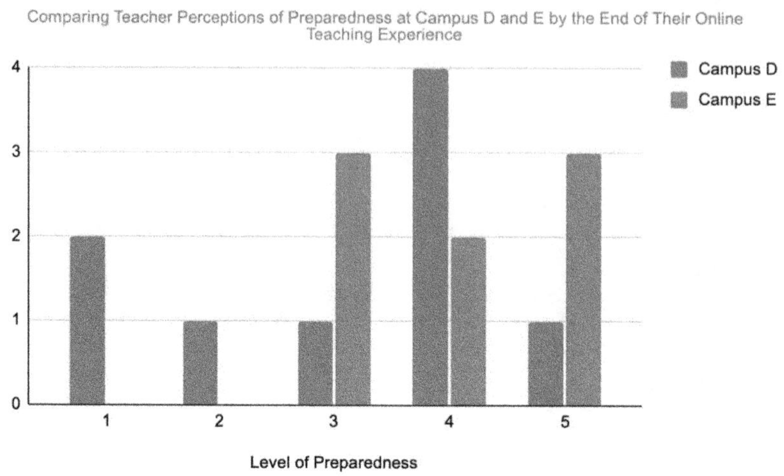
Comparing Teacher Perceptions of Preparedness at Campus D and E by the End of Their Online Teaching Experience
Campus D
Campus E
4
3
2
1
0
1
2
3
4
5
Level of Preparedness

Chapter 5
COVID and Counseling: A Shift to Telehealth During a Global Crisis

Bonnie Carter King
https://orcid.org/0000-0001-9694-1857
Texas Woman's University, USA

ABSTRACT

Due to the COVID-19 crisis, the mental health profession has shifted to online service provision, or telehealth. The aim of this chapter is to describe the COVID-19 crisis and subsequent changes that occurred to mental health service provision; the benefits and drawbacks to telehealth from practical, ethical, and cultural perspectives; and the learning opportunities that have come from this crisis. Finally, reflections on the future of the counseling profession and trends in service provision for serving an increasingly diverse population will be analyzed.

INTRODUCTION

The purpose of this book chapter is to describe the transition of mental health providers from practicing in person, to online video conferencing due to the COVID-19 global pandemic of 2020/2021. Telehealth mental health service provision is defined as the use of synchronous audio (and/or) video conferencing technology or telephone to provide mental health counseling services (Pierce et al., 2020). This chapter will provide a description of the COVID-19 pandemic and the reasons mental health providers moved their practices online, benefits, practical and ethical issues related to

DOI: 10.4018/978-1-7998-8028-8.ch005

telehealth, cultural considerations for the use of telehealth for therapy, and reflections on lessons learned from the COVID-19 crisis for mental health providers. Predictions of how this may impact the profession moving forward are also provided.

BACKGROUND

The COVID-19 virus, first detected in Wuhan, China in 2019, was declared a public international health emergency on January 30th, 2020, and a pandemic on March 11, 2020 by the World Health Organization (WHO) (Almeda et al., 2021; Lee et al., 2021). Spreading to other countries in the early months of 2020, the virus has impacted the lives of citizens around the globe. COVID-19 has taken the lives of millions, and has impacted economies worldwide (Almeda et al., 2021; Lee et al., 2021). Considered the greatest public health crisis of the century, the world has not experienced an illness that has been so deadly and transmissible in such a short period of time since the Spanish Influenza pandemic of 1918 (Sanders et al., 2020). As COVID-19 is a newly emerging virus and treatments are being created rapidly, until mass vaccination and better life-saving treatments are accessible to all who need it, many will continue to become ill with the potentially fatal virus. Until these life-saving prevention measures and treatments are available to all, the WHO and local governmental public health agencies' only strategies to help contain the spread of the virus has included hand-washing, social distancing, and mask-wearing. According to the Centers for Disease Control, these measures have been implemented at least to some extent, by governments world-wide (2020).

Since the DNA for a vaccine was sequenced in January of 2020 (WHO, 2020) Scientists have raced to create and test vaccines to prevent severe COVID illness. As of December, 2020 vaccines were approved in the United States by the Food Drug Administration (FDA), as they have been proven efficacious to prevent severe illness and death from COVID-19 (FDA, 2020). Since the first report of a cluster of cases of COVID-19 on December 31, 2019 (WHO, 2020) until December of 2020,the virus has spread all over the world, infecting over 104,600,000 million people, and taking the lives of over 2, 274,000 people (453,545 deaths in the United States) as of February, 4, 2021 (Center for Systems Science and Engineering (CSSE), 2021). In addition, many governments have closed their borders, implemented mandatory "stay at home" orders, and have utilized governmental authority to temporarily close businesses, and limit capacity of local establishments such as work-places, schools, restaurants, bars and other public establishments. Furthermore, supply chains have been disrupted due to governmental closures and viral outbreaks at production facilities that have led to quarantines that pause production (McKibben, 2020).

These disruptions have impacted the global economy in devastating ways. In addition to the physical threat of illness, people have been greatly impacted by financial hardship, and the loss of employment. As a solution, many businesses and entire industries have moved to online platforms, and many businesses with the ability to do so, have sent their employees to work from home. While many industries had an online presence before the pandemic, some have shifted almost entirely to online platforms. Businesses have started using video conferencing for meetings, and to provide services to people that were previously "face-to-face." Teachers, doctors and mental health professionals have largely switched exclusively to video conferencing to provide services to clients in order to decrease the transmission of COVID-19 in attempts to keep themselves and their clients safe and healthy.

In addition to physical health and economic concerns, a major increase in mental health challenges have emerged as a result of so much uncertainty, compounded by the isolation that so many are experiencing as a result of long-term social distancing, recommended by local, state and federal governments. Loneliness, which has increased for many due to social distancing, is a risk factor for a number of physical and mental health issues such as heart disease and stroke (Valtorta et al., 2016) and depression (Pitman et al., 2018). Further, research has emerged to indicate that younger adults have been more deeply impacted by the social distancing measures than older adults, with 69% of Canadians ages 18-24 having reported loneliness as a result of social distancing, with females more impacted than males (Bricker, 2020). International studies have also identified being female as a risk factor in addition to general decreases in positive feelings since the beginning of the pandemic (Li et al., 2020).

People in quarantine frequently experience feelings such as boredom, fear, sadness and alienation (Xiang et al., 2020), and while most have not had to strictly quarantine due to illness for extended periods, many people have been social distancing as recommended by local governments for months. For those that have been ill, some hospitalized COVID-19 patients experience symptoms of post-traumatic stress disorder (PTSD) before and after discharge from treatment (Liu et al., 2020). Healthcare workers are also experiencing PTSD, burn-out, and extreme levels of stress and anxiety related to fear of contracting the illness, and being overworked, as hospitals have been overwhelmed past capacity in many areas with COVID patients (Almeda, 2021). Due to the high mortality rate of this coronavirus, grief has become a prevalent mental health concern, as many people have lost loved ones to the disease. Mental health treatment options are more important than ever, as so many are experiencing the emotional and psychological havoc the virus has left in its wake. In addition, those already suffering with mental health issues have also experienced fear, depression and anxiety, which for many has overshadowed their original presenting issues (Perrin, Ry barczyk, et al., 2020). The traditional ways in which so many coped with the stressors of life outside of therapy such as socializing with friends, attending support groups, church, and other community services have no longer been accessible due to social distancing (Harandi, Taghinasab, & Nayeri, 2017) As a response, public health organizations world-wide have organized to provide mental health treatment recommendations and resources to citizens that can be accessed online and by phone (Alameda, 2021). Many governmental recommendations include the provision of mental health services primarily online or by telephone to decrease the transmission of COVID-19, with the exception of those with needs that require a higher level of care such as hospitalization (Almeda, 2021).

TELEHEALTH FOR THE USE OF MENTAL HEALTH SERVICE PROVISION

In order to meet the growing need for mental health services, while still caring for those who were accessing services before the pandemic, a monumental shift in the mental health profession has occurred. Service providers have moved their practices almost exclusively to synchronous online telehealth formats. According to a cross-sectional study conducted in the United States by Pierce et al. (2020) that polled 2,619 licensed psychologists, "Prior to the COVID-19 pandemic, psychologists performed 7.07% of their clinical work with telepsychology, which increased 12-fold to 85.53% during the pandemic, with 67.32% of psychologists conducting all of their clinical work with telepsychology" (p.14). What once was almost exclusively a face-to face profession, is now primarily provided through online video conferencing platforms. For client and clinician safety to stop the spread of the coronavirus, many Health Insurance Portability and Accountability Act (HIPAA) compliant software have emerged to meet the growing demand for video conferencing approaches to psychotherapy. This rapid shift has led to major changes in the ways in which the counseling relationship manifests in practice. Further, practitioners in the study predicted that after the pandemic ends that 34% would continue to provide services through telepsychology (Pierce et al., 2020).

To accommodate this shift in service provision, drastic measures have been taken by licensing boards, the government and third party payers to ease restrictions that were previously barriers to accessing telehealth services. Insurance companies, such as Blue Cross Blue Shield (BCBS, 2020) Medicare, Medicaid and Children's Health Insurance Plans have eliminated reimbursement restrictions for telehealth based counseling during the on-going state of national public health emergency declared in the United States (United States Centers for Medicare and Medicaid Services [CMS], 2020). Licensing boards for psychiatrists in some states even waived rules preventing psychologists from practicing between states (Pierce et al., 2020). HIPAA restrictions were eased to allow for clinicians to use software to support online video conferencing for their counseling practices (OCR, 2020a, 2020b). Before, some of these software were not compliant with the law due to confidentiality limitations. Before COVID, only HIPAA compliant video software was allowed for use in telehealth, but based on current needs, software such as Skype, Zoom etc. are now allowed to meet the huge demand for online counseling (Pierce et al., 2020).

There are many benefits to telehealth counseling practice, in addition to increased safety from COVID transmission that online counseling provides. Some of the benefits include: increased access to specialty providers for clients in more broad geographical regions, more confidentiality for stigmatized clients who feel uncomfortable going out in public to access services, ease of access and decreased transportation costs for clients, and cost effectiveness for clinicians who no longer need expensive office space. Further, both clinicians and clients have greater access to individuals throughout the state, as licensing extends state-wide. Geographical distance less of an obstacle to access services.

POSSIBLE ISSUES THAT EMERGE IN TELEHEALTH

Although telehealth has been proven to be effective (Varker et al., 2019), before the COVID crisis, many mental health providers were hesitant to use video conferencing with clients due to HIPAA law barriers, biases that the approach would be less effective, and a perceived lack of training available that could increase comfort levels of practitioners (Perry et al, 2020; Pierce et al., 2020). Many more training programs have emerged to meet the therapists needs during this crisis and, one can assume that the gap is closing in terms of self-efficacy for providers using a telehealth approach in their practices.

Further, insurance companies have historically limited reimbursement for telehealth for their clients' mental health treatment. Many insurance plans including Medicare, only allowed telehealth for rural clients (O'Reilly, 2019), or those with specific plans, thereby requiring most individuals to seek treatment in an office setting. With a lack of demand due to telehealth not being reimbursed as readily as face-to-face sessions, providers were incentivized to provide services primarily in an office setting. Because of these limitations, many group practices or treatment centers have avoided the use of telehealth for their services, and therefore clinicians who were employed in group practices practiced primarily in an in-person format.

Telehealth is not the most appropriate intervention for clients requiring hospitalization or a higher level of care than regular talk-therapy, although, the instant access could help get clients to a safer space, more quickly. There can be barriers to immediate safety is someone shares they are in crisis, but refuse to take themselves to an inpatient facility. Police or mobile crisis units would be needed is such situations, and clinicians may have less control to keep someone physically safe than if the client is physically present in in their office until they can be transported to a hospital setting.

In addition, there are some inconsistencies when it comes to guaranteeing confidentiality in home offices or the homes of clients, and technology glitches that can lead to a bad connection. The social distancing that has occurred since the beginning of the pandemic have left many without safe access to childcare. So distractions can be more prevalent, and privacy could be harder to achieve outside the boundaries of a designated office. Clinicians should make certain that their home offices or work spaces are completely confidential to protect client confidentiality, even if they cannot control the level of confidentiality in the homes of their clients.

Next, although services have been expanded during this on-going crisis, we cannot predict if third party payers will open their policies in the future to reimburse providers for telehealth services once the pandemic has passed. As mentioned, third-party payers have largely expanded their coverage to include continued coverage for telehealth services throughout the state of public health emergency, but while 34% of psychologists predict they will expand telehealth services even after the pandemic has passed (Pierce et al., 2020), we cannot predict if insurance providers and government provided healthcare plans in the United States (Medicare and Medicaid) will continue to reimburse providers for such services.

CULTURAL BARRIERS TO USE OF TELEHEALTH

There are also cultural concerns that might hinder certain populations from accessing online therapy services. Culture is multidimensional and is typically comprised of aspects of identity such as race, ethnic origin, gender identity and expression, age, sexual orientation, education, religion, geography, customs and traditions, and socio-economic status (Lim & Lu, 2008). Cross-sectional aspects of identity come together in unique ways when working with clients as a mental health provider. It is a priority, and an ethical standard to serve diverse populations with cultural competence as a mental health provider (ACA, 2014). Effective cultural competence leads to more participation from culturally diverse clients, more engagement with the process and increased trust Roncoroni et al., 2014). The shift to online counseling impacts certain cultures differently than others, primarily regarding access and preferences. Individuals whose first language is not English may not be as open and willing to engage in telehealth with a clinician that does not also speak their native language, and frequently preferred service providers that were from a similar cultural background (Hilty et al., 2019). Further, interpreters can help bridge the gap between non-bilingual providers and clients when organizations prioritize the need for interpreters. Therefore, bilingualism, interpreters and culturally similar therapists are important to help clients feel more comfortable accessing telehealth services (Hilty et al., 2020). Clinicians should have training in both telehealth competencies and cultural competence to engage effectively with diverse populations through online video conferencing platforms.

Clients with financial difficulties, or in more remote areas may have intermittent access to internet or devices on which to video conference, which can also become a barrier to service access. Organizations operating from online platforms can help clients without access by providing internet hotspots, or even devices on which to communicate. This support was employed through many school districts in order to help students access online learning during the COVID-19 crisis. Organizations such as Devices for Students (2020) supported this effort, and telephone and internet companies such as AT&T, Sprint, Verizon, Consumer Cellular, Google Fi, T-Mobile etc. temporarily expanded their plans to include more hotspot access for those staying home to stop the spread of COVID-19. Local school districts reached out to student families to provide internet hot spots and devices to help support access to online education services. Employing this type of support for mental health services can provide access to the resources required to effectively access telehealth services.

Another potential issue is lack of access to confidential space. Clients that share their homes with roommates or family members may find it difficult to find a confidential space in which to process confidential material. Further, different cultures have different needs or cultural attitudes related to living situations, and more communal cultures are more apt to multigenerational family living situations. Financial situations can also impact the amount of family members that live in a home, or the amount of space in a particular living situation, all of which can impact confidentiality. Further, cultural attitudes may impact access to psychotherapy services. Trusting healthcare providers can be difficult for historically marginalized populations, or those who have been discriminated against while accessing medical or mental health services. The lack of access to an in-person relationship with a therapist may make it more difficult to relate on a personal level required to build a relationship with a mental health provider.

Finally, as mentioned above, insurance companies and other third party payers control how services are provided by paying for service provision through healthcare plans. If insurance companies do not approve telehealth services, individuals will not have access to services covered in a face to face format that was previously covered through their health insurance plan. In a post-pandemic world, if more providers take their practices online only, there may be less providers who accept insurance, which could lead to a shortage of providers that individuals have access to for mental health services. This may make the service of mental health care a luxury provided to those who can pay “out of pocket” for fees that typically start at over 100.00 per hour. This puts a greater burden on individuals (as well as society) with a lower or even average socio-economic status, thereby preventing many from being able to access psychotherapy.

SOLUTIONS AND RECOMMENDATIONS

As mental health providers have become more comfortable with online video conferencing to provide telehealth services to diverse populations of clients, a variety of solutions can be employed to make telehealth more effective. Organizations and individual providers should focus on access, confidentiality, and cultural competence in their individual practices.

Accessibility

By partnering with non-profits and internet and telephone companies, increased accessibility could become a permanent part of practice moving forward past the COVID-19 crisis. Regardless, organizations can work with the specific needs of the family through grants, or as part of the pricing plan to meet the individual access needs of the client. Staying informed about third party reimbursement may be the most important consideration. As these services have been expanded as acceptable forms of psychotherapy for almost an entire year due to the COVID-19 crisis, it is likely that telehealth may become more recognized by third-party payers as a useful and legitimate format by which to receive mental health care. The benefits are numerous for those in rural areas, or with lack of reliable transportation. It is also beneficial for individuals with mental health issues that make interaction in the world outside of their homes more difficult. For those with diagnoses such as agoraphobia, post-traumatic stress disorder (PTSD), obsessive compulsive disorder (OCD), and even depression, sometimes staying home to the exclusion of functioning in many areas in life is a symptom of their mental illness. Further, those who have been stigmatized by their mental illness or for high profile community members for whom confidentiality is of great concern, online options may be the most acceptable and private option to protect their confidentiality or even physical safety for marginalized individuals in some communities. The first step could be accessing a mental health counselor or medical doctor online through e-visits to prescribe medication or begin psychotherapy.

Confidentiality

Confidentiality can be addressed in a variety of ways through on-going informed consent processes. Informed consent is the process by which the clinician informs the client about services, the expectations and boundaries of the therapeutic relationship, fees, limits to confidentiality and theoretical and practical approaches to psychotherapy. This process is ongoing, and is both written (through consent forms and paperwork) and verbal, for example reminding a client of the limits of confidentiality throughout the therapeutic relationship, or negotiating and co-creating interventions and goals with the client. In an online situation where privacy cannot be completely guaranteed on the client's end, the therapist and client can create solutions such as sitting in the yard away from the family, pausing the conversation when someone walks close by, and setting boundaries with family members to allow for confidential processing time, or other plans that work best based on the individual needs of the client. The therapist can assist the client in coming up with plans that work best for them, given their individual situation. Therapists can also utilize a flexible approach to the informed consent process and change the plan when necessary. The limits of confidentiality should be also addressed thoroughly in the informed consent paperwork that is reviewed and signed by the client.

For clinicians, it is an ethical duty to create a confidential space in which to practice when providing telehealth services. Any home office should be sound-proofed, boundaries with family members should be clearly set to avoid interruptions, and clinical software that is encrypted and HIPAA compliant should be the standard for providing long-term telehealth services. Many therapeutic practice provider software now offer such standards for their telehealth features, scheduling, and billing software (Simple Practice, 2021). These services make it easier than ever to provide confidential, high quality video conferencing for clients. While HIPAA laws have been eased in light of COVID-19, this law will likely revert back to a high level of protection for confidential information, therefore making specific HIPAA compliant software a necessary part of any online practice. Further, confidentiality training for interpreters who might support the counseling process for non-English speaking clients, must be completed to assure confidentiality of client information.

Cultural Competence

Cultural competence is an important part of any psychotherapeutic practice, and training programs and ethical standards promote non-discrimination, and the ability to work with diverse groups of individuals. For example, marriage and family therapists, licensed professional counselors, psychologists, and licensed clinical social workers, are all required to adhere to their respective ethical codes (AAMFT, 2015, ACA,

2014; APA, 2017; NASW, 2017). Going beyond anti-discrimination, the movement for more inclusivity in mental health professions has mirrored social movements in society, but gained more traction in the 1990's (Sue et al., 1992). Clinicians are encouraged to examine their own privilege, oppression, and specific cultural identities in order to be reflective about how their personal identity can impact the personal identities of the client in the counseling relationship (Ratts et. al, 2016). This self-awareness must extend to openness, curiosity, and an examination of personal biases that could potentially influence the personal values of the therapist, and become a barrier to nonjudgmental and culturally sensitive treatment of clients in session.

This awareness is imperative to extend to the telehealth counseling relationship to build trust, to meet access needs, and overcome any language barriers, or other barriers to treatment that might require cultural sensitivity. Confidentiality becomes even more important if the client is exploring something about their identity that might conflict with the beliefs of their family of origin, or cultural heritage. Awareness of cultural identities can allow the therapist to incorporate important aspects of the client's culture into the counseling process so that important ideologies are not overlooked or undervalued. An awareness on the part of the therapist about ways in which the specific populations they are working with have been discriminated against can help the therapist understand any hesitance to trust the service provider, as well as areas of exploration in session about how discrimination has impacted the client's life. It is the therapists duty to research specific issues that can emerge for the populations with which they work, and various aspects of the specific culture. Therapists should not lean too heavily on stereotypes however, but should instead operate from an understanding that culture is multidimensional, complex and defined ultimately by the client.

FUTURE RESEARCH RECOMMENDATIONS

The use of Telehealth for mental health services in the U.S. as a response to the COVID-19 pandemic of 2019-2021 has grown substantially to slow the spread to the virus as a safe alternative to meeting clients in face-to-face formats. The use of telehealth has been proven as an effective means to treat a variety of mental health disorders (Varker et al., 2019), and more clinicians than ever before are using this approach with clients (Pierce et al., 2020). As clinicians have adjusted over the past year, the likelihood that the use of telehealth will become a standard approach to the ways in which mental healthcare is delivered is high. Research is emerging almost as quickly as the use of the format itself, but more research will be needed to continue to support more widespread use of telehealth for treatment programs and practitioners alike. More research regarding effectiveness could also provide incentive to third-party payers to continue to provide reimbursement for services. More research on specific issues related to cultural competence as it specifically pertains to telehealth could also help populations who are less served in general by mental health services, as they can be more easily accessed through telehealth than in-person if living in rural areas. This massive shift of services will certainly impact the profession for many years to come. As this chapter is written, the pandemic rages on, and there is no definite timeline for when it will end. It may be months or longer before life looks anything like it used to before the pandemic. In that time it is likely that standards, training, products to support the practice, and research will all improve to help give practitioners the ability to provide even more effective delivery of telehealth services to clients. It is likely that telehealth is here to stay, and the benefits of that can be helpful to all those who find the service more convenient and time efficient. If therapy proves to be as effective through telehealth as in-person services, the whole format of the therapeutic relationship could change, as well as the structure, ethical standards and standard approaches to the practice.

CONCLUSION

Telehealth is an approach to psychotherapy that has been present for many years, albeit as an outlier service typically reserved for those in very rural areas. Prior to the pandemic, few clinicians felt comfortable providing telehealth services due to lack of training, lack of comfort, and biases related to telehealth as an effective approach to psychotherapy. Further, HIPAA compliance, and a lack of third-party reimbursement have been barriers to incorporating telehealth into their practices (Perry et al, 2020; Pierce et al., 2020). Due to the COVID-19 pandemic, governments and insurance entities have eased such restrictions to allow for desperately needed, safe forms of psychotherapy to treat those with mental health conditions during this uncertain time in history. Governmental social distancing recommendations, as well as fear and unease generated by the pandemic itself, has increased stress, loneliness and mental health concerns in populations that did not experience such struggles previously. The need for online mental health services is at an all-time high, and at this moment in time we have the hardware and software to meet the demand. The many benefits include accessibility, convenience for clients, the lack of office space needed for clinicians, ability to access specialists outside of a client's geographical location, and lack of stigma attached to attending sessions at an office space. For rural clients who did not have access to a provider in their area, telehealth is a great alternative to not accessing services at all, or committing to driving long distances to access a provider.

There are also potential drawbacks to the approach such as access, confidentiality concerns, cultural considerations, and reimbursement issues that clinicians must attend to in order for telehealth to be the most accessible and beneficial approach for mental health service provision. Certainly during the COVID-19 pandemic, the accessibility of these services has been crucial to support diverse populations of individuals in need of psychotherapy during an unprecedented time of loss, fear, and instability for so many worldwide. With continued learning, training, and research, in addition to consideration for the issues that can emerge in this practice, telehealth is an effective treatment option for mental health counseling. With continued dedication to upholding best practices and cultural considerations for this type of practice, many clients may prefer to be served in the future through telehealth. Increased access to services and decreased stigma can contribute to more people accessing life-saving and life enhancing treatment more easily than ever before, making the telehealth trend a permanent fixture in the psychotherapy professions.

ACKNOWLEDGMENT

The author would like to acknowledge all the doctors, nurses, and other healthcare providers during the COVID-19 crisis who put their health and lives on the line to treat those who were hospitalized with the virus. This research received no specific grant from any funding agency in the public, commercial, or not-for-profit sectors.

REFERENCES

Almeda, N., García-Alonso, C., & Salvador-Carulla, L. (2021). Mental health planning at a very early stage of the COVID-19 crisis: A systematic review of online international strategies and recommendations. *BMC Psychiatry*, *21*(1), 1–15. doi:10.118612888-020-03015-y PMID:33451305

American Association of Marriage and Family Therapists. (2015). *Code of Ethics.* https://www.aamft.org/Legal_Ethics/Code_of_Ethics.aspx

American Counseling Association. (2014). *Code of Ethics.* https://www.counseling.org/resources/aca-code-of-ethics.pdf

American Psychological Association. (2017). *Clinical practice guideline for the treatment of post-traumatic stress disorder: What is psychotherapy.* https://www.apa.org/ptsd-guideline/patients-and-families/psychotherapy

Blue Cross Blue Shield. (2020) *Telehealth benefits continue in 2021.* https://www.bcbsil.com/provider/education/2020/2020_12_22.html

Bricker, D. (2020). *Majority (54%) of Canadians Say Physical Distancing has Left Them Feeling Lonely or Isolated.* Ipsos.

Center for Systems Science and Engineering (CSSE) at Johns Hopkins University (JHU) COVID-19 Dashboard. (2021). https://gisanddata.maps.arcgis.com/apps/opsdashboard/index.html#/bda7594740fd40299 423467b48e9ecf6

Centers for Disease Control and Prevention. (2020). *Coronavirus (COVID-19).* https://www.cdc.gov/coronavirus/2019-nCoV/index.html

Harandi, T. F., Taghinasab, M. M., & Nayeri, T. D. (2017). The correlation of social support with mental health: A meta-analysis. *Electronic Physician, 9*, 5212–5222. https://dx.doi.org/10.19082/5212

Hilty, D., Feliberti, J., Evangelatos, G., Lu, F., & Lim, R. (2019). Competent cultural telebehavioral healthcare to rural diverse populations: Administration, evaluation, and financing. *Journal of Technology in Behavioral Science., 4.* Advance online publication. doi:10.100741347-018-0076-0

Hilty, D. M., Gentry, M. T., McKean, A. J., Cowan, K. E., Lim, R. F., & Lu, F. G. (2020). Telehealth for rural diverse populations: Telebehavioral and cultural competencies, clinical outcomes and administrative approaches. *mHealth, 20*(6). https://doi.org/10.21037/mhealth.2019.10.04

Law, T. (2020). *What Makes COVID-19 Different From the Flu? Time.com.*

Lee, J. H., Lee, H., Kim, J. E., Moon, S. J., & Nam, E. W. (2021). Analysis of personal and national factors that influence depression in individuals during the COVID-19 pandemic: A web-based cross-sectional survey. *Globalization and Health, 17*(1), 1–12. https://doi-org.ezp.twu.edu/10.1186/s12992-020-00650-8

Li, S., Wang, Y., Xue, J., Zhao, N., & Zhu, T. (2020). The impact of COVID-19 epidemic declaration on psychological consequences: A study on active Weibo users. *International Journal of Environmental Research and Public Health*, 17.

Lim, R. F., & Lu, F. (2008). Culture and psychiatric education. *Academic Psychiatry*, (32), 269–271. doi:10.1176/appi.ap.32.4.269

Liu, N., Zhang, F., Wei, C., Jia, Y., Shang, Z., & Sun, L. (2020). Prevalence and predictors of PTSD during COVID-19 outbreak in China hardest-hit areas: Gender differences matter. *Psychiatry Research, 287*, 112921.

McKibben. (2020). The economic impact of COVID-19. In R. Baldwin & B. Weder di Mauro (Eds.), *Economics in the Time of COVID-19* (pp. 45–53). CEPR Press.

National Association of Social Workers. (2017). *Code of Ethics*. https://www.socialworkers.org/About/Ethics/Code-of-Ethics

O'Reilly, K. B. (2019). *Federal law cripples telehealth in Medicare. New bill changes that*. Retrieved from https://www.ama-assn.org/practice management/digital/federal-law-cripples-telehealth-medicare-new-bill changes

Perrin, P. B., Rybarczyk, B. D., Pierce, B. S., Jones, H. A., Shaffer, C., & Islam, L. (2020). Rapid telepsychology deployment during the COVID-19 pandemic: A special issue commentary and lessons from primary care psychology training. *Journal of Clinical Psychology*, *76*, 1173–1185. https://dx.doi.org/10.1002/jclp.22969

Perry, K., Gold, S., & Shearer, E. M. (2020). Identifying and addressing mental health providers' perceived barriers to clinical video telehealth utilization. *Journal of Clinical Psychology*, *76*, 1125–1134. http://dx.doi .org/10.1002/jclp.22770

Pierce, B. S., Perrin, P. B., Tyler, C. M., & McKee, G. B. (2021). The COVID-19 Telepsychology Revolution: A National Study of Pandemic-Based Changes in U.S. Mental Health Care Delivery. *The American Psychologist*, *76*(1), 14–25.

Pitman, A., Mann, F., & Johnson, S. (2018). Advancing our Understanding of Loneliness and Mental Health Problems in Young People. *The Lancet Psychiatry*. doi:https://doi.org/10.1016/S2215-0366(18)30436-X

Ratts, M., Singh, A., Nassar-McMillan, S., Butler, S., & McCullough, J. (2016). Multicultural and social justice counseling competencies: Guidelines for the counseling profession. *Journal of Multicultural Counseling and Development*, *44*, 28–48. doi:10.1002/jmcd.12035

Roncoroni, J., Tucker, C. M., & Wall, W. (2014). Patient perceived cultural sensitivity of clinic environment and its association with patient satisfaction with care and treatment adherence. *American Journal of Lifestyle Medicine*, (8), 421–429. doi:10.1177/1559827614521760

Sanders, J. M., Monogue, M. L., Jodlowski, T. Z., & Cutrell, J. B. (2020). Pharmacologic treatments for coronavirus disease 2019 (COVID-19): A Review. *Journal of the American Medical Association*, *323*(18), 1824–1836. doi:10.1001/jama.2020.6019

Simple Practice. (2021). *Support clients anywhere, at any time*. https://www.simplepractice.com/telehealth/?utm_source=google&utm_medium=cpc&utm_campaign=rc-search-non-brand-dsa&utm_term=&gclid=Cj0KCQiApY6BBhCsARIsAOI_GjYIrCOtvVuOQDGy86qC-JcaEnKDG7oRiL9cfFzxmNZYwzPbG5of4xQaAh_OEALw_wcB

Sue, D. W., Arredondo, P., & McDavis, R. J. (1992). Multicultural counseling competencies and standards: A call to the profession. *Journal of Multicultural Counseling and Development*, *20*(2), 64–88. https://doi.org/10.1002/j.2161-1912.1992.tb00563.x

United States Centers for Medicare and Medicaid Services. (2019). *Tele medicine*. Retrieved from https://www.medicaid.gov/medicaid/benefits/ telemedicine/index.html

United States Department of Health and Human Services' Office for Civil Rights. (2020a). *Bulletin: HIPAA Privacy and Novel Coronavirus*. Retrieved from https://www.hhs.gov/sites/default/files/february-2020- hipaa-and-novel-coronavirus.pdf

United States Department of Health and Human Services' Office for Civil Rights. (2020b). *OCR eases HIPAA telehealth enforcement for COVID-19 emergency*. Retrieved from https://www.aha.org/news/headline/2020-03-18-ocr-eases-hipaa-telehealth-enforcement-covid-19-emergency

U.S. Food and Drug Administration. (2020). *FDA Takes Additional Action in Fight Against COVID-19 By Issuing Emergency Use Authorization for Second COVID- 19 Vaccine*. https://www.fda.gov/news-events/press-announcements/fda-takes-additional- ction-fight-against-covid-19-issuing-emergency-use-authorization-second-covid

Valtorta, N. K., Kanaan, M., Gilbody, S., Ronzi, S., & Hanratty, B. (2016). Loneliness and social isolation as risk factors for coronary heart disease and stroke: Systematic review and meta-analysis of longitudinal observational studies. *Heart (British Cardiac Society)*, *102*, 1009–1016. https://doi.org/10.1136/heartjnl-2015-308790

ADDITIONAL READING

Almeda, N., García-Alonso, C., & Salvador-Carulla, L. (2021). Mental health planning at a very early stage of the COVID-19 crisis: A systematic review of online international strategies and recommendations. *BMC Psychiatry*, *21*(1), 1–15. doi:10.118612888-020-03015-y PMID:33451305

Hilty, D. M., Gentry, M. T., McKean, A. J., Cowan, K. E., Lim, R. F., & Lu, F. G. (2020). Telehealth for rural diverse populations: Telebehavioral and cultural competencies, clinical outcomes and administrative approaches. *mHealth, 20*(6), 20. doi:10.21037/mhealth.2019.10.04 PMID:32270012

Pierce, B. S., Perrin, P. B., Tyler, C. M., McKee, G. B., & Watson, J. D. (2021). The COVID-19 Telepsychology Revolution: A National Study of Pandemic-Based Changes in U.S. Mental Health Care Delivery. *The American Psychologist, 76*(1), 14–25. doi:10.1037/amp0000722 PMID:32816503

Ratts, M., Singh, A., Nassar-McMillan, S., Butler, S., & McCullough, J. (2016). Multicultural and social justice counseling competencies: Guidelines for the counseling profession. *Journal of Multicultural Counseling and Development, 44*(1), 28–48. doi:10.1002/jmcd.12035

Varker, T., Brand, R. M., Ward, J., Terhaag, S., & Phelps, A. (2019). Efficacy of synchronous telepsychology interventions for people with anxiety, depression, posttraumatic stress disorder, and adjustment disorder: A rapid evidence assessment. *Psychological Services, 16*(4), 621–635. doi:10.1037er0000239 PMID:29809025

KEY TERMS AND DEFINITIONS

COVID-19: The novel coronavirus emerged at the end of 2019 in Wuhan China and spread worldwide infecting and killing millions of people. It is classified as a global pandemic.

Cultural Competence: The ability to work effectively with diverse groups of people through the process of evaluating personal biases, researching the specific culture of the clients served, and serving with curiosity and a non-judgmental attitude.

HIPAA: The Health Insurance Portability Act is a federal law aimed at protecting the confidential health information of citizens of the United States.

Mental Health: The state of one's psychological well-being. If someone is feeling content and adjusting normally, then they have good mental health. If someone is struggling with their mental health, they may have a mental illness, large amounts of stress, or psychological distress that interferes with their ability to function healthily in relationships, occupational functioning or self-care.

Online Counseling: Online counseling services that are accessed online either by text message or video conferencing

Telehealth: Accessing a medical or mental health service by telephone or online video conference. This is a synchronous exchange as opposed to an email visit or online questionnaire to access services.

Third-Party Payers: Third party payers include insurance companies, federal and local government entities, and grant money that funds medical or mental health treatment provided by individual providers or group treatment centers.

Section 2

Authenticity: Creating and Maintaining Equitable Personal Connections

Chapter 6
Digital Access to Culturally Relevant Curricula:
The Impact on the Native and Indigenous Student

Tim Swagerty
Midwestern State University, USA

ABSTRACT

Native and Indigenous students in low SES urban, rural, and reservation areas have the lowest retention-to-graduation rate and college completion rate of any demographic in the United States. The primary contributing factor is lack of culturally relevant curricula in a culturally sustaining pedagogy to present their culture in a positive representation. The second contributing factor is a lack of access to computer and internet resources to foster and bolster online, remote, and distance learning.

INTRODUCTION

The educational achievement gaps and digital divide for Native and Indigenous students in the United States in the twenty-first century has never been more evident or expansive. Availability and access to culturally relevant curricula in a culturally sustaining pedagogy has been a paramount issue since the inception of American Indian education policies. The contemporary and compounding issue of availability and accessibility to computers and internet services further deceases the probability or possibility of success for this underserved and underrepresented subpopulation of American society. Whether inner city urban or rural- and reservation-based,

DOI: 10.4018/978-1-7998-8028-8.ch006

these Native and Indigenous students suffer under conditions more readily seen in third-world countries.

High school graduation rates of Native and Indigenous students rank the lowest of any demographic in America today, 67% compared with a national average of 80%. This fact is evident in statistics representing reported state graduation rates with a high concentration of Native and Indigenous students. These states include in descending order are Alaska (14%), New Mexico (9%), South Dakota (8%), Oklahoma (7%), Montana (6%) and Arizona (4%). For those that do receive a high school diploma, GED or HSE, the increasing percentage of those students who enroll in, but do not complete a two or four year degree is represented in Table 1 by ethnicity for years 1976 to 2006. (Camera, 2015, DeVoe & Darling-Churchill, 2008, Oliff, 2017).

Figure 1. Percentage distribution of students enrolled in public and private degree-granting institutions. By type of institution and race/ethnicity: Selected years 1976-2006

	1976			1980			1990			2000			2006		
Race/ethnicity	Total	2-year	4-year	Total	2-year	4-year	Total	2-year	4-year	Total	2-year	4-year	Total	2-year	4-year
White	84.3	80.2	86.6	83.5	79.8	85.7	79.9	76.4	82.0	70.8	64.9	74.6	67.4	61.8	70.8
Total minority	15.7	19.8	13.4	16.5	20.2	14.3	20.1	23.6	18.0	29.2	35.1	25.4	32.6	38.2	29.2
Black	9.6	11.2	8.7	9.4	10.6	8.7	9.3	10.1	8.8	11.7	12.5	11.2	13.3	14.3	12.7
Hispanic	3.6	5.5	2.5	4.0	5.7	3.0	5.8	8.2	4.3	9.9	14.4	6.9	11.4	15.8	8.8
Asian/Pacific Islander	1.8	2.1	1.7	2.4	2.8	2.2	4.3	4.2	4.3	6.6	6.9	6.5	6.8	6.9	6.7
American Indian/ Alaska Native	0.7	1.1	0.5	0.7	1.1	0.5	0.8	1.1	0.6	1.0	1.3	0.9	1.1	1.3	0.9

NOTE: Race categories exclude persons of Hispanic ethnicity. Data from 1976 to 1996 are for institutions of higher education that were accredited by an agency or association that was recognized by the U.S. Department of Education, or recognized directly by the Secretary of Education. Data from 1996 and later years are for degree-granting institutions. The new degree-granting classification is very similar to the earlier higher education classification, except that it includes some additional institutions, primarily 2-year colleges, and excludes a few higher education institutions that did not award associate's or higher degrees. Detail may not sum to totals because of rounding.
SOURCE: U.S. Department of Education, National Center for Education Statistics, *Digest of Education Statistics, 2007*, based on Higher Education General Information Survey (HEGIS), "Fall Enrollment in Colleges and Universities," 1976 and 1980; and Integrated Postsecondary Education Data System (IPEDS), "Fall Enrollment Survey," 1990, Spring 2001, and Spring 2007.

The literature in this area of study acknowledges the high percentage of both high school and college non-completion rates among Native and Indigenous populations, but offers less information into the impact of culturally relevant curriculum and access to technology at the elementary, middle school and high school levels. The primary obstacle to understanding this lack of content and resource available to this subpopulation is the lack of knowledge and understating of the unique relationship with the Native and Indigenous Nations of the Americas and their respective national governments. (Bissell, 2004, Donnellan, 2017, Douglas, 2021, Fish & Syed, 2018, Rodriguez, 2018, Wilson, 2018).

To select a beginning point in explaining this relationship between these Nations of North America and most specifically within the political and chronological boundaries of the United States of America, one must begin with the catalytic moment that framed U.S. Indian policy during the nineteenth century that persisted unabated until well into the twentieth century. Richard Henry Pratt first used a phrase in an official address that defined the exchange between White educators and Native students until very recently, **"...all the Indian there is in the race should be dead. Kill the Indian in him, and save the man.**" Though not a peer reviewed or scholarly article, it is the source where this discussion most obviously begins. These early attitudes are what all of the subsequent literature stems from into modernity (Carlisle Indian School Digital Resource Center, 2021, Official Report of the Nineteenth Annual Conference of Charities & Correction, 1892).

Pratt's previous experiences detailed in his autobiography of educating "Negroes" after his commission as a second lieutenant in the famed all-Black 10th Cavalry Regiment during America's Civil War (1861-1865) laid the groundwork for what he later applied to his Indian Boarding Schools. He also concurrently commanded units of Cherokee, Choctaw, Osage and Tonkawa scouts for reconnaissance purposes in that conflict and future engagements against Western Plains hostiles in the Indian Wars (1864-1890). He additionally credits his time as a jailor of Native prisoners-of-war at Fort Leavenworth, Kansas and Fort Sill, I.T. (later Oklahoma) and subsequent attachment as such to Fort Marion, Florida as his qualifying exposure to Native culture (1874-1903). (Pratt, Utley & Adams, 2003, Series I: General Correspondence and Official Papers, 1867-1924, 2021, TSHA Online, 2021).

It was in this autobiographical reflection that he touted his skill and efficacy at educating Africans on the battlefield and its application to Natives in the classroom. From these a posteriori observations of Natives in captivity through the lens of his sophistic experiences with transplanted and subjugated Africans on the battlefield, Pratt made seriously overreaching conjecture of the absence of empiric multifariousness of both groups. His acknowledgement of the cognitive diversity among dissimilarly enculturated ethnicities and the application of rigid and unfamiliar European education models to accomplish the acculturation of both into accepted "white" society left much to be desired. His belief was that the absolute eradication of aboriginal culture and complete assimilation of the Indian by the American people was the only way to insure the "civilization" of these conquered peoples. (Pratt, Utley & Adams, 2003, Series I: General Correspondence and Official Papers, 1867-1924, 2021, TSHA Online, 2021, Van der Vegt & Janssen, 2003).

The idea that Native and Indigenous students lacked cognitive diversity or possessed learning styles different from white, European students and previously overpowered peoples was another promotion that prompted much consideration in not what was presented, but how. Natives were identified as spatial, visual and auditory learners with a proclivity toward observational learning. There is little, if any, research to support any of these stereotyped suppositions. The adherence by some "modern" educators even today to this type of cultural generalization is why many Native and Indigenous students find little enjoyment in their educational experiences. The vein of thought represented in many of these works as written by non-Native authors as to the best ways to educate Native students is an issue presently under contemporary academic scrutiny (Kaulback, 1984, Kleinfeld & Nelson, 1991, MacKinnon, 1972, McCartin & Schill, 1977).

The most current promotion of scholarly work related to Native and Indigenous students is that bicultural and cross-cultural education is a means to combat the effects of classrooms as a tool of cultural genocide and reinforced by interviews from Native and Indigenous elders and parents, as well as educators of Native and Indigenous ancestry. Many of these persons were direct witness and victims of Pratt's Indian Boarding Schools in the twentieth century application and testify to the lack of culturally sensitive and culturally supportive representation in their educational experience. An idea not fully realized initially then, but under considerable twenty-first century examination now, is the benefit of culturally relevant pedagogical practice. This, in support of culturally relevant curricula, with bolster the overall academic learning of Native and Indigenous students through a greater involvement and investment by the student (Klug & Whitfield, 2012, Ladson-Billings, 1995, McCarthy & Lee, 2014, Van Leeuwen, 1978, Witkin & Berry, 1975).

Negative examples of culturally insensitive curricula to Native and Indigenous students are replete in the history of the educational exchange between the Nations and the institutionalized European educational model. References to sitting "Indian" style, the term "Indian giver", Thanksgiving themed activities where students make "Indian" vests from paper grocery sacks and/or "Indian" headbands of colored construction paper decorated with these "Indian" symbols are a few examples of oft-used and time-honored inappropriate representations of Native and Indigenous culture in American classrooms. The most egregious example to date is a coloring sheet presented to a fourth grade class in North Texas in 2015 depicting a Comanche warrior engaged in an act of depredation and despoliation. The information provided on the reverse of the page offers very little in the way of historically supported content, but instead perpetuates incorrect information and culturally insensitive misinformation about Comanche culture and lifeways (See Appendix A and B) (Castagno, McKinley & Brayboy, 2008, Hicklin-Hudson & Ahlquist, 2003, Pewewardy & Hammer 2003).

Another contributing factor to the negative impact of the lack of culturally relevant curricula in American education is the lack of full-time instructional faculty in elementary, middle and high school classes of Native and Indigenous origination or identification. The same dynamic persists into higher education with less than one percent of full-time instructional staff at degree granting institutions being of Native and Indigenous origination or identification, as represented in Table 2. The benefit to both Native and non-Native students of the influence of Native and Indigenous instructional staff is the presentation of a cultural democracy, a concept that pushes back against the idea of the monoculturalistic Anglo society that biculturalism refers to a deficient or disempowered minority. This educational philosophy protects and supports the Native and Indigenous students' right to be educated in a manner that acknowledges and maintains their cultural identity while integrating in a constructive manner into the accepted and institutionalized values of the dominant culture. The benefit to the non-Native student is the exposure to a critical pedagogy that empowers all students to develop the capacity to reflect, critique, act to transform and develop sensitivity to aspects of other cultures. (Castellanos & Leggio, 1983, Darder, 1991, Harris, 1994, Horse, 2005, Ramirez, Castaneda & Herold, 1974, Watahomigie & Yamamoto, 2019).

Figure 2. Percentage Distribution offull-time instructionalfaculty in degree-granting institutions, race/ethnicity and academic rank

Faculty/rank	White	Total minority faculty	Black	Hispanic	Asian/ Pacific Islander	American Indian/ Alaska Native
Total	**82.8**	**17.2**	**5.6**	**3.6**	**7.6**	**0.5**
Professors	87.5	12.5	3.3	2.3	6.6	0.3
Associate professors	83.4	16.6	5.5	3.2	7.5	0.4
Assistant professors	78.6	21.4	6.8	3.9	10.2	0.5
Instructors	80.6	19.4	7.9	5.6	5.0	1.0
Lecturers	82.9	17.1	5.1	4.9	6.8	0.4
Other faculty	81.9	18.1	5.6	3.5	8.4	0.6

NOTE: Detail may not sum to totals because of rounding. Race categories exclude persons of Hispanic ethnicity.
SOURCE: U.S. Department of Education, National Center for Education Statistics, *Digest of Education Statistics 2007*, based on 2005–06 Integrated Postsecondary Education Data System (IPEDS), Winter 2005–06.

A further complicating factor to negative cultural stereotypes in the curricula and the lack of full-time Native and Indigenous instructional staff in the classroom are issues of accessibility to basic resources to compete educationally in a digital world. As shown in Table 3, American Indians and Alaskan Natives rank highest in percentage of households with no computer or computer access, high-speed internet or both. Computer is defined as and includes desktops, laptops, tablets or smartphones. From the middle of the twentieth century until now, the lack of supportive infrastructure in the areas of highest Native and Indigenous student populations to support on-line and distance learning are woefully below the national average. Through the unique relationship between the respective Nations' governments and the United States government, these Nations have the autonomy to pursue the development of supportive infrastructure, but rarely have the internal funding or federal support to accomplish the measures (Bissell, 2004, Hayes, 2003, Population Reference Bureau, 2020, Ujifusa, 2020).

Figure 3. Economic and Digital Divides of Children Ages 5-17 by Race and Ethnicity 2018
PRB analysis of data from the U.S. Census Bureau's 2018 American Community Survey Microdata Sample.

Racial or Ethnic Group	Lacks Computer, High-Speed Internet Access, or Both	Family Income Below 100% of Poverty	Household Receives SNAP Benefits
American Indian/Alaska Native	50%	30%	35%
Black	36%	31%	39%
Hispanic or Latino	34%	25%	28%
White	20%	10%	12%
Asian/NHOPI	14%	11%	11%
Two+ Races	19%	16%	23%
All Children	25%	17%	21%

The situation is exacerbated by the conditions inflicted by the most recent and profound ecological calamity of the twenty-first century. The digital divide has widened considerably under conditions of quarantine for insular areas of the United States. While urban-based Native and Indigenous communities were impacted as well, it is the rural and reservation-based populations that are most devastated by the combination of lack of hardware and connectivity. The states of the United States with the highest concentrations of Native and Indigenous students are more profoundly impacted than other areas of the country, as seen in Table 4. Where access had been available, even at some distance, at reservation schools and tribal community centers, the potential hazard of exposure to COVID-19 and transmission to susceptible populations at home; i.e., advanced age and/or immunocompromised, hindered many from participating in on-line or distance learning (American Indian Policy Institute, 2021, Camera, 2015, DeVoe & Darling-Churchill, 2008, Oliff, 2017).

Efforts by the tribal governments in conjunction with federal initiatives have extended phone and internet services to more rural and reservation-based Native and Indigenous communities in pursuance of the goal of "universal service" as promoted in federal mandate. Programs such as the Link-Up America and Lifeline Assistance, through the expansion and execution of the measures set out in the Communications Act of 1934 and the Telecommunications Act of 1996, have provided digital access to many of the underserved communities. The 1996 Act mandates that consumers in all regions of the Nation, including low-income consumers and those in rural, insular, and high cost areas, should have access to telecommunications and information services. It establishes that these services are required to be reasonably and comparable to those services provided in urban areas and that are available at rates that are reasonably and comparable to rates charged for similar services in urban areas (Communications Act, 1934, FCC, 2003, Telecommunications Act, 1996).

Figure 4. U.S. School Age Children at Educational Risk Due to COVID-19 Pandemic
PRB analysis of the 2018 ACS Public Use Microdata Sample and the U.S. Department of Education's 2019 National Assessment of Education Progress (NAEP) database.

	Percent of all school-age children who belong to racial or ethnic minority	Without laptop, desktop, or tablet	Without paid high-speed internet access	Without computer, or paid high-speed internet access, or both	Family income is below 100% of Poverty	Household receives SNAP benefits
United States	44.5	15.3	29.4	33	26	30.9
Alaska	45.1	14	36.3	39.2	14.5	25.9
Arizona	57.9	17.5	33.8	37.1	26.5	30.3
Montana	20.6	14.9	31.7	33.9	23.2	25.8
New Mexico	74.3	22.2	39.2	42.7	30.5	36.2
Oklahoma	45	18.4	39.9	43.3	28.9	27.3
South Dakota	25.9	35	46.2	55.8	42.8	37.4

The Federal Communications Commission (2003) report on promoting deployment and subscribership in underserved areas, including tribal and insular areas indicated that:

- On Indian reservations, only 47% of households have telephone service.
- On the Navajo reservation, only 22% of households have telephone service.
- Number of tribal households – 453,269

Section 706 of the Telecommunications Act of 1996 requires that the Federal Communications Commission (FCC) report on the status of broadband internet access being "…deployed to all Americans in a reasonable and timely fashion…" The results of the 2015 reporting are:

- 63 percent of Americans living on Tribal lands (2.5 million people) lack access to 25 Mbps/3 Mbps broadband

- 85 percent living in rural areas of Tribal lands (1.7 million people) lack any access to broadband or internet

A similar reporting of 2019 in pursuance of the Repack Airwaves Yielding Better Access for Users of Modern Services Act of 2018 indicated little improvement. This data is based on and in conjunction with data from the 2010 U.S. Census. It is this most recent report seen in Table 5 that indicates only 21% of Tribal housing units are covered by two or more wired providers of 25/3 Mbps services, while two or more such providers cover 56% of non-tribal housing (Communication Act, 1934, FCC, 2003, FCC, 2015, FCC, 2019, Telecommunications Act, 1996).

Figure 5. Deployment of 25/3 Mbps Fixed Service to Non-Tribal Housing Units
Federal Communications Commission (2019) Report on Broadband Deployment in Indian Country, Pursuant to the Repack Airwaves Yielding Better Access for Users of Modern Services Act of 2018.

Deployment of 25/3 Mbps Fixed Service to Non-Tribal and Tribal Housing Units (HUs)												
	By Any Technology				By Any Terrestrial Technology				By Any Wired Technology			
# providers	Non-tribal HU (000s)	Tribal HU (000s)	*Non-tribal*	*Tribal*	Non-tribal HU (000s)	Tribal HU (000s)	*Non-tribal*	*Tribal*	Non-tribal HU (000s)	Tribal HU (000s)	*Non-tribal*	*Tribal*
0	98	28	0%	2%	9,418	550	7%	31%	11,734	628	8%	36%
1	1,615	195	1%	11%	42,023	644	30%	37%	48,975	762	36%	43%
2	13,578	503	10%	29%	58,640	362	43%	21%	62,578	335	46%	19%
3 or more	121,962	1,024	89%	58%	27,173	194	20%	11%	13,966	26	10%	2%
Total	137,254	1,750	100%	100%	137,254	1,750	100%	100%	137,254	1,750	100%	100%

The 2019 report also indicates the results presented in Table 6 that 25% of housing units on Tribal lands—about 431,000—have no wired option for 10/1 Mbps service, while only 6% of housing units on non-Tribal lands lack coverage by any wired provider of 10/1 Mbps. Of that 431, 000 housing units, 45% of those housing units on Tribal lands are limited to only one wired option, compared to only 22% of those on non-Tribal lands (FCC, 2019).

Figure 6. Deployment of 10/1 Mbps Fixed Service to Non-Tribal Housing Units
Federal Communications Commission (2019) Report on Broadband Deployment in Indian Country, Pursuant to the Repack Airwaves Yielding Better Access for Users of Modern Services Act of 2018.

Deployment of 10/1 Mbps Fixed Service to Non-Tribal and Tribal Housing Units (HUs)												
	By Any Technology				By Any Terrestrial Technology				By Any Wired Technology			
# providers	Non-tribal HU (000s)	Tribal HU (000s)	*Non-tribal*	*Tribal*	Non-tribal HU (000s)	Tribal HU (000s)	*Non-tribal*	*Tribal*	Non-tribal HU (000s)	Tribal HU (000s)	*Non-tribal*	*Tribal*
0	21	26	0%	2%	4,411	287	3%	16%	7,685	431	6%	25%
1	745	99	1%	5%	23,829	591	17%	34%	30,989	797	22%	45%
2	6,631	362	5%	21%	62,997	518	46%	30%	79,661	471	58%	27%
3 or more	129,857	1,264	94%	72%	46,016	355	34%	20%	18,918	51	14%	3%
Total	137,254	1,750	100%	100%	137,254	1,750	100%	100%	137,254	1,750	100%	100%

As seen in Table 7 from the FCC 2019 report, a very low 12% of Tribal housing units have a choice of more than one terrestrial provider of 100/10 Mbps service, as opposed to approximately 46% of non-Tribal housing units. An additional 12% of Tribal housing units do have a choice of more than one wired provider of such service, but compared to approximately 44% of non-Tribal housing units (FCC, 2019).

Figure 7. Deployment of 100/10 Mbps Fixed Service to Non-Tribal and Tribal Housing Units

Federal Communications Commission (2019) Report on Broadband Deployment in Indian Country, Pursuant to the Repack Airwaves Yielding Better Access for Users of Modern Services Act of 2018.

Deployment of 100/10 Mbps Fixed Service to Non-Tribal and Tribal Housing Units (HUs)												
	By Any Technology				By Any Terrestrial Technology				By Any Wired Technology			
# providers	Non-tribal HU (000s)	Tribal HU (000s)	*Non-tribal*	*Tribal*	Non-tribal HU (000s)	Tribal HU (000s)	*Non-tribal*	*Tribal*	Non-tribal HU (000s)	Tribal HU (000s)	*Non-tribal*	*Tribal*
0	17,505	782	13%	45%	17,505	782	13%	45%	18,127	794	13%	45%
1	56,427	765	41%	43%	56,427	765	41%	43%	59,162	757	43%	43%
2	48,587	187	35%	11%	48,587	187	35%	11%	48,769	183	36%	11%
3 or more	14,734	17	11%	1%	14,734	17	11%	1%	11,196	16	8%	1%
Total	137,254	1,750	100%	100%	137,254	1,750	100%	100%	137,254	1,750	100%	100%

Many locally based grass roots efforts at the tribal and reservation levels have manifested and are producing favorable results in the provision of broadband network and high-speed internet to their rural- and reservation based-populations. There are presently forty tribal owned broadband network companies covering sixty-five Native and Indigenous Nations in the United States. Another thirty-seven Nations are working in partnerships with private service providers to extend service to insular populations in previously unserved/underserved areas. The Federal Communication Commission, in support of these efforts, created the Rural Tribal Priority Window in 2020 to allow Nations to apply for 2.5 GHz spectrum licenses at no cost. This coverage is suitable for both mobile and fixed point-to-point use. Four-hundred license applications were received in the February 3, 2020 to September 2, 2020 window. The FCC will be reviewing the applications in 2021 and aiding Nations to embark on these valuable infrastructure projects. This will play an important role in the extension of broadband and communication services to Native and Indigenous communities on Tribal reservation lands in the United States. (Adcock, 2014, Donnellan, 2017, Douglas, 2021, Sottile, 2020,).

The identified issues of absence of culturally relevant educational content, nonexistence of critical pedagogy and deficiencies in existence, availability and access to digital resources for rural- and reservation-based Native and Indigenous students in the United States comprise the trifecta of impedances to the ability of these students to compete in a global, digital environment. While effort has been made and improvement seen, the continued focus and allocation of resources by federal, state, local and tribal authorities to narrow this obvious educational and digital divide is paramount to the continued academic success and achievement of this population, the Native and Indigenous students of the United States.

REFERENCES

Adcock, T. (2014). Technology Integration in American Indian Education: An Overview. *Journal of American Indian Education*, *53*(2), 104–121. Retrieved April 28, 2021, from https://www.jstor.org/stable/43610478es

American Indian Policy Institute. (n.d.). *Digital Divide*. Retrieved April 28, 2021, from https://aipi.asu.edu/content/digital-divide

Bissell, T. (2004). The Digital Divide Dilemma: Preserving Native American Culture While Increasing Access to Information Technology on the Reservations. *Journal of Law, Technology and Policy, 2004*(1), 129-150. Retrieved from http://illinoisjltp.com/journal/wp-content/uploads/2013/10/bissell.pdf

Camera, L. (n.d.). *Native American students left behind*. Retrieved March 31, 2021, from https://www.usnews.com/news/articles/2015/11/06/native-american-students-left-behind

Carlisle Indian School Digital Resource Center. (n.d.). Retrieved May 4, 2021 from https://carlisleindian.dickinson.edu/teach/kill-indian-and-save-man-capt-richard-h-pratt-education-native-americans

Castagno, A., & Brayboy, B. (2008). Culturally Responsive Schooling for Indigenous Youth: A Review of the Literature. *Review of Educational Research*, *78*(4), 941–993. doi:10.3102/0034654308323036

Castellanos, D., & Leggio, P. (1983). *The best of two worlds: Bilingual-bicultural education in the US*. Academic Press.

Communications Act of 1934, 48 Stat. 1064 (1934) (current version at 47 U.S.C. § 151 (2000)

Darder, A. (1991). *Culture and power in the classroom: A critical foundation for bicultural education*. Greenwood Publishing Group.

DeVoe, J. F., & Darling-Churchill, K. E. (2008). *Status and Trends in the Education of American Indians and Alaska Natives: 2008 (NCES 2008-084)*. National Center for Education Statistics, Institute of Education Sciences, U.S. Department of Education.

Donnellan, E. S. (2017). No Connection: The Issue of Internet on the Reservation. *American Indian Law Journal, 5*(2).

Douglas, J. (2021, February 4). *Bridging the digital divide in Indian country*. Retrieved April 29, 2021, from https://www.hcn.org/issues/53.4/indigenous-affairs-technology-bridging-the-digital-divide-in-indian-country

Federal Communications Commission. (2003). *Promoting Deployment/Subscribership in Underserved Areas, Including Tribal and Insular Areas*. Retrieved May 3, 2021 from https://www.fcc.gov/Bureaus/Wireless/Factsheets/tribal_lands_fact_sheet.txt

Federal Communications Commission. (2015). *Broadband Progress Report*. Author.

Federal Communications Commission. (2019). *Report on Broadband Deployment in Indian Country, Pursuant to the Repack Airwaves Yielding Better Access for Users of Modern Services Act of 2018*. Retrieved May 03, 2021, from https://aipi.clas.asu.edu/sites/default/files/05011019fccreport_on_broadband_deployment_in_indian_country_pursuant_to_the_repack_airwaves_yielding_better_access_for_users_of_modern_services_act_of_2018.pdf

Federal Communications Commission. (2021). *2.5 GHz Rural Tribal Window*. Retrieved May 03, 2021, from https://www.fcc.gov/25-ghz-rural-tribal-window

Fish, J., & Syed, M. (2018). Native Americans in Higher Education: An Ecological Systems Perspective. *Journal of College Student Development, 59*(4), 387–403. doi:10.1353/csd.2018.0038

Harris, S. (1994). "Soft" and "hard" domain theory for bicultural education in Indigenous groups. *Peabody Journal of Education, 69*(2), 140–153. doi:10.1080/01619569409538769

Hayes, S. (2003). *Digital Divide in Indian Country: Obstacles and Opportunities for Native American Communities*. Retrieved April 28, 2021, from http://students.washington.edu/aliss/silverfish/archive/april2003/hayes.pdf

Hickling-Hudson, A., & Ahlquist, R. (2003). Contesting the Curriculum in the Schooling of Indigenous Children in Australia and the United States: From Eurocentrism to Culturally Powerful Pedagogies. *Comparative Education Review, 47*(1), 64–89. doi:10.1086/345837

Horse, P. G. (2005). *Native American Identity*. Retrieved October 27, 2020, from http://convention.myacpa.org/houston2018/wp-content/uploads/2017/11/Horse_2005.pdf

Kaulback, B. (1984). Styles of learning among native children: A review of the research. *Canadian Journal of Native Education, 11*(3), 27–37.

Kleinfeld, J., & Nelson, P. (1991). Adapting instruction to Native Americans' learning styles: An iconoclastic view. *Journal of Cross-Cultural Psychology, 22*(2), 273–282. doi:10.1177/0022022191222007

Klug, B. & Whitfield, P. (2012). *Widening the Circle*. doi:10.4324/9780203616703

Ladson-Billings, G. (1995). Toward a theory of culturally relevant pedagogy. *American Educational Research Journal, 32*(3), 465–491. doi:10.3102/00028312032003465

MacKinnon, A. A. (1972). Eskimo and Caucasian: A discordant note on cognitive-perceptual abilities. In *Proceedings of the annual convention of the American Psychological Association*. American Psychological Association.

McCartin, R., & Schill, W. J. (1977, October). Three modes of instruction. *Journal of American Indian Education*, 14–20.

McCarty, T., & Lee, T. (2014, April). Critical Culturally Sustaining/Revitalizing Pedagogy and Indigenous Education Sovereignty. *Harvard Educational Review, 84*(1), 101–124. doi:10.17763/haer.84.1.q83746nl5pj34216

Official Report of the Nineteenth Annual Conference of Charities and Correction (1892), 46–59. Retrieved from http://historymatters.gmu.edu/d/4929/

Oliff, H. (2017, May 16). *Graduation rates and American Indian Education* [Web blog post]. Retrieved from http://nativepartnership.org/graduation-rates-american-indian-education/

Pewewardy, C. D., & Hammer, P. (2003). *Culturally Responsive Teaching for American Indian Students*. ERIC Digest.

Population Reference Bureau. (2020). *Children, coronavirus, and the digital divide: Native American, black, and Hispanic students at greater Educational risk during pandemic*. Retrieved April 27, 2021, from https://www.prb.org/coronavirus-digital-divide-education/

Pratt, R. H. (n.d.). *Series I: General Correspondence and Official Papers, 1867-1924*. https://archives.yale.edu/repositories/11/archival_objects/435399

Pratt, R. H., Utley, R. M., & Adams, D. W. (2003). *Battlefield and classroom: four decades with the American Indian, 1867-1904*. University of Oklahoma Press.

Ramirez, M., Castaneda, A., & Herold, P. L. (1974). The relationship of acculturation to cognitive style among Mexican Americans. *Journal of Cross-Cultural Psychology*, *5*(4), 424–433. doi:10.1177/002202217400500404

Rodriguez, K. (2018). *Native American Education: Building Stronger Families, Communities, and Youth through Cultural Education*. Retrieved from https://digital.sandiego.edu/solesmalscap/22

Sottile, C. (2020, October 16). *Native American Teachers, entrepreneurs seek new ways to close digital divide*. Retrieved April 28, 2021, from https://www.nbcnews.com/news/us-news/native-american-teachers-entrepreneurs-seek-new-ways-close-digital-divide-n1243746

Telecommunications Act of 1996, Pub. LA. No. 104-104, 110 Stat.

TSHA. (n.d.). *Pratt, Richard Henry (1840–1924).* https://www.tshaonline.org/handbook/entries/pratt-richard-henry#:~:text=With%20the%20outbreak%20of%20the,Rosecrans%20and%20George%20H

Ujifusa, A. (2020, December 4). *1 in 3 American Indian, black, and Latino children fall into digital divide, study says*. Retrieved April 28, 2021, from https://www.edweek.org/education/1-in-3-american-indian-black-and-latino-children-fall-into-digital-divide-study-says/2020/07

Van der Vegt, G. S., & Janssen, O. (2003). Joint impact of interdependence and group diversity on innovation. *Journal of Management*, *29*(5), 729–751. doi:10.1016/S0149-2063_03_00033-3

Van Leeuwen, M. S. (1978). A cross-cultural examination of psychological differentiation in males and females 1. *International Journal of Psychology*, *13*(2), 87–122. doi:10.1080/00207597808246615

Watahomigie, L. J., & Yamamoto, A. Y. (2019). Linguistics in action: The Hualapai bilingual/bicultural education program. In *Collaborative research and social change* (pp. 77–98). Routledge. doi:10.4324/9780429040573-6

Wilson, A. T. (2018). *Native American Male Perspectives on Successful Higher Educational Attainment: Primary Motives and Practices in Overcoming Hindering Barriers in Degree Attainment*. UCLA. Retrieved from https://escholarship.org/uc/item/3qv9d1sw

Witkin, H. A., & Berry, J. W. (1975). Psychological differentiation in cross-cultural perspective. *ETS Research Bulletin Series*, *1975*(1), i-100. doi:10.1002/j.2333-8504.1975.tb01043.x

APPENDIX A

Figure 8.

APPENDIX B

Figure 9.

HORSEMEN OF THE PLAINS

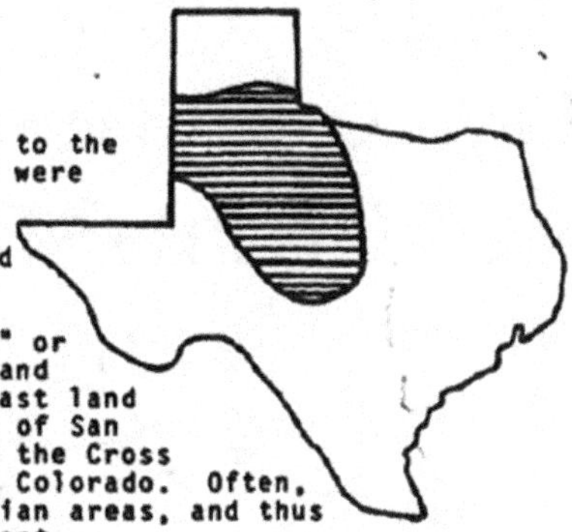

Their name meant "enemy." True to the meaning of their name, the Comanches were the determined enemy of the white settlers in Texas. From the early 1700s until 1875, the Comanches would roam and terrorize the Texas plains and plateaus. The Comanche Indians considered the lands of "Comancheria" or "Comanche Country" to be their home and hunting ground. Comancheria was a vast land whose boundaries extended from south of San Antonio northward to Kansas, east to the Cross Timbers and west into New Mexico and Colorado. Often, the Comancheria overlapped other Indian areas, and thus the threat of battle was always present.

The Comanches, like the Apaches, were divided into several different groups or bands. There were the Honeyeaters who lived mainly in the area between the Brazos and Colorado Rivers in northwest Texas. There were the Wanderers who roamed the lands of northwest Texas between the Red and Brazos Rivers, and finally, there were the Antelopes who considered the Texas Panhandle and beyond as their hunting grounds.

The Comanche Indians were certainly the true "horsemen of the plains," learning to ride a horse before they could walk! Both men and women of the Comanche tribe were excellent riders. The Comanche boy, for example, was not only riding a pony by the age of four or five, he was also learning to pick up objects from the ground while his mount traveled at a fast speed. Eventually, a young brave would be strong enough to pick up a fallen warrior. The horse, as a result, was an important influence in the Comanche culture. This was true of the buffalo as well.

Buffalo hunts usually involved most of the Comanche tribesmen. The mounted hunters would try to surround the buffalo and move them into a circle where the braves would kill them with bows and arrows or spears. Once the buffalo had been killed and skinned, some of the meat was prepared by drying it into jerky. The rest was either boiled, broiled, or eaten raw! The buffalo also provided water bags that were made from the stomach, and ornaments, cups, and utensils that were made from the horns, hooves, and bones. Strings, thread, and ropes were made from buffalo hair. Tanned buffalo hides were used for clothing, bedding, tepee coverings, saddles, and robes. Moreover, the Comanches also hunted antelope, longhorn cattle, bear, and elk.

The Comanches were also mighty warriors. Not only did war demonstrate bravery, but it was necessary to protect the Comancheria from other neighboring tribes and white settlers who threatened to attack at any time. Comanche braves on the warpath were an impressive site. Many wore feather headdresses mounted

20

Chapter 7
Is Telehealth an Effective Way to Increase Accessibility to Counseling Services?

Tara Renee Fox
Midwestern State University, USA

ABSTRACT

Providing telehealth is often a means to increase the accessibility to and availability of clinical mental healthcare services. Due to the COVID-19 pandemic, telehealth has been globally implemented into healthcare systems. Today, almost 390 million individuals have at least one mental illness. There are many challenges to seeking clinical mental healthcare, including availability and accessibility, anonymity, finances and insurance, stigma, and travel and transportation. Due to these barriers, many individuals have untreated mental health conditions, which can burden healthcare systems. By utilizing innovative delivery models such as telehealth technologies, the disparities experienced by individuals when attempting to seek clinical mental healthcare services can decrease.

INTRODUCTION

Telehealth services are increasingly becoming a preferable delivery model for healthcare. Having the ability to utilize telehealth services often means increasing access to mental healthcare services. Today, the integration of telehealth mental healthcare proves necessary to decrease barriers to seeking clinical mental healthcare services. In 2016, The National Institute of Mental Health stated that one in five American adults lived with a mental illness (National Institute of Mental Health,

DOI: 10.4018/978-1-7998-8028-8.ch007

2017). The prevalence of mental illness fluctuates significantly depending on age, gender, life experiences, race, and socioeconomic status. Mental illness is defined as a mental, behavioral, or emotional disorder that causes impairment (Aaronson et al., 2018; National Institute of Mental Health, 2017). Stress can be a factor that can increase symptoms of mental illness. During the COVID-19 global pandemic, stress, isolation, grief, and loss influenced individuals' mental health. Despite one in five individuals dealing with mental health issues, clinical mental healthcare services can be challenging to navigate. In recent decades, mental and behavioral disorders have been the fifth most costly diseases in the United States (Diederich & Song, 2014; Langarizadeh et al., 2017). Due to the prevalence of mental illness, telehealth is a viable platform to create better pathways for an otherwise complicated mental healthcare system. When thinking about telehealth and how viable it is to treat mental health issues, it is important to consider the effectiveness of telehealth for common mental health diagnoses, the advantages and disadvantages of telehealth, and the future of clinical mental healthcare treatment through telehealth services.

MENTAL HEALTH A GLOBAL ISSUE

Research has suggested that mental health is a global issue (Acharya et al., 2017; Handley et al., 2014b; Monteiro et al., 2014; Robinson et al., 2012; Sinclair et al., 2013; Zubatsky et al., 2018), significantly impacting one in 10 individuals globally and accounting for 30% of the global non-fatal disease burden (Mnookin, 2016). Over the last decade, global mental health has attracted the attention of health policymakers and researchers. Global health has been defined as an area of study, research, and practice that emphasizes improving health and achieving evenhandedness in health for individuals worldwide (Patel, 2014; Koplan et al., 2009). Global health has three key aspects, including the priorities are determined by the burden of disease, striving for fairness in the delivery of health between populations, and a global scope concerning knowledge and actions to benefit individuals globally (Patel, 2014). Due to the growing burden of global mental health, actions to promote global health are necessary. An estimated 5% of the worldwide population is impacted by mental health disorders, including psychosis, intellectual disabilities, dementia, drug or alcohol dependence, and severe depression (Patel, 2014). Five percent of the global population translates to an average of almost 390 million individuals impacted globally (Patel, 2014; U.S. Census Bureau et al., 2021).

The Impact of Mental Health Issues

Over the last decade, there has been a resilient international undertaking to address clinical mental healthcare needs and decrease disparities worldwide, specifically for low socioeconomic populations (Cooper, 2016; Patel & Prince, 2010; Rathod et al., 2017). Mental health issues, including depression, anxiety, and substance abuse, are some of the most common and incapacitating health conditions internationally. The crucial issue, from an international viewpoint, is not only the enormous burden and growing numbers of individuals impacted by mental disorders, but the fact that many of these persons are not receiving the services we know can improve their lives (Patel, 2014; World Health Organization, 2011; Zubatsky et al., 2018). The limited access to clinical mental health services globally discourages individuals from seeking help for their mental illness. Global mental health is defined as promoting healthy quality of life for individuals with mental disorders worldwide, including those who live in lower socioeconomic countries who only have access to a fraction of mental health resources (Patel, 2014). Today, global mental health has a wide range of stakeholders, including consumers, civil society groups, policymakers, donors, and development agencies. Making a difference in mental health trenches requires a change in delivery methods to improve global mental health practice. Clinical mental health providers are responsible for reducing the disparities of mental healthcare services to individuals worldwide.

The Future of Global Mental Health

Due to the COVID-19 pandemic, global health resources are needed to address psychosocial stressors, adverse outcomes from isolation, and fear among international populations (Torales et al., 2020). There is a need for a worldwide response addressing the impact that COVID-19 had on individual's mental health. When mental health issues are left untreated, it creates a burden not only on the individual but the individual's family, friends, or place of occupation. When individuals are battling mental health issues, it impacts their ability to take care of themselves, maintain medication compliance, and decrease their occupational productivity. There is a great need to increase global resources to address international mental health concerns. Individuals living in low-socioeconomic areas are at high risk for developing disorders that cause significant impairment. If we continue to ignore the burden of untreated mental health issues, personal consequences, social consequences, and economic loss will continue to grow (Andrade et al., 2014).

COVID-19

In 2020 the World Health Organization named COVID-19 the greatest global pandemic experienced in over a century (Li et al., 2020; Lyne et al., 2020; World Health Organization, 2020). COVID-19 has affected physical, mental, and social health issues. Due to the global pandemic, many individuals have experienced grief from the death of friends and family while being socially isolated. An individual cannot plan for death; it is difficult and unwanted (Butler, 2002; Crunk et al., 2017; Dyer & Hagedorn, 2013; Neimeyer & Burke, 2012; Neimeyer et al., 2010). The first case of COVID-19 within the United States was reported on January 21, 2020, and the first death on February 6, 2020 (Hauck et al., 2020). As of summer 2021, nearly 180 million cases of COVID-19 have been confirmed, totaling over 3.8 million deaths (World Health Organization, 2021).

COVID-19 and Mental Health

Due to the COVID-19 global pandemic, the delivery of mental health services has been disrupted. The pandemic has caused the mental healthcare system, an already fragmented and challenging to navigate system, to buckle under current clinical mental healthcare demands. Virtual care, otherwise known as telehealth, has begun to be globally adopted to reduce the risk of exposure for healthcare workers (Kola, 2020; Webster, 2020). As a result of the current focus of the pandemic being the transmission rate globally, the psychosocial consequences, including long-lasting problems, isolation, stigma, emotional trauma, and grief, are getting overlooked (Brooks et al., 2020; Lyne et al., 2020; Torales et al., 2020). Despite many countries beginning to open back up, the demand for clinical mental healthcare services may continue to increase. Due to this demand, clinical mental healthcare providers need to quickly adapt delivery models to accommodate the rapidly changing mental health landscape. The use of telehealth, phone, and video technologies, for assessments, has dramatically increased (Lyne et al., 2020). There is a need for leadership within community mental healthcare systems to battle the challenges created by COVID-19 on a local and global level.

Moving Past the Pandemic

As the impact of COVID-19 continues to evolve, there is much to consider from a clinical mental healthcare perspective. The psychosocial drain will become progressively apparent as the effect of physical distancing, loneliness, grief, and unemployment begins to manifest. Telehealth services need to continue to develop in order to treat the long-term effects of the global pandemic. The COVID-19 crisis has emphasized the need to re-examine policies to increase the accessibility to clinical mental healthcare services. Due to the long-term effects experienced, support and follow-up need to be provided for extended periods. "Psychological crisis interventions include three key points: 1) understanding the mental health status in different populations influenced by the COVID-19 outbreak; 2) identifying people who are at high risk of suicide and aggression, and 3) providing appropriate psychological interventions for those in need" (Li et al., 2020, p. 1735). The clinical mental healthcare profession needs researchers willing to build on current literature to explore innovative ways to decrease mental healthcare disparities worldwide.

IMPACTING URBAN AND RURAL AREAS

Mental health issues impact urban and rural areas; however, research has shown unique challenges faced by rural individuals compared to their urban counterparts. For rural communities, barriers such as negative social and cultural influences, lack of network ties, the concentration of poverty, stigma, dual relationships, lack of accessibility, and geographic isolation are factors that increase disparities when seeking clinical mental healthcare services (Bischoff et al., 2014; Breslau et al., 2014; Fox, 2019; Galea et al., 2007; Marsella, 1998; Procter et al., 2015). Due to the difficulty experienced when accessing mental healthcare, individuals often do not seek treatment. In the United States, roughly 91 million adults live in communities with a deficiency of mental healthcare professionals (Fox, 2019; Health Resources and Services Administration, 2016; Soltis-Jarrett, 2019). Without a system-wide change to improve internet access and increase economic opportunities, clinical mental healthcare disparities will persist in rural areas. Advocacy for the clinical mental healthcare profession is necessary to fill the educational gaps where myths and biases about mental healthcare are heavily present.

Barriers and Challenges

There is a strong correlation between poverty levels and mental health issues for individuals in urban and rural areas (Cidade et al., 2012; Fox, 2019; Inder et al., 2011; Nepomuceno et al., 2015). Researchers have identified low socioeconomic status as a multidimensional factor. The stressors that low socioeconomic individuals are exposed to place individuals at a greater danger of poor mental health outcomes. Individual's challenges when seeking clinical mental healthcare are unique; therefore, they require innovative healthcare delivery methods. Without the aid of policy changes, individuals may continue to struggle to access quality mental healthcare services.

OVERWHELMED MENTAL HEALTH SYSTEMS

Researchers concluded that in the United States, public health systems are overcome by primary mental healthcare needs (Brezenski, 2014; Cook et al., 2007; Fox, 2019). Traditional healthcare systems are becoming overwhelmed by the increasing number of individuals presenting with mental health issues. An overburdened mental healthcare system produces a substantial overload on clinical mental health providers serving underserved communities. This strain was only compounded during the global pandemic, where mental health systems were stunned by the number of individuals needing mental healthcare services. The level of compassion and empathy shown by healthcare providers, emergency services, and community resource professionals has proven psychically, mentally, and economically costly (Cocker & Joss, 2016). In order to increase accessibility to individuals living with common mental health problems and provide some relief to clinical mental healthcare providers, exploring innovative delivery models such as telemental healthcare services is necessary.

BARRIERS TO SEEKING CLINICAL MENTAL HEALTH SERVICES

Research has shown that many individuals who might benefit from mental healthcare do not receive treatment or do not complete treatment (Andrade et al., 2014). Having a working understanding of the barriers to seeking clinical mental health treatment is essential when planning for mental healthcare services, prioritizing the distribution of resources, and reducing the burden created by untreated mental health issues. When seeking clinical mental healthcare services, individuals may face significant barriers. Researchers described barriers such as lack of availability and accessibility of mental health providers (Aisbett et al., 2007; Fox, 2019; Hall & Gjesfjeld, 2013; Inder et al., 2011; Jackson et al., 2007; Jameson & Blank, 2007; Monteiro et al., 2014; Pelling & Butler, 2015; Robinson et al., 2012), lack of anonymity for individuals living in rural areas (Aisbett et al., 2007; Fox, 2019; Hall & Gjesfjeld, 2013; Jameson & Blank, 2007; Procter et al., 2015), mental health stigma or fear of exclusion (Aisbett et al., 2007; American Academy of Family Physicians, 2011; Brezenski, 2014; Fox, 2019; Jameson & Blank, 2007; Monteiro et al., 2014; Procter et al., 2015; Robinson et al., 2012), lack of transportation or substantial traveling distance to a mental health specialist (Aisbett et al., 2007; Colon-Gonzalez et al., 2013; Fox, 2019; Jackson et al., 2007; Procter et al., 2015), and lack of adequate finances or insurance coverage for clinical mental health counseling services (Arora et al., 2017; Fox, 2019; Inder et al., 2011; Jameson & Blank, 2007; Robinson et al., 2012). These barriers contribute to individuals not receiving the mental health assistance they need.

Availability and Accessibility

The first barrier to seeking clinical mental healthcare services identified by the research is availability and accessibility to mental healthcare providers. Rural areas, specifically, struggle with a lack of availability of and accessibility to mental healthcare professionals. When the availability of and accessibility to clinical mental healthcare services are limited, individuals living in rural communities are then required to seek out specialty services or increase their traveling distances (Aisbett et al., 2007; Fox, 2019; Hall & Gjesfjeld, 2013; Inder et al., 2011; Jackson et al., 2007). The deficiency in access to clinical mental health providers contributes to the increasing rate of mental health issues being treated in primary care medical facilities. Due to the limited access to clinical mental healthcare providers, individuals are more likely to neglect their mental healthcare needs.

Anonymity and Confidentiality

The second barrier identified by the research is a lack of anonymity and confidentiality. In rural communities specifically, when a clinical mental health counselor is practicing in a rural area, privacy and remaining anonymous may be concerning (Corrigan & Shapiro, 2010; Fox, 2019; Paulson, 2010; Pescosolido et al., 2010). The culture of everyone knowing everyone in rural communities provides a challenge to anonymity and confidentiality. For example, having a vehicle parked in front of a counseling office may increase client vulnerability and increase perceived mental health stigma. A feeling of lack of anonymity can severely hinder an individual's attitudes towards clinical mental healthcare services.

Mental Health Stigma

Mental health stigma is the third barrier identified that might prevent individuals from seeking clinical mental health services. Mental health stigma is a common roadblock that prevents individuals from seeking clinical mental health services (Aaronson et al., 2018; Brezenski, 2014; Corrigan et al., 2014; Corrigan & Penn, 2015; Fox, 2019; Procter et al., 2015; Robinson et al., 2012). During the ages of 18-24 years old, acceptance from peers is highly sought out. Due to this trait, mental health stigma can greatly impact a young adult's desire to seek clinical mental healthcare services, even when necessary. Due to the complexity of mental health stigma, many individuals choose not to seek clinical mental healthcare services to avoid any negative consequences. These negative consequences include appearing weak, lack of support from friends or family, or mental healthcare not being accepted by the individual's culture (Aaronson et al., 2018; Corrigan et al., 2014; Corrigan & Shapiro, 2010; Fox, 2019; Pescosolido et al., 2010). Stigma impacts not only individuals who are seeking clinical mental healthcare services but also their support systems. Mental healthcare providers need to advocate for their clients, educate the public on mental healthcare services, and promote the importance of caring for one's mental health.

Travel and Transportation

Fourth, researchers identify lack of transportation or the need to travel substantial distances to meet with a clinical mental health provider as a barrier. Due to the travel and transportation barrier, individuals may find it more convenient to disregard their mental healthcare needs. Lack of transportation differs from the need to travel distances to clinical mental healthcare services, limiting individuals living in the city as the provider. The financial burden of finding consistent transportation to sustain clinical mental healthcare treatment may prove too expensive for individuals (Correll et al., 2011; Fox, 2019). Individuals may find it challenging to acquire transportation or take multiple hours off to travel to clinical mental health services.

Finances and Insurance

Lack of ability to pay for specialty services is the fifth barrier to seeking clinical mental healthcare services. This barrier could present as a lack of finances or insurance to retain clinical mental healthcare services. Lack of finances or lack of insurance can complicate navigating mental healthcare systems. Most individuals living in low-socioeconomic settings lack appropriate health insurance (Aguirre & Carrion, 2013; Fox, 2019). When individuals lack the financial ability to seek clinical mental health services, this dramatically reduces the likelihood of reaching out to mental health resources.

Removing Barriers

Due to the immense barriers identified, there is a significant need for utilizing innovative mental health delivery methods. Research has yielded five possible solutions to remove barriers when seeking clinical mental healthcare services. These solutions included tailored mental health advertising, integrating mental health into formal education, education through semiformal support services, accessible mental healthcare, and making new meaning (Lynch et al., 2018). Communities creating partnerships with clinical mental healthcare treatment centers may reduce stigma when seeking clinical mental healthcare services. When dealing with university students, targeting students with more educational programs regarding positive outcomes when receiving clinical mental healthcare services may prove beneficial (Vidourek et al., 2014).

TELEHEALTH: THE FUTURE OF CLINICAL MENTAL HEALTH SERVICES

Telehealth has been defined as delivering healthcare services where healthcare providers use information and communication technologies to exchange information for the diagnosis, treatment, prevention strategies, research, evaluation, and continuing education (Kruse et al., 2017; World Health Organization, 2010). Due to the number of individuals with Internet access, the viability of utilizing telehealth technology continues to grow. Advances in telehealth platforms have begun to change the delivery of clinical mental healthcare services. Telehealth provides a delivery method that expands access and offers high-quality care at a lower cost (Shigekawa et al., 2018). With the access to telehealth technology expanding, telehealth services can decrease mental healthcare delivery service disparities. Utilizing telehealth platforms to treat mental health issues can strengthen healthcare systems. Despite the barriers providers and clients may face when utilizing telehealth services, telehealth technology remains an innovative delivery method for clinical mental healthcare services (Fox, 2019; Guerin, 2014; Kramer et al., 2015; Mackie, 2015). The number of individuals able to use telehealth services continues to grow by millions.

History of Telehealth

When the Telecommunication Act of 1996 was passed, it removed specific economic and legal barriers for usage. This Act assisted in paving the way for telehealth technology to be utilized as a treatment modality. In the past, the Rural Health Care Program of the Universal Service Fund has aided rural communities to receive the technology needed to receive telehealth services (Benavides-Vaello et al., 2013). The Balanced Budget Act of 1997 presented the first authorization for using Medicare funds for telehealth. Previously, reimbursement was only allowed in Health Professional Shortage Areas and not in Metropolitan Statistical Areas (Benavides-Vaello et al., 2013). Since the 1990s, research has provided significant evidence of the effectiveness and acceptability of telehealth service delivery models when treating children, adolescents, and adults.

Mainstream America

The usage of telehealth technologies to deliver clinical mental health services continues to expand. Mental health services are now widely offered through telehealth modalities. Telehealth has become a popular delivery model to provide information, communication, support, and healthcare services. Telemental healthcare is an innovative multidisciplinary approach that enables clinical mental healthcare services to be rendered. With the broadening of internet access to rural areas, telemental healthcare seeks to facilitate clinical mental healthcare services to remote locations. Research has shown that provider satisfaction utilizing telehealth for clinical mental healthcare has been largely positive (Connolly et al., 2020).

Benefits of Telehealth

Telemental healthcare services offer an option to seek clinical mental healthcare services at the client's desired location. For individuals who are short on time, this provides zero travel time, making a clinical hour session possible within an already busy schedule. Additional benefits when utilizing telehealth services include increasing access to clinical mental healthcare providers, decreasing missed appointments, enhancing emergency response, increasing client satisfaction, improving the efficiency of mental health services, increasing provider satisfaction, improving treatment adherence, and reducing the cost of care (Benavides-Vaello et al., 2013; Connolly et al., 2020; Langarizadeh et al., 2017; Reay et al., 2020; Smith et al., 2020). Telehealth services can be stand-alone or complementary to existing in-person services. Research has shown that when utilizing telehealth services, it is equally as effective as in-person treatment when treating adults with common mental health issues, including adjustment disorder, anxiety, depression, and post-traumatic stress disorder (Reay et al., 2020; Varker et al., 2019). In order to expand the beneficial services of telehealth technology to more clinical mental healthcare clients, more support from national, state, and local policymakers is needed to impact nationwide change.

Challenges of Telehealth

Researchers have highlighted challenges to implementing telehealth technologies, including difficulty building client rapport, feeling undertrained, reimbursement issues, lack of broadband access, lack of technological skill, lack of provider or client acceptance, and no hands-on capabilities (Benavides-Vaello et al., 2013; Dinesen et al., 2016; Leite et al., 2020; Reay et al., 2020; Smith et al., 2020; Summers-Gabr, 2020). If there is a lack of training providing on telehealth platforms, providers may feel undertrained to utilize the telehealth equipment to provide services. Due to this lack of training, providers may find utilizing telehealth technologies challenging. Clinical mental healthcare providers may decide that the equipment or subscription to software needed to provide telemental healthcare services is too expensive. Providers have voiced concerns, including how telehealth services might negatively impact client's therapeutic relationships and privacy and safety concerns when utilizing telehealth technologies (Reay et al., 2020). Even though the growth of telehealth rapidly expanded during the pandemic, private insurance, Medicare, and Medicaid reimbursement services are inconsistent.

Overcoming Challenges

Many clinical mental healthcare service platforms have been constructed; however, unified reimbursement tactics, nationwide management, and coordination policies are still inadequate. If more federal policies supported telehealth services, it could serve as a practical solution to the lack of mental health providers (Colon-Gonzalez et al., 2013; Fox, 2019; Handley et al., 2014a). Due to the pandemic, traditional clinical mental healthcare services have rapidly transitioned to telehealth delivery models to ensure continuity of care. Lack of training for clinical mental healthcare providers is a challenge to ensuring the ethical utilization of telehealth technologies. Providers need to be guided on best practices, developing therapeutic relationships, and innovative ways to deliver traditional treatments within a non-traditional setting (Reay et al., 2020). Training clinical mental healthcare providers increases confidence in utilizing technologies, and technical support is established. Without proper training, providers may struggle to assess individual's appropriateness for telehealth services, including the ability to access technology, the severity of mental health symptoms, and analyzing client risk of harm (Reay et al., 2020).

Rural Outreach

Clinical mental healthcare providers have been overwhelmed throughout the pandemic, especially those working in isolated rural areas. It is common for geographically isolated areas to have insufficient access to clinical mental healthcare providers. When seeking clinical mental healthcare services, significant challenges for rural communities include disparities between social classes, distinct cultural groups, geographical distance to services, lack of funding, language barriers, and low socioeconomic status (Benavides-Vaello et al., 2013). Due to the economic climate in rural areas, rural communities struggle to recruit and retain qualified clinical mental healthcare providers. Despite the demand for clinical mental healthcare services increasing, accessibility and availability issues create challenges when rural individuals attempt to seek clinical mental healthcare services. Utilizing telehealth services can reduce disparities in delivering mental healthcare services to rural communities (Benavides-Vaello et al., 2013; Langarizadeh et al., 2017). The continued growth of the clinical mental health services disparity in rural areas demands attention.

TYPES OF TELEHEALTH SERVICES

Telehealth technologies provide virtual platforms to deliver health information, improve prevention strategies, monitor clients, and perform video consultations. There are several modalities for telehealth services, including video conference, telephone, and messaging systems, web-based interventions, mobile phone technology, networking via social media and group discussions, and simulated people and places (Langarizadeh et al., 2017). Today, Telehealth is the fastest growing division of the healthcare system. Through the growth of telehealth technologies, telemental healthcare services can utilize synchronous, asynchronous, remote monitoring, and clinical mental health provider consultation amenities. Synchronous, asynchronous, and remote monitoring telehealth services are becoming more and more accepted by clients and providers alike.

Synchronous

Synchronous approaches to telehealth are characterized by a real-time approach where the service occurs concurrently between the client and the professional provider via video conference or audio conversations (Costa et al., 2021; Mechanic et al., 2020). Synchronous methods allow for live interactions between client and provider. Synchronous technologies are the closest comparison to face-to-face client and provider interactions. Synchronous communication assists in providing face-to-face contact, image and voice, via the usage of videoconferencing equipment between provider and client (Reay et al., 2020; Verhoeven et al., 2010). Synchronous approaches remain what are commonly understood when thinking about telehealth services.

Asynchronous

Unlike synchronous interactions, asynchronous methods include email, text, and fax. Asynchronous modalities have long been implemented within healthcare services. Many providers use asynchronous interventions to remind clients of appointments, provide client check-ins, share support documents, and disperse online assessments to consumers. An asynchronous approach provides an outlet for communication without the need for real-time involvement. Asynchronous telehealth communication assists providers in monitoring clients, delivering feedback to clients, and automated messaging systems without face-to-face contact (Verhoeven et al., 2010). Asynchronous approaches are widely used throughout the healthcare system.

Remote Monitoring

Throughout the pandemic, remote monitoring services were extensively utilized within the healthcare profession. Remote monitoring comprises continuous evaluation of a client's clinical status, whether through direct video monitoring or reviewing tests and images collected remotely (Mechanic et al., 2020; Vegesna et al., 2017). Remote monitoring can keep clients engaged in their care between face-to-face or telehealth visits. Research has shown that remote monitoring enhances a provider's ability to monitor and manage clients in non-traditional healthcare settings (Vegesna et al., 2017). Telehealth technologies including synchronous, asynchronous, and remote monitoring modalities have proved impactful throughout the global pandemic. Despite technological advancements within telehealth services, obstacles, such as reimbursement problems, lack of client belief in the validity of telehealth, and lack of adequate training for practitioners, remain. (Volterrani & Sposato, 2019).

THE FUTURE OF TELEHEALTH

The utilization of telehealth technology has increased rapidly since spring 2020. Due to the pandemic, new technology innovations, more evidenced-based research, client and provider acceptance, and efficient software platforms have led to telehealth mental healthcare service growth. For telemental healthcare services to continue to develop, focused research directed at outcomes and effectiveness within clinical mental healthcare services is needed. In order to solidify telehealth's effectiveness outside of the COVID-19 pandemic, telehealth must continue to be integrated into current healthcare systems assimilating into traditional healthcare models (Smith et al., 2020). Telemental healthcare services are practical, feasible, and appropriate to support clients, family members, and mental health service providers. Specific professional training addressing competencies and providing care will continue to be needed (Reay et al., 2020).

Innovation in the World of Telehealth

Technologies that support telehealth services are thriving. Innovative technologies, including wearable devices, smartphones, and smart homes, can be utilized to monitor client health and send notifications to providers when emergencies are detected (Blandford et al., 2020). Previously, pioneering adopters of telemental health services have been isolated communities with limited access to clinical mental healthcare providers. Today, telemental healthcare provides access to services to a culturally diverse client population. The telehealth industry needs to collaborate with clients and providers to ensure digital inclusion, data security, and tailoring their platforms to the user's needs (Blandford et al., 2020). Telehealth platforms not only improve existing mental healthcare systems but can create a unified g

Global Telehealth

Telehealth technologies have been used within healthcare professions since the 19th century (Dinesen et al., 2016). The importance of telehealth as a modality for delivering care over distance is increasingly relevant due to the global pandemic of COVID-19 experienced worldwide. By utilizing telehealth systems, services can effectively be provided between providers and clients within the same town, state, across the country, or different parts of the world. According to research, telehealth services can play a critical role in the global response to mental health issues (Smith et al., 2020). For telemental healthcare to be successful globally, the following criteria need to be met.

- Ensure that all mental healthcare providers receive applicable telehealth platform education and best-practice training;
- Create and introduce telemental healthcare certification for providers;
- Establish sustainable reimbursement policies and procedures to cover the cost of providing telehealth adequately;
- Redesign clinical healthcare models to support telemental health;
- Enhance communication strategies with all stakeholders about resources, telehealth services, and treatment options;
- Provide quarterly surveys for clients and providers to manage the effectiveness of telemental healthcare services and identify areas for improvement.

As the prevalence of telehealth usage worldwide continues to grow, the importance of matching appropriate telehealth platforms with specific populations to meet client needs is vital. Telehealth platforms and app developers are typically not healthcare experts, nor do they have clinical licenses to work with clients. With the usage of telehealth technologies in global healthcare delivery becoming common, it is important to develop evidence-based, successful, and innovative telehealth solutions that can provide sustainable telehealth services (Dinesen et al., 2016). There is a need for researchers worldwide to focus their attention on utilizing telemental healthcare services to treat mental health issues to create a framework for pinpointing best practices.

ANALYSIS

The COVID-19 pandemic had distressing consequences for individual and collective well-being, emotional health, and social functioning (Pfefferbaum and North, 2020). Due to the rapid change in clinical mental healthcare service delivery, providers experienced periods of confusion, followed by scrambling to figure out innovative ways to serve their clients still. Many clinical mental healthcare agencies were not previously prepared to provide telemental health services. This lack of preparation created chaos when navigating new mental healthcare delivery models to ensure continuity of care for clients. Telemental health services have proven themselves a viable alternative to face-to-face clinical mental healthcare interactions. Providers and clients became more proficient in telehealth technology due to recurrent usage.

Research supports that telemental health can help to overcome barriers when accessing clinical mental healthcare services to diminish the existing treatment gap. A multitude of studies has established the effectiveness of utilizing telehealth services to treat mental health issues. Telemental health is flexible and can be utilized across the continuum of care to deliver and improve clinical mental healthcare services (Mace et al., 2018). Essentially, anyone with an internet connection can access telemental health treatment. Telemental healthcare increases convenience when seeking clinical mental healthcare services and can reach populations that were previously reluctant to seek face-to-face services. Telemental health provides accessible specialty mental healthcare services for mental health issues that might otherwise go untreated.

It is vital to continue to developing telehealth services, given the potential for the COVID-19 pandemic to remain a long-term risk and plan for any future global crisis (Lyne et al., 2020). Advancements in telehealth technologies are ongoing and can create a broader scope of telehealth opportunities. Evidence supports that by overcoming funding and reimbursement challenges, governing and compliance challenges, and training challenges, telemental health can provide an ethical modality to increase accessibility to clinical mental healthcare services (Mace et al., 2018). Before COVID-19, supporters have fought for the expansion of telemental healthcare for decades. Despite the overwhelming success of utilizing telemental health to seek clinical mental healthcare services, many critical orders that allowed for the expansion of telemental health services during the pandemic will expire (Warren & Smalley, 2020).

REFERENCES

Aaronson, A. L., Backes, K., Agarwal, G., Goldstein, J. L., & Anzia, J. (2018). Mental Health During Residency Training: Assessing the Barriers to Seeking Care. *Academic Psychiatry, 42*(4), 469–472. doi:10.1007/s40596-017-0881-3

Acharya, B., Maru, D., Schwarz, R., Citrin, D., Tenpa, J., Hirachan, S., & Ekstrand, M. (2017). Partnerships in mental healthcare service delivery in low-resource settings: Developing an innovative network in rural Nepal. *Globalization and Health, 13*(2), 2. Advance online publication. doi:10.118612992-016-0226-0 PMID:28086925

Aguirre, J., & Carrion, V. (2013). Integrated behavioral health services. *Clinical Pediatrics, 52*(12), 1178–1180. doi:10.1177/0009922812470744 PMID:23299835

Aisbett, D., Boyd, C., Francis, K., & Newnham, K. (2007). Understanding barriers to mental health service utilization for adolescents in rural Australia. *Rural and Remote Health, 7*(1), 624. PMID:17302509

American Academy of Family Physicians. (2011). *Mental health care services by family physicians* [Position paper]. Retrieved from https://www.aafp.org/online/en/home/policy/policies/m/mentalhealthcareservices.html

Andrade, L. H., Alonso, J., Mneimneh, Z., Wells, J. E., Al-Hamzawi, A., Borges, G., Bromet, E., Bruffaerts, R., de Girolamo, G., de Graaf, R., Florescu, S., Gureje, O., Hinkov, H. R., Hu, C., Huang, Y., Hwang, I., Jin, R., Karam, E. G., Kovess-Masfety, V., ... Kessler, R. C. (2014). Barriers to mental health treatment: Results from the WHO World Mental Health surveys. *Psychological Medicine, 44*(6), 1303–1317. doi:10.1017/S0033291713001943 PMID:23931656

Arora, P. G., Godoy, L., & Hodgkinson, S. (2017). Serving the underserved: Cultural considerations in behavioral health integration in pediatric primary care. *Professional Psychology, Research and Practice, 48*(3), 139–148. doi:10.1037/pro0000131

Benavides-Vaello, S., Strode, A., & Sheeran, B. C. (2013). Using Technology in the Delivery of Mental Health and Substance Abuse Treatment in Rural Communities: *A Review. The Journal of Behavioral Health Services & Research, 40*(1), 111–120. doi:10.100711414-012-9299-6 PMID:23093443

Bischoff, R., Reisbig, J., Springer, A., Schultz, M., Robinson, P., & Olson, R. (2014). Succeeding in rural mental health practice: Being sensitive to culture by fitting in and collaborating. *Contemporary Family Therapy*, *36*(1), 1–16. doi:10.100710591-013-9287-x

Blandford, A., Wesson, J., Amalberti, R., AlHazme, R., & Allwihan, R. (2020). Opportunities and challenges for telehealth within, and beyond, a pandemic. *The Lancet. Global Health*, *8*(11), e1364–e1365. Advance online publication. doi:10.1016/S2214-109X(20)30362-4 PMID:32791119

Breslau, J., Marshall, G. N., Pincus, H. A., & Brown, R. A. (2014). Are mental disorders more common in urban than rural areas of the United States? *Journal of Psychiatric Research*, *56*, 50–55. doi:10.1016/j.jpsychires.2014.05.004 PMID:24857610

Brezenski, T. F. (2014). Interest group formation and mental health care consumers in the United States: The effects of stigma, fiscal policy, and the Affordable Care Act. *Journal of Multidisciplinary Research, 6*(3), 39–52.

Brooks, S. K., Webster, R. K., Smith, L. E., Woodland, L., Wessely, S., Greenberg, N., & Rubin, G. J. (2020). The psychological impact of quarantine and how to reduce it: Rapid review of the evidence. *Lancet*, *395*(10227), 912–920. doi:10.1016/S0140-6736(20)30460-8 PMID:32112714

Butler, R. N. (2002). Age, death, and life review. In K. J. Doka (Ed.), *Living with grief: Loss in later life. Hospice Foundation of American.*

Cidade, E. C., Junior, J. F. M., & Ximenes, V. M. (2012). Implicações psicológicas da pobreza na vida do povo latinoamericano. [Psychological implications of poverty in the life of Latin-American people]. *Psicologia Argumento*, *30*(68), 87–98. doi:10.7213/psicol.argum.5886

Cocker, F., & Joss, N. (2016). Compassion Fatigue among Healthcare, Emergency and Community Service Workers: A Systematic Review. *International Journal of Environmental Research and Public Health*, *13*(6), 618. doi:10.3390/ijerph13060618 PMID:27338436

Colon-Gonzalez, M. C., McCall-Hosenfeld, J. S., Weisman, C. S., Hillemeier, M. M., Perry, A. N., & Chuang, C. H. (2013). "Someone's got to do it" – Primary care providers (PCPs) describe caring for rural women with mental health problems. *Mental Health in Family Medicine*, *10*(4), 191–202. PMID:25632302

Connolly, S. L., Miller, C. J., Lindsay, J. A., & Bauer, M. S. (2020). A systematic review of providers' attitudes toward telemental health via videoconferencing. *Clinical Psychology: Science and Practice*, *27*(2), e12311. doi:10.1111/cpsp.12311

Cook, J. A., Ruggiero, K., Shore, S., Daggett, P., & Butler, S. B. (2007). Public-academic collaboration in the application of evidence-based practice in Texas mental health system redesign. *International Journal of Mental Health, 36*(2), 36–49. doi:10.2753/IMH0020-7411360204

Cooper, S. (2016). Global mental health and its critics: Moving beyond the impasse. *Critical Public Health, 26*(4), 355–358. doi:10.1080/09581596.2016.1161730

Correll, J. A., Cantrell, P., & Dalton, W. M. (2011). Integration of behavioral health services in a primary care clinic serving rural Appalachia: Reflections on a clinical experience. *Families, Systems & Health, 29*(4), 291–302. doi:10.1037/a0026303 PMID:22214296

Corrigan, P. W., Druss, B. G., & Perlick, D. A. (2014). The impact of mental illness stigma on seeking and participating in mental health care. *Psychological Science in the Public Interest, 15*(2), 37–70. doi:10.1177/1529100614531398 PMID:26171956

Corrigan, P. W., & Penn, D. L. (2015). Lessons from social psychology on discrediting psychiatric stigma. *Stigma and Health, 1*(S), 2–17. doi:10.1037/2376-6972.1.S.2

Corrigan, P. W., & Shapiro, J. R. (2010). Measuring the impact of programs that challenge the public stigma of mental illness. *Clinical Psychology Review, 30*(8), 907–922. doi:.2010.06.004 doi:10.1016/j.cpr

Costa, R. R. G., Dorneles, J. R., Veloso, J. H., Gonçalves, C. W., & Neto, F. R. (2021). Synchronous and asynchronous tele-exercise during the coronavirus disease 2019 pandemic: Comparisons of implementation and training load in individuals with spinal cord injury. *Journal of Telemedicine and Telecare*, 1357633X2098273. Advance online publication. doi:10.1177/1357633X20982732 PMID:33461399

Crunk, A. E., Burke, L. A., & Robinson, E. H. M. III. (2017). Complicated grief: An evolving theoretical landscape. *Journal of Counseling and Development, 95*(2), 226–233. doi:10.1002/jcad.12134

Diederich, J., & Song, I. (2014). Mental health informatics: current approaches. Mental Health Informatics, 1-16.

Dinesen, B., Nonnecke, B., Lindeman, D., Toft, E., Kidholm, K., Jethwani, K., Young, H. M., Spindler, H., Oestergaard, C. U., Southard, J. A., Gutierrez, M., Anderson, N., Albert, N. M., Han, J. J., & Nesbitt, T. (2016). Personalized Telehealth in the Future: A Global Research Agenda. *Journal of Medical Internet Research, 18*(3), e53. doi:10.2196/jmir.5257 PMID:26932229

Dyer, J. E. T., & Hagedorn, W. B. (2013). Navigating bereavement with spirituality-based interventions: Implications for non-faith-based counselors. *Counseling and Values*, *58*(1), 69–84. doi:10.1002/j.2161-007X.2013.00026.x

Fox, T. (2019). *Integrated behavioral healthcare: The future of rural mental health.* Retrieved from https://ttu-ir.tdl.org/handle/2346/85520

Galea, S., Ahern, J., Nandi, A., Tracy, M., Beard, J., & Vlahov, D. (2007). Urban neighborhood poverty and the incidence of depression in a Population-Based cohort study. *Annals of Epidemiology*, *17*(3), 171–179. doi:10.1016/j.annepidem.2006.07.008 PMID:17320784

Guerin, E. (2014). Rural Americans have inferior internet access: Does it matter that broadband quality varies so greatly? *High Country News*. Retrieved from https://www.hcn.org/issues/46.2/rural-americans-have-inferior-internet-access

Hall, S. A., & Gjesfjeld, C. D. (2013). Clergy: A partner in rural mental health? *Rural Mental Health*, *37*(1), 50–57. doi:10.1037/rmh0000006

Handley, T., Kay-Lambkin, F., Inder, K., Attia, J., Lewin, T., & Kelly, B. (2014a). Feasibility of internet-delivered mental health treatments for rural populations. *Social Psychiatry and Psychiatric Epidemiology*, *49*(2), 275–282. doi:10.100700127-013-0708-9 PMID:23689825

Handley, T., Kay-Lambkin, F., Inder, K., Lewin, T., Attia, J., Fuller, J., & Kelly, B. (2014b). Self-reported contacts for mental health problems by rural residents: Predicted service needs, facilitators and barriers. *BMC Psychiatry*, *14*(1), 1–23. doi:10.118612888-014-0249-0 PMID:25193400

Hauck, G., Gelles, K., Bravo, V., & Thorson, M. (2020). Five months in: A timeline of how COVID-19 has unfolded in the U.S. *USA Today*. https://www.usatoday.com/indepth/news/nation/2020/04/21/coronavirus-updates-how-covid-19-unfolded-u-stimeline/2990956001/

Health Resources and Services Administration. (2016). *Health Professional Shortage Areas (HPSAs)*. Retrieved from https://bhw.hrsa.gov/shortage-designation/hpsas

Inder, K. J., Berry, H., & Kelly, B. J. (2011). Using cohort studies to investigate rural and remote mental health. *The Australian Journal of Rural Health*, *19*(4), 171–178. doi:10.1111/j.1440-1584.2011.01208.x PMID:21771157

Jackson, H., Judd, F., Komiti, A., Fraser, C., Murray, G., Robins, G., & Wearing, A. (2007). Mental health problems in rural contexts: What are the barriers to seeking help from professional providers? *Australian Psychologist*, *42*(2), 147–160. doi:10.1080/00050060701299532

Jameson, J. P., & Blank, M. B. (2007). The role of clinical psychology in rural mental health services: Defining problems and developing solutions. *Clinical Psychology: Science and Practice*, *14*(3), 283–298. doi:10.1111/j.1468-2850.2007.00089.x

Kola, L. (2020). Global mental health and COVID-19. *The Lancet. Psychiatry*, *7*(8), 655–657. doi:10.1016/S2215-0366(20)30235-2 PMID:32502468

Koplan, J. P., Bond, T. C., Merson, M. H., Reddy, K. S., Rodriguez, M. H., Sewankambo, N. K., & Wasserheit, J. N. (2009). Towards a common definition of global health. *Lancet*, *373*(9679), 1993–1995. doi:10.1016/S0140-6736(09)60332-9 PMID:19493564

Kramer, G. M., Kinn, J. T., & Mishkind, M. C. (2015). Legal, regulatory, and risk management issues in the use of technology to deliver mental health care. *Cognitive and Behavioral Practice*, *22*(3), 258–268. doi:10.1016/j.cbpra.2014.04.008

Kruse, C. S., Krowski, N., Rodriguez, B., Tran, L., Vela, J., & Brooks, M. (2017). Telehealth and patient satisfaction: A systematic review and narrative analysis. *BMJ Open*, *7*(8), e016242. doi:10.1136/bmjopen-2017-016242 PMID:28775188

Langarizadeh, M., Tabatabaei, M. S., Tavakol, K., Naghipour, M., Rostami, A., & Moghbeli, F. (2017). Telemental Health Care, an Effective Alternative to Conventional Mental Care: A Systematic Review. *Acta Informatica Medica*, *25*(4), 240–246. doi:10.5455/aim.2017.25.240-246 PMID:29284913

Leite, H., Hodgkinson, I. R., & Gruber, T. (2020). New development: 'Healing at a distance'—Telemedicine and COVID-19. *Public Money & Management*, *40*(6), 483–485. Advance online publication. doi:10.1080/09540962.2020.1748855

Li, W., Yang, Y., Liu, Z. H., Zhao, Y. J., Zhang, Q., Zhang, L., Cheung, T., & Xiang, Y. T. (2020). Progression of Mental Health Services during the COVID-19 Outbreak in China. *International Journal of Biological Sciences*, *16*(10), 1732–1738. doi:10.7150/ijbs.45120 PMID:32226291

Lynch, L., Long, M., & Moorhead, A. (2018). Young Men, Help-Seeking, and Mental Health Services: Exploring Barriers and Solutions. *American Journal of Men's Health*, *12*(1), 138–149. doi:10.1177/1557988315619469 PMID:27365212

Lyne, J., Roche, E., Kamali, M., & Feeney, L. (2020). COVID-19 from the perspective of urban and rural general adult mental health services. *Irish Journal of Psychological Medicine*, *37*(3), 181–186. doi:10.1017/ipm.2020.62 PMID:32434599

Mace, S., Boccanelli, A., & Dormond, M. (2018). *The Use of Telehealth Within Behavioral Health Settings: Utilization, Opportunities, and Challenges*. Behavioral Health Workforce Research Center. Retrieved from https://behavioralhealthworkforce.org/wp-content/uploads/2018/05/Telehealth-Full-Paper_5.17.18-clean.pdf

Mackie, P. F. (2015). Technology in rural behavioral health care practice: Policy concerns and solution suggestions. *Rural Mental Health*, *39*(1), 5–12. doi:10.1037/rmh0000027

Marsella, A. J. (1998). Urbanization, mental health and social deviancy – a review of issues and research. *The American Psychologist*, *53*(6), 624–634. doi:10.1037/0003-066X.53.6.624 PMID:9633263

Mechanic, O. J., Persaud, Y., & Kimball, A. B. (2020). Telehealth Systems. In *StatPearls*. StatPearls Publishing.

Mnookin, S. (2016). *Out of the shadows: Making mental health a global development priority*. World Bank Group. Retrieved from https://documents.worldbank.org/curated/en/2016/04/26281016/out-shadowsmaking-mental-health-global-development-priority

Monteiro, N. M., Ndiaye, Y., Blanas, D., & Ba, I. (2014). Policy perspectives and attitudes towards mental health treatment in rural Senegal. *International Journal of Mental Health Systems*, *8*(1), 1–16. doi:10.1186/1752-4458-8-9 PMID:24646335

National Institute of Mental Health. (2017). *Mental illness*. Retrieved from https://www.nimh.nih.gov/health/statistics/mental-illness.shtml

Neimeyer, R. A., & Burke, L. A. (2012). Complicated grief and the end-of-life: Risk factors and treatment considerations. In J. T. Werth (Ed.), *Counseling clients near the end-of-life: A practical guide for mental health professions* (pp. 205–228). Springer.

Neimeyer, R. A., Burke, L. A., Mackay, M. M., & Stringer, J. G. (2010). Grief therapy and the reconstruction of meaning: From principles to practice. *Journal of Contemporary Psychotherapy*, *40*(2), 73–78. doi:10.100710879-009-9135-3

Nepomuceno, B., Cardoso, A., Ximenes, V., Barros, J., & Leite, J. (2016). Mental health, well-being, and poverty: A study in urban and rural communities in Northeastern Brazil. *Journal of Prevention & Intervention in the Community*, *44*(1), 63–75. doi:10.1080/10852352.2016.1102590 PMID:26697905

Patel, V. (2014). Why mental health matters to global health. *Transcultural Psychiatry, 51*(6), 777–789. doi:10.1177/1363461514524473 PMID:24595266

Patel, V., & Prince, M. (2010). Global mental health: A new global health field comes of age. *Journal of the American Medical Association, 303*(19), 1976–1977. doi:10.1001/jama.2010.616 PMID:20483977

Paulson, L. (2010). *The lived experience of supervisors in rural settings.* Retrieved from https://www.counseling.org/docs/default-source/vistas/the-lived-experience-ofsupervisors-in-rural-settings.pdf?sfvrsn=1

Pelling, N., & Butler, B. (2015). Rural counselors and psychologists in Australia: A descriptive survey of professionals and practice. *International Journal of Mental Health, 44*(1/2), 94–106. doi:10.1080/00207411.2015.1009782

Pescosolido, B. A., Martin, J. K., Long, J. S., Medina, T. R., Phelan, J. C., & Link, B. G. (2010). "A disease like any other"? A decade of change in public reactions to schizophrenia, depression, and alcohol dependence. *The American Journal of Psychiatry, 167*(11), 1321–1330. doi:10.1176/appi.ajp.2010.09121743 PMID:20843872

Pfefferbaum, B., & North, C. S. (2020). Mental Health and the Covid-19 Pandemic. *The New England Journal of Medicine, 383*(6), 510–512. doi:10.1056/NEJMp2008017 PMID:32283003

Procter, N., Ferguson, M., Backhouse, J., Cother, I., Jackson, A., Murison, J., & Reilly, J. (2015). Face to face, person to person: Skills and attributes deployed by rural mental health clinicians when engaging with consumers. *The Australian Journal of Rural Health, 23*(6), 352–358. doi:10.1111/ajr.12204 PMID:26683718

Rathod, S., Pinninti, N., Irfan, M., Gorczynski, P., Rathod, P., Gega, L., & Naeem, F. (2017). Mental Health Service Provision in Low- and Middle-Income Countries. *Health Services Insights, 10.* Advance online publication. doi:10.1177/1178632917694350 PMID:28469456

Reay, R. E., Looi, J. C., & Keightley, P. (2020). Telehealth mental health services during COVID-19: Summary of evidence and clinical practice. *Australasian Psychiatry, 28*(5), 514–516. doi:10.1177/1039856220943032 PMID:32722963

Robinson, W. D., Geske, J., Backer, E., Jarzynka, K., Springer, P. R., Bischoff, R., & Swinton, J. (2012). Rural experiences with mental illness: Through the eyes of patients and their families. *Families, Systems & Health, 30*(4), 308–321. doi:10.1037/a0030171 PMID:23088821

Shigekawa, E., Fix, M., Corbett, G., Roby, D. H., & Coffman, J. (2018). The Current State Of Telehealth Evidence: A Rapid Review. *Health Affairs (Project Hope)*, *37*(12), 1975–1982. doi:10.1377/hlthaff.2018.05132 PMID:30633674

Sinclair, C. C., Holloway, K., Riley, A. M. G., Auret, K., & Eysenbach, G. (2013). Online mental health resources in rural Australia: Clinician perceptions of acceptability. *Journal of Medical Internet Research*, *15*(9), e193. Advance online publication. doi:10.2196/jmir.2772 PMID:24007949

Smith, A. C., Thomas, E., Snoswell, C. L., Haydon, H., Mehrotra, A., Clemensen, J., & Caffery, L. J. (2020). Telehealth for global emergencies: Implications for coronavirus disease 2019 (COVID-19). *Journal of Telemedicine and Telecare*, *26*(5), 309–313. doi:10.1177/1357633X20916567 PMID:32196391

Soltis-Jarrett, V. (2019). The TANDEM3-PC: The Foundation for an Innovative, Integrated Behavioral Health NP-led Model of Practice in Rural Primary Care. *Archives of Psychiatric Nursing*, *33*(1), 2–10. doi:10.1016/j.apnu.2018.08.007 PMID:30663620

Summers-Gabr, N. M. (2020). Rural–urban mental health disparities in the United States during COVID-19. *Psychological Trauma: Theory, Research, Practice, and Policy*, *12*(S1), S222–S224. doi:10.1037/tra0000871 PMID:32478541

Torales, J., O'Higgins, M., Castaldelli-Maia, J. M., & Ventriglio, A. (2020). The outbreak of COVID-19 coronavirus and its impact on global mental health. *The International Journal of Social Psychiatry*, *66*(4), 317–320. doi:10.1177/0020764020915212 PMID:32233719

U.S. Census Bureau, International Data Base, & USA Trade Online. (2021). *U.S. and World Population Clock*. U.S. Census Bureau. Retrieved from https://www.census.gov/popclock/world

Varker, T., Brand, R. M., Ward, J., Terhaag, S., & Phelps, A. (2019). Efficacy of synchronous telepsychology interventions for people with anxiety, depression, posttraumatic stress disorder, and adjustment disorder: A rapid evidence assessment. *Psychological Services*, *16*(4), 621–635. doi:10.1037er0000239 PMID:29809025

Vegesna, A., Tran, M., Angelaccio, M., & Arcona, S. (2017). Remote Patient Monitoring via Non-Invasive Digital Technologies: A Systematic Review. *Telemedicine Journal and E-Health*, *23*(1), 3–17. doi:10.1089/tmj.2016.0051

Verhoeven, F., Tanja-Dijkstra, K., Nijland, N., Eysenbach, G., & van Gemert-Pijnen, L. (2010). Asynchronous and synchronous teleconsultation for diabetes care: A systematic literature review. *Journal of Diabetes Science and Technology*, *4*(3), 666–684. doi:10.1177/193229681000400323 PMID:20513335

Vidourek, R. A., King, K. A., Nabors, L. A., & Merianos, A. L. (2014). Students' benefits and barriers to mental health help-seeking. *Health Psychology and Behavioral Medicine*, *2*(1), 1009–1022. doi:10.1080/21642850.2014.963586 PMID:25750831

Volterrani, M. & Sposato, B. (2019). Remote monitoring and telemedicine. *European Heart Journal Supplements*, *21*(Supplement M), M54-M56. doi:10.1093/eurheartj/suz266

Warren, J. C., & Smalley, K. B. (2020). *Using Telehealth to Meet Mental Health Needs During the COVID-19 Crisis*. Commonwealth Fund. doi:10.26099/qb81-6c84

Webster, P. (2020). Virtual health care in the era of COVID-19. *Lancet*, *395*(10231), 1180–1181. doi:10.1016/S0140-6736(20)30818-7 PMID:32278374

World Health Organization. (2010). *Telemedicine: Opportunities and developments in Member States: report on the second global survey on eHealth.* World Health Organization.

World Health Organization. (2011). *Mental health atlas 2011*. World Health Organization.

World Health Organization. (2020). *Mental Health and Psychosocial Considerations During the COVID-19 Outbreak*. World Health Organization.

World Health Organization. (2021). *WHO Coronavirus disease (COVID-19) dashboard.* Retrieved from https://covid19.who.int/

Zubatsky, M., Edwards, T. M., Wakabayashi, H., & Ivbijaro, G. (2018). Integrated behavioral health in primary care across the world: Three countries, three perspectives. *Family Practice*, *35*(6), 645–648. doi:10.1093/fampra/cmy034 PMID:29741628

Chapter 8
The Equity of Remote Teaching During COVID-19

Allison Victoria Wilson
https://orcid.org/0000-0003-2387-0482
Midwestern State University, USA

ABSTRACT

This chapter is written by a second-grade teacher who gives a personal account for remote teaching during quarantine and the 2020 - 2021 school year. The equity and diversity of various situations involved during the past year are discussed throughout the chapter: from the beginning of quarantine 2020 to the end of the school year in May 2021. Remote teaching, social and emotional factors, and the diversity of families are also addressed. The chapter is dedicated to Jordan Lea Darnell, a teacher who lost her battle to COVID-19 in the Spring of 2021.

INTRODUCTION

The Equity of Remote Teaching During COVID-19

In the spring of 2020, our world turned upside down. COVID-19 stormed through the United States, completely rerouting our ordinary lives to virtual ones. Toilet paper, hand sanitizer, and Lysol seemed to disappear from the face of the earth. Schools closed their doors to students, and teachers conquered their duties from their dinner tables. Google Meet became our classroom, and Google Classroom became our whiteboard. Teachers did not stop teaching as they continued to do their jobs amidst the pandemic continuing to control our daily lives.

DOI: 10.4018/978-1-7998-8028-8.ch008

On Friday, March 13, 2020, our students saw their classroom and school hallways for the very last time for the 2019-2020 school year. It was the start of spring break and the unknown ending of an in-person school year, unseen by educators across the United States. I remember walking into our local Sam's Club that Friday and seeing the shelves stripped of bottled water and toilet paper. It was panic-causing panic amongst consumers in the store. As spring break continued, our school district closed school the following week for curriculum instructors and administration from all the campuses to come together and come up with a game plan for how and what needed to happen to continue teaching students during this time of shutdown.

Teachers across my school district called the parents of their students to determine who needed what to continue learning; either a Chromebook for virtual instruction or paper packets for students that did not have internet access. The equity of students given a Chromebook had somewhat of an advantage compared to those who received paper packets during this time. Students that had the opportunity to receive virtual instruction during quarantine had the chance to see their peers and me, their teacher, through check-in times on Google Meet Monday through Friday. Social and emotional factors played a significant role in this learning format, allowing students to see their peers and teachers. My students that only received weekly paper packets did not get the opportunity to see their peers daily. The diversity and equity in the factors of social and emotional learning differentiated in both instruction types. I had one parent that I sent personal teaching YouTube videos assisting with instruction throughout our time during quarantine. I also remember talking on the phone with a parent who did not quite understand a phonics lesson her student had to complete, so I ended up talking on the phone for an hour with her. The conversation I had with that mom went in five million different directions, as I believe she honestly needed someone to talk to during this challenging time that the entire world saw.

Other equity matters arose when students had to submit work into Google Classroom or parents had to send me pictures of their child's work through email or text message. Students had to be accounted for in some format regarding attendance purposes, making this part of the solution. All material assigned to students was a review from the school year as teachers did not teach new content due to the nature of our shutdown situation state-wide (quarantine). Considering the impact this shutdown had on families and students, having the last nine weeks as a review of second-grade was the best decision, as some of my students' parents lost their jobs. I believe that during quarantine in the spring of 2020, schools were in survival mode and did the best they could with what they had.

As an educator, equity and diversity amongst the families were the hardest things to manage. One of my students did not qualify for special education, and she was a remote student (her undercover name for this will be Ana). Her mom worked nights, and she did not have someone to help her with her school work during this time consistently. Ana was intelligent with technology but was a Kindergarten reader and could not complete her assignments on Google Slides without an adult assisting her. I had her log into Google Meet, and I would read her slides to her almost daily. I would then help her with the assignment aspect of it on Google Classroom, telling her how to manipulate the slides and scaffold questions. As an educator, we want all of our students to succeed and be successful in the classroom. The diversity of this family's situation made my job as her teacher vital in her upside-down world. Equity of education for students like her made the meaning and duty of our remote job greater than ever before. At the end of the school year, my school had a "drive-thru" end-of-the-year celebration. Ana gave me the biggest hug, and her mother had tears in her eyes and thanked me for assisting her with her school work. I will never forget that day and the feeling of reward I got from that family.

After the school year ended, I felt a sense of incompleteness that many educators felt as well. From a personal account, the mental fatigue in comparison to an ordinary in-person school year was evident. Remote teaching takes a lot out of a person, as I would much rather be teaching in a classroom. My students, along with many other students worldwide, missed an entire nine weeks of learning. I remember my students asking me almost every day on Google Meet, "Ms. Wilson, when are we going back to school?". The truth behind that question was hard to answer to an eight-year-old, but to me, that question meant that they missed being at school, and that made me so eternally joyful. I missed school as well; nine weeks of remote teaching during a pandemic was enough for a lifetime. But, we educators did it amid a chaotic world.

When the Fall of 2020 came around, my school planned to have two remote teachers in my grade level; one being myself. Our lesson plans became daily Google Slides instead of a Google Doc with TEKS and information regarding our weekly lessons. I started the 2020-2021 school year with eight students in person and eight remote students, as parents had the option as to what format of education their child could receive. My daily schedule involved me teaching my in-person students all day until 2:25 PM. At 2:25 PM, I would teach my remote students that day's lesson. I would answer any questions they had and try my best to engage them. I slowly had my virtual students come back to in-person school throughout the school year for various reasons, predominantly because they were struggling with remote learning at home. The equity in the decision to leave a child at home for remote learning comes back on the parents for this age of students. Eight-year-olds are not fully capable high school students; they are young kids. In order for students to learn at their full potential that they can with remote learning, they need parents to assist them with their lessons daily.

As of May 2021, I can say that all my students are back to in-person learning. I felt this sense of relief when all my students came back to school this spring. It is lovely to have all of them together, healthy, in my classroom with their peers. The amount of academic growth I have seen this school year from all my students has been unreal. One of my remote students that came back to in-person learning jumped eleven reading levels this school year. All she needed to fill her gap in reading was small group interventions and practice. It is amazing what students can accomplish when they have a predictable environment, such as school, that directs and supports learning. The benefit of students who attend in-person school is evident not just in academics but also in social and emotional aspects.

Education continued despite a pandemic; educators served students during a season of unknown and time of foreseen normalcy. My heart will forever hold a special place for all students who went through quarantine and remote learning, as it was hard on many of them for many reasons.

Ana, my past student, stops by my classroom to this day and hugs me and smiles her beautiful smile before school starts; along with other students from my quarantine group. Many challenges and triumphs happened in the world of education within the past year. Something that remained the same was my love for teaching, regardless of instruction format. Education will always be able to continue, and teachers will forever be a need in our world, as students will always need to learn.

I dedicate this chapter to Jordan Lea Darnell. A teacher, friend, sister, and daughter who worked at a high school in north Texas as a Special Education Teacher. Jordan lost her battle to COVID-19 in the Spring of 2021. She was a light to all who knew her and gave her students 110% in every aspect. I dedicate this to her and all she did for her students throughout her years as an educator.

LITERATURE REVIEW

This literature review will analyze a variety of topics and research findings regarding remote teaching during the COVID-19 pandemic. The focus of my literature review is the equity behind teaching students with remote instruction and students learning throughout the pandemic. According to OECD (2018), “equity in education means that schools and education systems provide equal learning opportunities to all students” (p.22). This proposed definition took educators for a whirlwind during remote instruction, as there are so many different circumstances behind every student’s home life that can make or break the meaning of equity.

For my literature search I used Google Scholar as my database to find all articles. I chose to search only articles from 2018 to 2021. The chosen written language for my articles is English. Some words and phrases I used to search for my articles are listed as the following: remote teaching, equity in remote teaching, and equity during COVID-19. I found a plethora amount of articles in relation to COVID-19, but used total of ten articles on the related topic of study involving equity.

The context of this literature review is to gain a greater understanding of the relationship that equity had during remote teaching in the 2020 and 2021 school years. This literature review would benefit teachers, administrators, district professionals, college professors or relatable career in the context of using remote instruction. It would also benefit anyone who is wanting to learn more about equity in a virtual world. My views regarding remote teaching and personal experiences are discussed throughout the literature review, as I give personal connections and accounts from my own teaching experiences.

The literature review will discuss the following ideas: equity in remote teaching, device and internet access, physical education equity, teacher support with emergency remote learning, remote teacher self-efficacy, emotional and social support for students, learning equity and lost learning, education after COVID-19, and a conclusion will be stated at the end.

EQUITY IN REMOTE TEACHING

The all-in question comes down to equity in the world of remote teaching and student learning. William et al. (2021) conducted a study consisting of open-responses and Likert-type questions involving teachers, counselors, and administrators. Research from William et al. (2021) found the following three common themes emerge from their focus group discussions:

(a) when educators put in extra work, students benefit, (b) the pandemic forced a shift in mindset that inspired new behaviors and resulted in positively surprising outcomes, and (c) teachers want the lessons learned from the experience to lead to lasting change (p.11).

The extra work that educators put into remote teaching was unreal. The amount of time spent assisting parents with technology issues or teaching them how to log their students into Google Classroom was no joke. William et al (2021) also stated that, "educators were overwhelmed and exhausted with the amount of uncompensated time and energy they had to put in during the pandemic, the investment of extra time and energy resulted in positive outcomes for them and their students" (p.11). The true light that kept me going through the pandemic was seeing my students through Google Meet and assisting them as much as I could. Due to the curveball that the pandemic threw at everyone, the need for a sense of normalcy was necessary to obtain the sense of fulfillment that our daily in-person job gave us.

The conclusion from William et al. (2021) suggests that technology is still a problem in the world of education and serves as a "digital divide" (p.17). William et al. (2021) stated that, "the teachers, counselors, and administrators in the study learned the importance of community building. Effective community-building actions included those that connected the school and home environments, provided encouragement and support for students and parents, and allowed for flexibility" (p.18). Community-building was true for my personal school as well. During the pandemic my school had uplifting virtual staff meetings and a few meals provided to us by our PTO. Parents of students in my class were supportive of my job, as I showed them as much grace as I could during that scary time. Community-building was a something that benefited everyone during the pandemic at my school, as it allowed for an opportunity to feel supported and appreciated.

DEVICES AND INTERNET ACCESS

According to a report from The School Superintendents Association, 25% of superintendents reported that nearly all their students (91-100%) had internet access at home and that 26% of students had devices to connect to the internet at home. (Rogers and Ellerson, 2020). This reality led schools to come up with a plan regarding how students would continue learning even without a device or internet access. According to Ali and Herrera (2020), "some school districts have opted to create packets for students and families to pick up which include learning materials and exercises" (p.1). This was the alternate option my school district went with but instead of parents picking them up our administration delivered them to their front porch.

As the pandemic continued in the spring of 2020, my personal students who received paper packets depreciated as families opted for devices to help with instruction. Different areas around the United States evolved ideas and solutions to the Internet and Wi-Fi access dilemmas for families. Ali and Herrera (2020) found the following:

In South Carolina, Charleston County School District is deploying 10 buses equipped with Wi-Fi to provide additional online coverage for students. In Texas, Dallas County Independent School District Board of Trustees approved the purchase of 12,000 Wi-Fi hotspots to support distance learning. In Caddo Parish Public Schools, Louisiana, the NAACP and Comcast have joined forces to provide two months of free internet to help families without internet access engage in virtual learning activities. (p.2)

In relation to devices and internet access, equipment rentals and distribution was an area that school districts setup as well. From my class in the spring of 2020, about five out of my 18 students needed a device to use during remote instruction. These devices were picked-up outside of our school building with administration passing the devices out to parents. Ali and Herrera (2020) found the following about equipment rental and distribution:

The School District of Palm Beach in Florida is distributing devices on loan for students. Families are allowed up to one device, allowing for families with more than one student to share the device. Montgomery County Public Schools, Maryland is distributing Chromebooks to students and families. Chromebooks are distributed at elementary and high schools across the country. Families without internet access will be able to get free service through Comcast (p.2).

Another area of need in some school districts was print materials and instructional television. Print materials was something my campus needed for sure, as not all students at my Title 1 school had a device or internet access. Print materials were printed off at the school by administration and delivered to the students' houses. Ali and Herrera (2020) stated the following in regard to print materials and instructional television:

Charlottesville City Schools, Virginia is enabling families to check out laptops and is also providing low-tech options like printouts of educational materials to ensure continued learning for students at home. In Hamilton County Schools, Tennessee, office supply company Staples has offered to print and distribute the district's continued learning packets at no cost for families that do not have the ability to print at home. Jackson Public Schools, Mississippi is developing a range of learning materials that are paper based, online, and shared via their instructional Television Channels (p.3).

All the areas of need in relation to devices, internet access and student learning materials all relate back to the equity of teaching during remote instruction. If students do not have what they need to be successful, they cannot continue to learn.

PHYSICAL EDUCATION EQUITY

When looking at various classes during the pandemic, physical education stood out as an extra-curricular that was completely lost for some students. According to research done by Pinotti (2021), "equity issues have made it difficult or nearly impossible for some students to access their online physical education course or have hindered their ability to engage in regular physical activity in a safe environment" (p.17). Not only did students have their physical activities taken from them, with gyms being shut down, teachers did as well. The impact of having something ripped from your daily schedule effects your mental state as a human being.

Pinotti's (2021) research was conducted at a Title 1 school composed of a high number of socioeconomically disadvantaged students. After interviewing two teachers and 16 students three themes arose from their answers regarding what they thought could be done to improve the equity issues affecting students' physical activity during online P.E. classes. Those themes are stated as the following: access to more resources, technology assistance, and family support (Pinotti, 2021, p.18). A theme that stood out to me as an educator was family support. Family support is the backbone to remote learning, especially at the elementary level. From personal experience, if it was not for my parents support growing up, I would not want to go exercise my body daily, as they were the ones who motivated me outside of school. Pinotti (2021) stated the following:

As an avid runner, active lifestyle enthusiast, and with a current education emphasis in Health & Wellness, I know first-hand the benefits that physical activity can provide to oneself. The skills to become physically active and the ability to sustain an active lifestyle were established through learned motor skills and confidence building throughout my public schooling P.E. classes. Maintaining an active lifestyle has positively affected my mental health, my academic performance, and has helped reduce any potential health risks. The COVID-19 outbreak, however, has had a significant impact on P.E. and, therefore, students' physical activity levels. These new equity issues are out of children's control, yet dictate their abilities to further develop the knowledge, skills, and confidence to be physically active (p.14).

Being someone who has lost a significant amount of weight during my undergrad, the mental aspect of moving my body, lifting weight and running are huge mental breaks for me. When I go to the gym every morning, it relieves stress and prepares me for my day. Students need that time in P.E. class to run, play and socialize with their peers. Elementary P.E. class was filled with so much joy for me as a kid during my younger days. The pandemic took that time from students because of our shutdown situation and led to equity issues in the aspect of students getting that time to move their bodies at least five days per week. The equity of having students move their bodies was put on hold if they did not have that support and guidance in their own household. Pinotti (2021) stated that "having in-person help and encouragement can significantly increase the likelihood of children participating in physical education" (p.27). Kids need positive examples to lead them to healthy lifestyles. From my experience, during the shutdown, my mental state was not where it normally was with a normal daily schedule. It was hard to adjust not having somewhere to go to work out or have an environment to conduct your actual job. Everything was shifted to the context of your home, a place that is meant to be the opposite of that. I had to adjust my daily schedule regarding physical activity as a teacher. Since gyms were completely closed, the next best thing was to be outside, so I drove to a nearby walking trail and would walk or run two to five miles per day before my first Google Meet with my students every morning. Having some sort of structure, as an adult, made the day-to-day requirements of my job, easier to tackle. When looking at this from a student perspective, kids need that support at home to allow for safe physical activities outside of the classroom and physical education classes. Students need the opportunity to move their bodies outside of a remote instruction or paper packet assignments, as these are very sedentary actions. I know for me the aspect of sitting for most of the day, staring at a screen was tiresome in the mental aspect as a teacher. During the start of the in-person fall of 2020 schoolyear, I made it mandatory in my classroom to have multiple GoNoddle breaks throughout the school day. These breaks allowed students time to social distance, take their mask off, move their bodies for three or four minutes to a fun song, and not think about academics for a little while. These breaks helped with classroom behaviors and were a great reset button for students throughout the day. Movement and physical activities are very important to for all human beings. It allows us an escape and is virtuous for your one and only body you are given.

TEACHER SUPPORT WITH EMERGENCY REMOTE LEARNING

Teachers took the pandemic day by day, as alterations to our contracted job happened with a split-second email notification. The reality of school shutdown was like your job being rerouted into a whole new platform. Life was hard to adjust to, as our in-person students and our classroom were not a part of our daily routine. Teachers had to adjust to using a computer to instruct students and arrange their daily schedules at home to become their virtual classrooms. Teachers who are moms and dads had to balance their personal lives at home along with teaching in their own home environments. The following is research by Trust and Whalen (2020):

Overall, participants felt overwhelmed and unprepared to use online or remote teaching strategies and tools and they struggled to adapt their pedagogy to fluctuating situations, such as students' unreliable Internet access, changing personal needs, and unclear or shifting educational or governmental directives. Participants reported needing significant support with shifting their practice and, as result, mainly relied on informal, self-directed learning with their professional learning networks for assistance. (p.191)

Overwhelmed was a feeling common amongst my grade-level team as we conquered this pandemic. Results from the online survey suggest that 61% of educators felt overwhelmed with the online learning resources (Trust and Whalen, 2020, p.191). When addressing the administration aspect regarding support of their teachers, I felt like this was an exceptional category for my school. Only 16% of the participants questioned during the online survey stated that they felt a lack of support by administrators (Trust and Whalen, 2020, p.191). Being blindsided by COVID-19, this low number suggest the leadership in education during this crazy time was supportive of their teachers.

When addressing the aspect of teacher preparedness for remote teaching, Trust and Whalen (2020) research stated that "out of the 256 recorded comments, 169 (66%) agreed that educators should have more training, 47 (18%) disagreed, and 15 (6%) were unsure" (p.192). Before COVID-19 hit, my students were familiar with using Google Classroom for activities at school. This was very helpful, as I did not have to do so much groundwork in preparation for remote learning. My district offered multiple Google Meets regarding instruction and informational meetings about technology and applications during shutdown. My district also had a back-to-school in-person meeting with district leaders in reference to the expectations of creating online lessons through Google Slides. Educators were transformable human beings during this time of remote teaching. Teachers did teach during the pandemic even though everything was new and overwhelming. Teachers led parents and students through unchartered waters, just like administration did to their own staff of educators. All areas in education were transformed as our world did the same.

REMOTE TEACHING SELF-EFFICACY

When viewing remote teaching, there are variations of advantages, disadvantages and challenges presented to teachers and students. Cardullo at el. (2021) presented qualitative findings involving 141 responses from teachers; disadvantages, advantages and challenges were found after evaluating the results. The themes evident from the research in the category of challenges of remote teaching are listed as the following: access and equity, emotional uncertainty, interactions, at-risks learners, teacher proficiency, and assessment and integrity. Cardullo at el. (2021) stated the following regarding the challenges of student access and equity:

Student access and equity emerged as 23 participants listed this as a concern challenge in the open-ended response: 'some students do not have enough access to [a] devices and Internet reliability during class time. And no technology support from family' (q28). Many participants positioned this issue of equity under several umbrellas: not all of my students have Internet access or devices that they can use, conflicting priorities within households for device/Internet use, sharing technology with multiple school-age siblings or unreliable WIFI connections which impact audio and video needed for remote learning (p.38).

When analyzing the advantages of remote teaching, Cardullo at al. (2021) found one theme out of ten that stood out, flexibility. Cardullo at el. (2021) stated that, "the overall analysis of the theme flexibility revealed that teachers and students could work at their own pace, creating more time to practice and less time to plan" (p.40). The equity of this advantage is beneficial for students who had parents who had to work and could not assist with schoolwork until they got off. This advantage could also be a disadvantage for students who needed assistance logging onto scheduled Google Meets for virtual instruction time and could not because they did not have anyone to help them at home. Flexibility in the world of remote teaching for students can be both an advantage and disadvantage, depending on their situation and preferred learning type.

When looking at the disadvantages of remote teaching the following was found: inequity in technology, student motivation and engagement and at-risk learners (Cardullo at el., 2021, p.42). Cardullo at el. (2021) stated the following about inequality:

Remote teaching by nature exacerbates the inequality gap in society without solving every student's access to devices/Internet and knowledge and skills, which can create broader gaps in instruction for at-risk youth. Participants often felt that students lack preparation for remote learning, as 35 participants stated that students often lack motivation and engagement (p.42).

The significance of the study completed by Cardullo at el. (2021) is stated the following:

During COVID-19, many teachers were learning about the platforms simultaneously as they were instructing students. To ensure quality remote instruction and to ensure that students receive the support to make instruction equitable, teachers need to perceive that their instructional technology needs are met to focus on the teaching, learning and needs of their students (p.43).

EMOTIONAL AND SOCIAL SUPPORT FOR STUDENTS

Academics is not the only thing that is taught in an in-person classroom. Social and emotional learning are enormous skills experienced by students in a classroom with their peers. Schools provide a sense of predictability for students and allow for socialization with kids their own age. As a classroom teacher, the emotional aspect was evident during remote teaching and during the return of students to an in-person school year in the fall of 2020. I found that in fall 2020 school year after quarantine teaching in the spring of 2020, the students seemed more emotional than from previous years in the classroom. The following is from According to Garcia and Weiss (2020):

For children, going to school is not just about learning reading and math: it's about developing the social and emotional skills critical to succeeding in life. School closures eliminated some of these critically important aspects of school beyond academic activity, such as the development that occurs through personal relationships among students and between students and teachers, after-school activities that support children's mental and emotional well-being and skills development, and a sense of routine (p.14).

The shutdown of schools brought trauma to students across the world. It was such a sudden and unknown event that brought fear to students, as none have ever experienced anything like this before. Teachers felt the trauma through the screens of their devices when talking to students remotely. The endless questions from students about if we were going to be back in school soon, were heartbreaking. I felt this trauma continue in the emotionally aspects for students when they returned in the fall of 2020. I had more students than ever before exhibit anxiousness and emotional outbursts. The fear of going home for spring break in the spring of 2021 also showed, as a couple of my students showed fear of leaving school. Trauma was a result for some students because of school shutdown due to COVID-19. Garcia and Weiss (2020) stated the following regarding trauma:

The sudden need for children across the board to adapt to uncertain and rapidly changing circumstances and to cope with new levels of trauma make it all the more urgent to address this disparity between what parents and teachers understand about the breadth of skills critical to child development and systems that focus on testing a narrow set of cognitive skills (p.15).

Although trauma was a result of school closures for millions across the world, some skills were learned from the result of the pandemic. Resilience, the ability to adapt to change and thrive, was an important life skill learned as students transitioned to online learning (Garcia and Weiss, 2020). The resilience skill developed depended upon the environment in which the student lived. Some students transitioned easier into remote learning because of their homelife structure and support. Other students took it harder as their daily schedule and routines were only known and felt within the context of an in-person classroom environment. Every student took remote learning with a different grain of salt, as no two students' environments were the same.

Emotional and social skills are represented in the category of noncognitive skills. Garcia and Weiss (2020) stated the following regarding noncognitive skills:

Noncognitive skills are demonstrably as important as other cognitive skills when it comes to ensuring that children will thrive both in school and later in life. Moreover, since academics and socioemotional skills develop in tandem, and in recognition of the added challenges during the pandemic, it will be more critical to approach skills development holistically and make teaching and nurturing the whole child central, rather than marginal (p.15).

An in-person classroom holds such a vital playing ground for students to make friends and to satisfy those noncognitive skills. When we returned to in-person teaching in the fall of 2020, students were so excited to be back at school with their peers. The emotion level of my students from the 2020-2021 schoolyear was higher than from my previous years in the classroom; I believe students felt the sense of unknown from adults that they lived with at home and from their peers in class.

LEARNING EQUITY AND LOST LEARNING

Teacher concerns regarding those who taught from their homes was evident in the research conducted by Bradbury et al. (2020). Out of the 976 educator responses nine common themes emerged in relation to the biggest concerns while teaching online. The nine themes are listed as the following: mental health/well-being, student participation, student support, technology, social interaction, assessment, lost learning, learning equity, and special needs (Bradbury et al., 2020, p.43). The following is a response from Bradbury et al. (2020) regarding learning equity:

Closely associated with the theme of lost learning is the theme of learning equity. Teachers were clear that they could not see how learning could be equitable when the levels of parental support vary. The argument could be made during normal educational times as well, but as with lost learning, the lack of support in the home simply magnifies the concerns connected to learning equity. In addition to concerns of parental support, there were concerns expressed regarding the online format of teaching. Teachers were concerned whether distance education could even produce equitable learning, especially in some subjects (p.47).

Parental support was hit and miss from my perspective teaching during the pandemic. Some students were very successful with remote learning due to their support at home and others not so much. One other theme that went together with equity was lost learning. If a student does not have the support, they need at home to be successful with remote learning, then their learning would suffer. Bradbury et al. (2020) stated the following regarding lost learning:

The theme of lost learning is associated with the themes of student support and student participation. Teachers expressed concerns that students would regress; that is, lost already obtained knowledge and skills. Moreover, teachers identified potential learning gaps that would occur, in large part due to the differences in support structures in the home. For those who have engaged parents, the concern with learning gaps was less than those who have little support in the home. Although this concern is not unique to education in general, it is amplified when teachers have limited contact with students. In other words, teachers simply can't provide enough to offset a lack of support in the home when they don't have regular contact with their students (p.47).

Both lost learning and learning equity were huge concerns for teachers during the COVID-19 pandemic. Students have no control over their environments and learning tools they are given in the home setting. Teachers hold a structure daily in their in-person classrooms, providing students with the best possible learning environment. That structure is the gateway of success for most students, as having a predictable and safe environment promotes the highest sense of learning.

From a personal account, the learning that was lost during the COVID-19 shutdown of spring 2020 allowed for amazing growth for the 2020-2021 schoolyear at my campus. I saw reading levels skyrocket over the academic year, along with MAP (Measure of Academic Progress) testing scores. My students did show a gap in their learning from the beginning of the 2020 schoolyear, but with in-person instruction, I believe those gaps were identified and interventions were used in the best was possible for my students.

EDUCATION AFTER COVID-19

Education has grown without a choice in the aspect of technology during the 2020 and 2021 schoolyears. Educators across the world have been resilient for their students throughout difficult times of change and the uncomfortable unknown. The role of education will forever be dissimilar due to the experience with the novel Coronavirus. According to the Colao et al. (2020), "the COVID-19 crisis highlights that school fulfils not only an educational mission of knowledge acquisitions, but it also satisfies the socialization needs of young people" (p.370). Socialization needs are essential for mental well-being of all humans and create prime learning opportunities for students. The equity behind socialization was limited for some students during the pandemic, as not all students had access to virtual class meetings.

Bradbury et al. (2020) stated the following in regard to the pre and post- COVID-19 times:

COVID-19 has created a demarcation line; that is, pre- and post-COVID-19. At present, we are living in a tumultuous time of transition. Uncertainty is omnipresent, and COVID- 19 is driving most aspects of our lives, including education. There will be a new normal in a post-COVID-19 era, and education will operate within the new normal in ways different than that which teachers and learners encounter in a pre-COVID-19 world (p.49).

Looking into the future, inevitably equity for the sake of students and teachers needs to be addressed in the precinct of remote learning. Colao et al. (2020) stated that, "the COVID-19 crisis gives us the opportunity to re-assess what type of school we want for the future" (p.370). The future of education needs to take advantage of in-person learning opportunities, as COVID-19 showed us what the opposite of that looked like. The future of education also needs more in-depth plans for shutdowns and quarantine possibilities. Having a plan that is equitable for all students and teachers needs to be in place to help create the best possible situation.

CONCLUSION

After examining the ten articles, the equity of remote learning mostly led to the access of internet and devices amongst students in schools across the United States. If students were not provided a learning device due to a lack of internet access, they had to complete schoolwork in the context of a paper packet. New ideas and backup plans for situations such as the COVID-19 pandemic and shutdown, have made it necessary for educational systems to come up with ideas regarding how education will continue. Maybe a new normal would be to give out devices and internet to those who need it at the beginning of each schoolyear in case we ever need them because of a situation that causes us to shut down again. Or possibly, school districts will provide parents with trainings at the beginning of each school year to learn how to assist students with their schoolwork at home. The main sub issues regarding equity is the environment and structure the student has at home; this being the saddest matter that we cannot control as educators.

There are multiple factors that were evaluated from various sources of literature in the realm of equity during remote learning and teaching. We visited the following topics throughout this literature review: equity in remote teaching, device and internet access, physical education equity, teacher support with emergency remote learning, remote teacher self-efficacy, emotional and social support for students, learning equity and lost learning, and education after COVID-19. All topics had equity prevalent in them and personal discussions were evaluated within the topics as well.

From personal experience, I can say that the 2020 and 2021 schoolyear was one for the books. From shutdown to remote teaching and to a full academic school year of teaching with a mask on, education prevailed. The COVID-19 pandemic is an event in my lifetime that I will never forget, as it made us grateful for those in-person moments in education and in our personal lives. Equity in the world of remote learning can be enriched within schools if students are given the correct resources at home for remote learning. Something we will never be able to control as educators is the structure students have at home. What we can do as educators is give students tools in the classroom and serve them the best we can when they are with us. Giving students a community and culture enriched classroom when they are with us, provides them with life experiences and moments to never forget throughout their lifetime. Students will never forget the teacher who was there for them in a time of need or who did not give up on them. Teachers are powerful forces that make an impact on students year after year and for years to come.

REFERENCES

Bradbury, B. L., Suarez-Sousa, X. P., Coquyt, M., Bockelmann, T. L., & Pahl, A. L. (2020). Teaching Under Crisis: Impact and Implications of the COVID-19 Pandemic on Education in Minnesota. *The Interactive Journal of Global Leadership and Learning*, *1*(2), 2.

Cardullo, V., Wang, C. H., Burton, M., & Dong, J. (2021). K-12 teachers' remote teaching self-efficacy during the pandemic. *Journal of Research in Innovative Teaching & Learning*.

Colao, A., Piscitelli, P., Pulimeno, M., Colazzo, S., Miani, A., & Giannini, S. (2020). Rethinking the role of the school after COVID-19. *The Lancet. Public Health*, *5*(7), e370. doi:10.1016/S2468-2667(20)30124-9 PMID:32464100

García, E., & Weiss, E. (2020). *COVID-19 and Student Performance, Equity, and US Education Policy: Lessons from Pre-Pandemic Research to Inform Relief, Recovery, and Rebuilding*. Economic Policy Institute.

OECD. (2018), *Equity in Education: Breaking Down Barriers to Social Mobility*. PISA, OECD Publishing. doi:10.1787/9789264073234-en

Pinotti, G. (2021). *Equal Education for All? Equity Issues Impacting Physical Education and Physical Activity During COVID-19 Pandemic*. Academic Press.

Rogers, C., & Ng, N. E. (2020). *Memorandum: Report of Initial Findings: COVID-19 Impact on Public Schools*. AASA: The School Superintendents Association.

Tinubu Ali, T., & Herrera, M. (2020). *Distance Learning During COVID-19: Seven Equity Considerations for Schools and Districts*. Southern Education Foundation. https://www.southerneducation.org/covid-19-digital-equity/

Whalen, J. (2020). Should teachers be trained in emergency remote teaching? Lessons learned from the COVID-19 pandemic. *Journal of Technology and Teacher Education*, *28*(2), 189–199.

Williams, T. K., McIntosh, R. W., & Russell, W. B. (2021). Equity in Distance Education During COVID-19. *Research in Social Sciences and Technology*, *6*(1), 1–24. doi:10.46303/ressat.2021.1

Chapter 9
Analyzing Diversity and Equity in Virtual Classrooms

Antoniece Bailey
Independent Researcher, USA

Shatara Sims
Independent Researcher, USA

ABSTRACT

Around the country, school systems are trying to navigate through the ever-changing impact COVID-19 has caused. School districts scramble to equip teachers and students with technology as the world transitions to a mostly virtual setting. This document gives an overview of challenges educators encountered shifting from traditional face-to-face learning to virtual learning. This chapter also analyzes how differentiation and the use of multiple communication platforms help to address equity and diversity in the virtual setting.

INTRODUCTION

The year 2020 has brought about many changes that have caused all professions to, not only rethink the way we conduct business, but also rethink the ways in which we communicate with our clientele. Companies, schools, hospitals, and restaurants have transitioned from in-person professional meetings to an all virtual setting. This change has also caused companies to modify how they protect their employees, as well as their consumers and stakeholders. The everyday commute to work has now become a short walk to home offices and makeshift desktops in the comforts of our own homes. With 75% of the nation encountering some form of virtual reality, we

DOI: 10.4018/978-1-7998-8028-8.ch009

have gone from knowing the basics to being forced into learning the ins and outs of technology. Communication, shopping, dining, teaching, learning, etc., have all changed. Many professions have been forced to create new ways to communicate and perform by utilizing various media platforms. Many organizations have implemented platforms such as: Zoom, Microsoft Teams, WebEx, and other educational portals such as Blackboard and Canvas as a means to communicate effectively and keep communication open and safe. This chapter will discuss how Coronavirus impacted our nation and brought on changes that have impacted teaching and learning.

The Virtual Shift

Across the globe education has been affected greatly by this new "virtual shift". The traditional style of teaching has been modified to accommodate a society that was, unfortunately, unprepared for COVID19. As a teacher in one of the largest school districts in Georgia, I knew changes would have to be made. With approximately 32,000 students attending school in urban Georgia, plans had to be made to ensure all students would be afforded the opportunity to continue to a receive proper education. Accordingly, students were given the opportunity to "learn from home". While some were very excited about this option, others were skeptical and look forward to in-school learning. As a result, approximately 65% of the district opted to do virtual learning while the other 35% returned to face-to-face instruction. The district was then faced with the challenge of supplying technology to many homes that did not have and possibly could not afford to purchase technology devices for their children. With many meetings and budget modifications, the district was able to devise a plan that would create a shaky but temporary fix. Laptops that were usually meant for in-school use only, were now being prepared to travel to homes in local neighborhoods. Along with this, mobile hotspots were purchased and distributed. Further, buses that contained wi-fi hotspots were positioned in surrounding neighborhoods for those who needed internet access. The lack of technical knowledge hampered student interaction in the first phase of the course; with students needing to quickly learn how best to use nonverbal communication tools. A common complaint early in proceedings was the limitations of students being able to communicate as effectively as they wanted to with their peers (eg, ICQ becoming a one line, short answer advice session, rather than consolidated information sharing session).

Meeting the needs of technology for the students was just the beginning, the true task was soliciting certified educators who would be willing to transition from face-to-face teaching to virtual teaching. One of the biggest disadvantages faced when transitioning from face-to-face learning to virtual learning was the lack of basic computer knowledge between teachers, parents, and students. Student technical knowledge also forced faculty to re-evaluate team learning objectives early in the course. All students had extensive knowledge of basic computing packages, however, few had used information and communication technologies (ICQ) or web-based chat facilities. This limitation required a focus on operational learning. When Coronavirus impacted the world we were forced into a "lifestyle" that none of us were accustomed to, and educators were mandated to learn all new resources within a months' time. Many educators had limited exposure to communication platforms and were reluctant to implement them into their daily planning. Such platforms like Microsoft Teams, offer a workspace chat and videoconferencing option, which are valuable and useful assets to any work setting; however, coupled with fear and lack of computer skills, many were beginning to panic. To add to the fear and anxiety, high school teachers were asked to teach both virtual and face–to-face lessons sequentially.

When it came to teaching in an online environment, one of the biggest fears for most educators was the "unknown". This situation was something many had never experienced before and the fear of not knowing was not only daunting but caused mass hysteria among many educators. The study preparing students for the virtual organization: an evaluation of learning with virtual learning technologies shows that learning behaviors are closely related to team performance, and that positive team management contributes to the development of on-line learning behaviors. Poor team management skills and difficulties with technology were major elements in limited development of on-line learning behaviors (Andrews & Schwarz, 2002).

In some schools, educators were faced with leaving their schools to work in a hub with new administration and students from all over the district, with the possibility of not being able to return to their home school. Such changes made teachers reluctant to buy into this new learning model. This reluctance of the educators and the worry that educators would vacate their positions, prompted the district to offer an incentive for applying for these positions. Unfortunately, this offer was rescinded based on the idea that compensation was not necessary for duties that were already detailed in the educator's job requirements. Subsequently, all 56 schools had to find and provide virtual teachers per grade level.

The Virtual Boot Camp

Once the teachers were chosen they had to complete an intense professional development "boot camp" to learn a new education database that was introduced over the summer. Many teachers were overwhelmed with learning a completely new education program (in less than a month) and faced with teaching in a way they have never experienced before. Prior to the Coronavirus, the need for online resources was slim to none, as online courses were few and far between. Therefore, resources for virtual learning were scarce across the district. The heart of a teacher would not stop them from meeting the challenge of engaging, inspiring and motivating their students to learn. They had to be innovative in order to create activities that would engage, inspire and motivate students in a virtual setting. Overwhelmed and nervous teachers were then prompted with the task of educating the families on how to transition from in-person learning to online learning, with only a short period of time to do so.

In the past students have solely relied on their teachers to educate them in a face-to-face setting. Now with parents transitioning into being the at-home co-teacher, following an online schedule, learning online resources and familiarizing oneself with a plethora of platforms, they too became overwhelmed. The students have been fortunate enough to be given resources such as online textbooks, a virtual library, virtual manipulatives, and access to interactive sites to assist them in their learning. Virtual classrooms allow students and instructors to communicate synchronously using features such as audio, video, text chat, interactive whiteboard, application sharing, instant polling, emoticons, and breakout rooms. Adobe Connect, Blackboard Collaborate, WebEx, and Saba Centra are synchronous virtual classrooms prevalent in higher education (Martin & Parker 2014). Canvas is the platform purchased by the district for virtual learning, it simplifies course management, communication, student connection and it makes video learning more engaging and interactive. Students are guided by their teacher and parent on how to utilize each of these resources so that they can master each of the standards being taught. These online resources coupled with lecture teaching, help to bridge the gap between the lack of face-to-face learning the students are missing. Although people believe virtual teaching is a "walk in a park" most do not understand what all it entails. Teachers are now faced with not only teaching multiple classes of students a day (each class being 2-hrs) but they also have to teach the parents how to support their child in order for them to be successful. Teaching adults how to log-in to multiple sites and platforms has been just as challenging as teaching it to the students. Teachers, parents, and students were astounded by the amount of commitment it takes to be successful in an online platform.

Addressing Diversity and Equity

When discussing diversity and equity and how it is addressed in the virtual classroom the main concern was how will teachers address accommodating Special Education (SPED) students (Gifted students and students with Individual Education Plans (IEP)) in a virtual setting. It is mandated by federal and state law that these specific students must receive their services outlined in their IEP. Teachers had to differentiate their lessons to meet the needs of all the students. The general education teacher and special education teacher collaborated to differentiate the lesson to ensure the multiple learning levels of the students were incorporated throughout the lesson. Many special education students would be lost without the SPED teacher pushing into the classroom assisting them with their assignments. In the face-to-face setting, the proximity control between the student and teacher helps to keep the student focused and on task. In the virtual setting this obstacle is almost impossible to control because the sad truth is most students are home alone.

Most virtual classroom technologies have a content frame to share the instructor's files, an electronic/interactive whiteboard for instructors and students to write or draw, breakout rooms for group activities, text chat to interact using words and emoticons, and audio chat to talk via a microphone or telephone with the instructor and other students (Martin & Parker 2014). One resource teachers introduced was the use of breakout rooms to help give students the small grouping they would have received in class. During an interactive synchronous lecture, students can type questions and comments without interrupting the presenter. These questions benefit the students asking them as well as the entire class because every student can see the questions. This builds critical thinking skills by causing them to reflect on the questions and posit answers to them for themselves. It can also draw their attention to material they missed and provide information when the question is answered (Martin & Parker 2014). In the SPED breakout room, the SPED teacher reiterates the lesson taught by the general teacher, then she assigns them work on their level to meet the needs of their IEP. The general teacher accommodates the gifted students in the breakout rooms by giving them more rigorous and independent assignments that will engage them in the learning while maintaining the task aligned to the standard. The gifted breakout room is normally student led due to the students' academic ability and maturity level to handle independent/small group tasks. Last but not least, the students that do not fall into a SPED category are in a teacher lead breakout room with the general teacher who scaffolds their work for them. The accommodations continue throughout the lesson, as the lesson comes to an end the teacher and student use individual whiteboards so that the teacher can communicate with the student on a one to one basis to give them guidance and feedback on the activity. In the world of education, differentiation is the key to students learning and making academic growth because it addresses their specific deficits. Even in the year of "Covid" teachers have still gone above and beyond to make sure students are being caught up "academically" as much as possible. The table in Figure 1 presents a schedule of virtual learning.

Table 1.

Time	Monday	Tuesday	Wednesday	Thursday	Friday
7:00-7:30	Morning Duty	Morning Duty	Morning Duty	Morning Duty	Morning Duty
7:30-8:00	Planning	Planning	Planning	Planning	Planning
8:00-9:00	Co-Teach Math with the face to face classes	Co-Teach Math with the face to face classes	Co-Teach Math with the face to face classes	Co-Teach Math with the face to face classes	Co-Teach Math with the face to face classes
9:00-9:30	Word Study (Vocabulary)	Guided Reading Instruction	Word Study (Vocabulary)	Guide Reading instruction	**RTI's Intervention PLS Parent Conference as needed Every Friday**
9:30-9:50	Writing	Math	Writing	Math	
9:50-10:00	Movement Break	Movement Break	Movement Break	Movement Break	
10:00-10:30	Math	Daily Read Aloud	Math	Daily Read Aloud	
10:30-11:00	Science & PE/Health Integrated Activities	Social Studies & PE/Health Integrated Activities	Science & PE/Health Integrated Activities	Social Studies & PE/Health Integrated Activities	
11:00-11:50	Co-Teach Math with the face to face classes	Co-Teach Math with the face to face classes	Co-Teach Math with the face to face classes	Co-Teach Math with the face to face classes	Co-Teach Math with the face to face classes
12:00-12:30	Tutoring	Tutoring	Tutoring	Tutoring	
12:30-1:00	Lunch Break	Lunch Break	Lunch Break	Lunch Break	
PM Schedule	**PM Schedule**	**PM Schedule**	**PM Schedule**	**PM Schedule**	**PM Schedule**
1:00-1:30	Word Study (Vocabulary)	Guided Reading Instruction	Word Study (Vocabulary)	Guide Reading instruction	**RTI's Intervention PLS Parent Conference as needed Every Friday.**
1:30-1:50	Writing	Math	Writing	Math	
1:50-2:00	Movement Break	Movement Break	Movement Break	Movement Break	
2:00-2:30	Math	Daily Read Aloud	Math	Daily Read Aloud	
2:30-3:00	Science & PE/Health Integrated Activities	Social Studies & PE/Health Integrated Activities	Science & PE/Health Integrated Activities	Social Studies & PE/Health Integrated Activities	

The Do's and Don'ts of Virtual Learning

Teaching virtually to elementary students as stated previously has had many challenges. Difficulties encountered within online learning were participation, work completion, and attendance. When it comes to participation it's always the same students engaged in the lesson. Some students have their cameras and microphones turned off and others are sleeping or paying attention to something else in their background. The students are afraid to participate due to below grade level comprehension, peer judgement, or the fear of being incorrect. Not only is there an absence in student participation but the lack of preparation plays a role in the student's lack of success as well. Many parents feel that they don't have to provide their child with supplies due to all learning being online. Having the necessary school supplies and tools needed to stay organized is essential to successful virtual learning. Although class participation and supply preparation is a very big issue in virtual learning, the abundance of backgrounds noises limits all the students from learning the most. Once the learning atmosphere has been interrupted it takes several minutes to get everyone back on track, which limits instructional time and learning. Every minute of instruction is essential in any classroom setting, there is no room for unnecessary distractions. Once class is dismissed students are expected to do asynchronous lessons to assess their knowledge of the standard that was taught. Many students feel that if they attend the instructional portion of the class they shouldn't have to complete the asynchronous assignments that follow after the 2 hours of instruction. Even though the guidelines stated that the asynchronous work was a requirement of virtual learning, many students still refuse to complete the assignments. The lack of attendance in the virtual classrooms was a huge issue the district was not prepared for. There are students who've attended virtual class very little or not at all. Even with completing social worker referrals many parents and students have still not been held accountable for the excessive amount of absences. Participation, work completion, and attendance all play a role in student achievement, without all of these components the learning deficit only becomes greater. The development of learning behaviors in a team based on-line learning environment where students are required to use information and communication technologies ICTs for a central part of the learning activities contrasts with Salmon's (2000) model of on-line learning which identified five stages of learning behaviors. As outlined earlier, Salmon considered that given the appropriate conditions, nearly all participants will progress through these stages (2000). In the current study, few teams demonstrated all five of Salmon's stages. Low performing teams demonstrated little on-line interaction, had few bulletin board postings, and had very limited social interaction. Research relating to other virtual learning environments (such as video-conferencing environments), strongly indicate that students feel much more comfortable with the learning environment and have more successful learning experiences if they know other students and lecturing staff before the commencement of the course. This observation would indicate that including or developing social learning processes in on-line learning environments may be critical to promoting effective learning behaviors, particularly for low and middle performing teams.

CONCLUSION

Coronavirus has caused the world to revamp how we perform on a daily basis. The education field took one of the biggest hits due to the pandemic forcing everyone to work from home and continue lessons virtually. Virtual learning was never a factor in the K-12 setting and now every parent, teacher, and student is expected to make a seamless transition into a whole new way of educating students. In an effort to prepare for the future the district purchased a new database (Canvas) that can be used virtually or face-to-face to maintain continuous learning in the event of an extended absence from in school learning. Also, they have invested millions of dollars into purchasing new technology so that they may become a one to one (1 device per student) district in an effort to make learning accessible for all students. The way things are now may be our new norm for a while or possibly forever. If we are going to make sure that our children still receive a quality education, we must be willing to adapt. On top of adapting to a technological society, parents and students must take online learning seriously in order to continue to close the learning gap.

REFERENCES

Andrews, T., & Schwarz, G. (2002). Preparing students for the virtual organization: An evaluation of learning with virtual learning technologies. *Journal of Educational Technology & Society*, *5*(3), 54–65. Retrieved June 30, 2021, from http://www.jstor.org/stable/jeductechsoci.5.3.54

Martin, F., & Parker, M. A. (2014). Use of Synchronous Virtual Classrooms: Why, Who, and How? *Journal of Online Learning and Teaching*, *10*(2). Retrieved June 30, 2021, from https://www.researchgate.net/profile/Florence-Martin-7/publication/265335172

Chapter 10
Diversity and Equity in Healthcare:
Are We Doing Enough?

Alexis S. K. Bain
Independent Researcher, USA

ABSTRACT

Overcoming diversity and equity challenges in healthcare is no small feat for healthcare facilities, professionals, and the healthcare industry at large. This chapter will bring awareness to healthcare disparities that transpire because of the lack of equity which can occur when diversity is finite. It will identify and explore how diversity and equity have been omitted from medical facilities and look at the impact of the presence of diversity and equity on both consumers and medical personnel. In addition, it will uncover factors that contribute to negative outcomes at the virtual level and uncover ways to reduce inefficiencies and negative outcomes caused by the lack of diversity and equity in healthcare.

INTRODUCTION

"Primum non nocere", "First, do no harm." In healthcare, this phrase rings throughout every facility, educational institution, and even in the ears of healthcare professionals and civilians. This is an ethical mandate that originated from a translation of the Greek Hippocratic oath that encourages justice and the provision of quality care practices because the people being treated are human beings, not their diseases (Lee & Young, 2018). Medicine was not created and implemented for physicians past and present to harm their patients; intentional or not. However, medical practices

DOI: 10.4018/978-1-7998-8028-8.ch010

that occurred between the 1700's and the present seem to tell a different tale. Over the years, the death rate due to medical errors ranges anywhere from approximately 210,000 to 440,000 lives annually as stated in a 2013 report (Carver et al., 2021), with the latter number indicating that one third of deaths in the United States can occur accidentally. Medical errors are the presentation of a preventable occurrence (Carver et al., 2021). Many of the recurrent sources of medical errors are listed by "... misdiagnoses, adverse drug events, improper transfusions, misdiagnosis, under and over treatment, surgical injuries and wrong-site surgery, suicides, restraint-related injuries or death, falls, burns, pressure ulcers, and mistaken patient identities," (Carver et al., 2021). However, statistically speaking, minorities are known to experience higher rates of medical errors that have had significant impacts on their safety in medical (mainly hospitals) facilities (Chauhan et al., 2020).

Black people have been scrutinized through inhumane medical practices that ultimately crippled the black community for centuries (Nuriddin et al., 2020). Blaack people have been poked, prodded, overutilized and under recognized for their impact in medicine. Nuriddin et al., (2020) uses the term introduced by Paul Farmer and his colleagues at the time as "structural violence." The term structural violence is the intentional harm caused to an individual without their consent or knowledge of what they were being subjected to. Like Farmer and colleagues, Nuriddin et al., (2020) classifies structural violence as the reason behind the objectification of people of color who turn to healthcare institutions for medical treatment and prevention that are harmed in the process.

For example, in 1865 black people, most commonly black women were subjected to vicious treatment practices and experimental procedures against their will (Nuriddin et al., 2020). In 1880-1919 medical schools used a black man by the name of Baker to dig up black bodies (to reduce the possibility of exposure) for dissection practices in white medical schools. From the 1930's to the 1970's, the Tuskegee experiment gave black men the Syphilis virus knowingly and left them untreated; those men later died (Nuriddin et al., 2020). Lastly, in 1951 Ms. Henrietta Lacks and her HeLa cells made remarkable contributions to healthcare fields such as Oncology, Immunology, Histology, etc. (Nuriddin et al., 2020). Unfortunately she too became an experiment, having gone in for treatment for cervical cancer, but was ultimately made to contribute lindly to science. Ms Lacks went to John Hopins medical facility for treatment where she was kept in an alluded section of the hospital. They took her blood against her will and used it as the foundation for the treatment of others (Nuriddin et al., 2020). Throughout history, people of color made major contributions to medicine, overwhelming against their will or without their knowledge or consent, only to become the people who do not experience the full benefits of those contributions made by those who were violently, ignorantly, and unethically coerced into becoming test subjects.

The examples listed here are only a few of the biggest medical advances that black minorities have contributed to involuntarily. Despite the advances that black minorities have contributed to medicine, they and the remainder of the minority population in the United States continue to experience high levels of harm and a lower quality of patient safety in medical institutions. In lieu of the negatively associated outcomes of medical practices with minorities, there have been many efforts to reverse inequity through the ethnic and racial diversification of professionals providing healthcare to minorities. Diversity has become the staple for many healthcare facilities to aid in the reduction of healthcare disparity rates.

Overcoming diversity and equity challenges in healthcare is no small feat for healthcare facilities, professionals, and the healthcare industry at large. The history of racially inflicted disparities in the quality of patient care has placed the healthcare industry at the forefront of scrutiny, and constant industrial evolution is required to accommodate their ever-changing consumer dynamics. In 2010 Peek et al., (2010) wrote that there are two root causes for the racial disparities encountered in healthcare. These causes are; 1. healthcare systems and 2. discrimination at both or either the patient or provider level. Health care systems hold many different definitions in the medical profession (Woolf & Aron, 2013). It can stem from medical facilities to the legal branches of medicine that have a major impact on the operations of healthcare practices (Woolf & Aron, 2013).

Innately, the discrimination felt by both providers and patients would be the voluntary action of delaying treatment (provision and reception) based on perception and past occurrences (Rivenbark & Ichou, 2020). The segregation felt by patients and providers alike is a reflection of what medical facilities and professionals have had to endure since before the 1960s. Leading up to the 1960s, prior to the implementation of the civil rights act, there were major conflicts between the National Medical Association (NMA) and the American Medical Association (AMA). The events that transpired because of the segregation experienced between the two organizations led to the NMA and AMA not having the ability to collectively co-exist. The staggered views of each organization were also inherited by their patients. This resulted in minorities being handed the poorer end of the medical stick so to speak. Minorities were more susceptible to inequity or being treated with limited supplies because their medical facilities lacked appropriate materials to properly treat their conditions. The NMA was a predominantly black physician organization where the AMA was predominantly white and viewed as the superior organization. Many of the derailed views and discretions found in medicine were derived from the animosity that stemmed between both the NMA and AMA organizations.

Historically, the NMA wanted to join with the AMA, but the American Medical Association combatted this at every chance they were given. Some have accredited the segregation between the two organizations to the decisions made by Dr. Nathan

Smith Davis. Davis claimed himself to be the father of the American Medical Association, however, that fact is still under investigation. The AMA's views of the National Medical Association were biased and skewed as they were based on their views of the black physicians that were within the organization. Ultimately, this has led to major differentiation in the treatment of minorities and their majority counterparts. Many of the chronic and medically significant diseases that differ between races and ethnicities can be traced back to this According to Hall et al., (2015) many of the challenges faced in healthcare span from the racial and ethnic divide that minorities are exposed to within the profession. Hall et al., (2015) found that this divide both racially and ethnically has resulted in these groups of individuals having poorer health statuses and substandard health outcomes for chronic diseases.

Unfortunately, the issue of race in medicine is still recurring and is actively inhibiting the effectiveness and efficacy needed to enhance the purpose of medicine. Taylor (2019) specified that the apartheid felt by the minority community has resulted in limited physician availability and a lack of medical facilities that would be necessary for their treatment. Taylor (2019) also associated the difference in races that seek medical attention with a difference in how they are covered through their healthcare policy coverage. She has found that minorities and their white counterparts have significant differences in mortality and treatment which could be linked to the type of healthcare policy coverage that they have. In traditional medicine, the negligence met by minorities has been seen as inequity and a lack of diversity for consumers. Within healthcare, inequity and the lack of diversity experienced have progressed from traditional medicine to virtual medicine. As the 21st century has become more complex and is constantly evolving, medicine can experience a lag that inhibits the necessary integration methods needed to properly acclimate to the needs of everyone involved.

EQUITY IN HEALTHCARE

Providing equity within healthcare institutions has been the overall goal in aiding the healthcare needs of patients. Before discussing equity it is imperative to know and understand what equity is. Equity has been defined as "the quality of being fair and impartial," (Oxford Dictionary). In healthcare, this means being able to provide the appropriate care that a patient or consumer needs in a timely manner (Braveman & Gruskin, 2003). Braveman & Gruskin (2003) further explained that equity is often misunderstood for equality, but there are stark differences between the two. Equality speaks to distributing resources equally between patients. Meanwhile, equity speaks to distributing the appropriate resources to patients to provide the best care possible (Braveman & Gruskin, 2003). In medicine, providing equity has become a major factor in producing the necessary standard of care for each patient. Unfortunately, measuring the appropriateness of care for each individual is difficult to determine as every person's needs differ.

DIVERSITY IN HEALTHCARE

Diversity means more than just racial differences. Diversity looks at cultural, religious, behavioral, socioeconomic status, education, and so many more factors. Achieving a medically diverse environment is no easy task for medical facilities and the industry alike. However, it is a task that must be undertaken to improve the health status of individuals and their overall reception of healthcare treatment. Imagine walking into a clinic because you need care, but no one in the clinic looks like you. No one has the same beliefs as you, or even understands the needs you may have. Or imagine going to a clinic whose staff and systems do not value or account for the difference, and because you are different, you are not treated appropriately and ultimately do not receive the standard of care that you rightfully deserve. The biases faced by people of color versus their white counterparts are no small occurrences in healthcare. The actions that healthcare professionals have had towards their patients of the opposite race, culture, belief systems, socio-economic status, etc. have changed from blatant outward expressions of disdain to subtle, but distinct measures of medical bias.

VIRTUAL HEALTHCARE

Virtual medicine has been in healthcare since the late 1800s (Institute of Medicine (U.S.) & Lustig, 2012). The first form of virtual healthcare was what is now termed as "telemedicine". In those days physicians were using telephone consultations to lower the number of patient office visits (Institute of Medicine (U.S.) & Lustig, 2012). In another time period, the 1920's, there is a record of a physician diagnosing a patient over the radio (Institute of Medicine (U.S.) & Lustig, 2012). This was done in hopes that future medical practitioners would expand their reach, multiply the ways that patients can receive care, and change the overall trajectory of medicine. Virtual care has been defined by Jamieson et al., (2015) as "Any interaction between patients and/or members of their circle of care, occurring remotely, using any forms of communication or information technologies, with the aim of facilitating or maximizing the quality and effectiveness of patient care," (pg. 4).

CHAPTER OUTLINE

This chapter will bring awareness to healthcare disparities that transpire because of the lack of equity which can occur when diversity is finite. It will identify and explore how diversity and equity have been omitted from healthcare facilities, and look at the impact of the presence of diversity and equity on both consumers and hospital personnel. In addition, it will uncover factors that contribute to negative outcomes at the virtual level and uncover ways to reduce inefficiencies and negative outcomes caused by the lack of diversity and equity in healthcare.

CAUSES OF RACIAL DISPARITIES IN HEALTHCARE

Can equity still be maintained with conscious and unconscious biases present?

NO! However, in many medical facilities people of all races, ethnicities, social and educational classes have experienced some form of prejudice in medicine (Nong et al., 2020). Whether they have experienced unwanted stares, snide and demeaning remarks, lack of attention or been ignored, or been denied service, because of personal disregard for their natural state. Unfortunately, we see many people try to cope with prejudiced acts every day while attempting to procure equal treatment from their healthcare provider.

There is a disproportionate representation of healthcare disparities in minorities in the United States (Sorkin et al., 2010). In 2006 Egede (2006) stated that even with the many technological advances that occurred during that time, minorities (ie. people of color) were still receiving below standard care (Hall et al., 2015; Nong et al., 2020). The quality and efficacy that they were subjected to were not the same as their majority counterparts. It was further explained that due to the misuse of technology and the inadequate treatment measures for minorities, their comorbidity and mortality rates were significantly higher than non-minorities (Egede, 2006). Even in terms of maternal treatment and harmful events post giving birth, women of color are approximately 254% more likely to die or experience harmful post birthing events (Centers for Disease Control and Prevention, 2019)

The existence of healthcare disparities on the basis of race and ethnicity is not invisible. The IOM report of 2002 identified, "... disparities in health care as "racial or ethnic differences in the quality of health care that are not due to access-related factors or clinical needs, preferences, and appropriateness of intervention." Bakke et al., (2014) endorsed this by confirming the fact that colored people and white people were treated differently (in their terms less aggressively) which promoted the significant gap in health disparities between minorities and nonminorities. Examples of this discrimination in healthcare can be seen by people of color experiencing a lower quality of care (treatment decision), negatively influence patient satisfaction, delayed referrals if any at all, and limited communication between patients and physicians (Bakke et al., 2014). Uncovering the causes of racial disparities in healthcare institutions is not simple. They can often be missed because of the subtlety of their occurrences.

Approximately 1 out of every 5 adults in the United States has experienced a form of racial discrimination while seeking medical intervention (Nong et al., 2020). They were either discriminated against because of their race, ethnicity, age, sex, financial ability, or level of education (Nong et al., 2020). The discrimination faced by minorities in the United States has become one of the main reasons behind the substantial gap in healthcare disparities and medical outcomes (Evans et al., 2020). The type of experience that an individual has received from a physician, nurse, administrator, et cetera, can impede the patient's judgment of whether to seek treatment immediately or delay. Their experiences can further diminish a patient's trust in healthcare providers, causing a strain on the patient-provider relationship. Historically, the patient-provider relationship has been known to create remarkable outcomes in the course of treatment for a patient (Evans et al., 2020). With the patient-provider relationship, once fostered correctly and appropriately, patients are more inclined to follow the instructions of their providers and are more susceptible to having better treatment outcomes (Evans et al., 2020).

In a report written in 2015 by Representative Kelly it was implied that the 5 areas that needed to be addressed for healthcare disparities in the black community were:

- Access
- Workforce diversity
- Innovation and research
- Community engagement
- Federal action

The areas that are discussed in their report are not new cause agents for racial disparities in the healthcare arena. These areas have been a challenge for medicine for decades. Sorkin et al., (2010) described the impact that something as simple as a poor attitude from a healthcare professional towards a minority patient had a significant impact on their quality of care. It has been confirmed in a study by Stepanikova & Oates (2017) that poor physician-patient communication and patient discerned prejudice can lead to their omission of treatment and a delay in their retrieval of care. In comparison to their counterparts, a survey conducted by Nong et al., (2020) specified that race/ ethnic discrimination was the most frequent form of discrimination experienced by patients with respect to education, finances, sex, weight, and age (Hall et al., 2015).

Another factor that can contribute to racial and ethnic health disparities is the perceptions of race and subtle biases held between patients and providers (Hall et al., 2015). In healthcare, patients are not the only parties that face discrimination. However, in terms of medical care, the attitudes towards minorities are astonishingly different from nonminorities (or Caucasians). Hall et al., (2015) acknowledged that the attitudes and subtle perception (implicit bias) of physicians towards people of color explains why they receive lower standards of treatment, and gaps can be seen in healthcare outcomes between minorities and nonminorities. ***Implicit bias*** is the unconscious negative perception of one individual based on that person's ideology (Penner et al., 2016). When a physician has an implicit bias they impose significant ramifications on healthcare institutions and their ability to provide equitable standards of care (Penner et al., 2014; Penner et al., 2016). It has been proposed that physicians do not provide opioid treatment regimes for people of color, because their belief is that these people can withstand the pain that they are feeling and that they do not need it (Hoffman et al., 2016). People of color are also neglected or given reduced amounts of opioid treatment, because of the negative images that have been painted of them in the news or television shows, social media et cetera (Hoffman et al., 2016). People of color, especially black people are treated according to the perception of their provider (Hoffman et al., 2016; Penner et al., 2014; Penner et al., 2016). Therefore, if the provider has a negative opinion of people of color, their treatment, diagnosis, and prognosis may not be based upon what the patient is experiencing but more so on what the physician feels is necessary (Penner et al., 2016). Hoffman et al., (2016) confirm that the treatment, especially pain management for people of color (minorities) are dependent upon the ideology that black people are not as fragile as their non minority counterparts and that their anatomy is viewed differently. This leads to the inclining growth rates of racial and ethnic disparities in the United States.

Figure 1 shows the statistics of the diversity rate of physicians.

As seen in the image the ratio of diversity is subpar in comparison to the increasing growth rate of minorities in this country. In an article written by Evans, et al., (2020) many of the racially inclined disparities are associated with the lack of diversity in medical physicians in medical facilities . The limited diversity seen in healthcare professionals obstructs their ability to competently understand and properly approach the cultural aspects or diverse patient needs to provide equitable healthcare. This accounted for the fact that in previous research the success of black male treatment relied on the fact that they were treated by other black males (Evans et al., 2020). As well as the utilization of medical facilities can be contingent upon a patient's ability to recognize healthcare professionals that resemble who they are. Having a diverse staff aids in the physician, nurse, administrator, etc. ability to properly communicate and comprehend the needs of the patient. Diversification of providers would also aid in understanding the prevalence of chronic diseases or the rarity of them. But unfortunately, based on this rate of black to white physicians, or asian to hispanic physicians, there are not enough active minority physicians to treat the needs of their communities. Jackson & Garcia (2014) suggested in accordance with the results of Evans et al., (2020) that having a more diverse staff would be a necessary tool to change the disparities felt in healthcare. Diversity should be used to change the competence and quality of care provided to the disproportionate racial and ethnic race mix within the United States. It has been hypothesized that by the year 2050 the amount of minorities in the United States will succeed their counterparts (Nair & Adetayo, 2019). Therefore, medical facilities that do not take into account the changing nature of their communities will do more harm than good.

Figure 1.

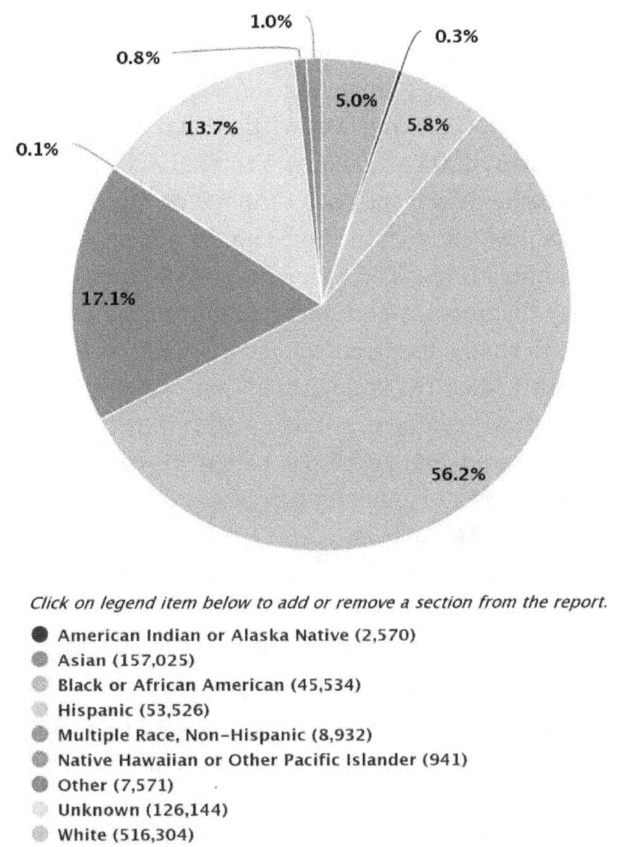

Note: Figure 18 shows the percentage of active physicians by race and ethnicity as of July 1, 2019.

Source: Race and ethnicity are obtained from a variety of sources including DBS, ERAS, APP, MCAT, SMDEP, GQ, MSQ, PMQ, FACULTY, GME, STUDENT with priority given to the most recent self-reported source.

DIFFERENT FORMS OF DIGITAL/VIRTUAL HEALTHCARE

Did you know that the phone you are holding, the tablet and computer that you use daily, and that smart watch that alerts you are all examples of devices that can transmit healthcare data and medical information? Yes, all of the aforementioned devices and more aid in the deliverance of medicine through virtual healthcare better known as digital health. **Digital health** as defined by Ronquillo et al., (2020) is the "...use of information and communications technologies in medicine and other health professions to manage illnesses and health risks and to promote wellness." The reach that digital health has achieved was obtained through the use of technological devices such as mobile phones, watches, telephones, and Health Information Technology (HIT) (Ronquillo et al., 2020). Digital health (Virtual healthcare) can be seen in different forms. Below are descriptions of common virtual/ digital health tools.

- **Health Information Technology (HIT)** is the use of technological devices and hardware that utilize applications along with specifically designed software to aid in storing, retrieving, sharing, and recording healthcare data to execute decision making, knowledge, and communication processes (Thompson & Brailer, 2004 as cited by Yen et al., 2017).
- **Mobile health (m-health)** is the use of mobile devices such as smartphones, smart watches, tablets, etc. that manage and decode any form of changes that occur with the human body. The mobile device shares information with the recording facility that records pertinent health information data and uses it to improve the health status of current and future patients (Park, 2016).
- **Electronic health (e-health)** is the utilization of recorded health data from technological devices to inform, educate and reinforce healthcare measures and decisions to the advantage of the patient (Ossebaard & Van Gemert-Pijnen, 2016). Examples of e-health devices are *Electronic Health Records (EHR) and Electronic Medical Records (EMR).*
- **Telehealth** is the receiving and decoding of medical information to ameliorate the patient's health through electronic communication interchanged between different sites (Kichloo et al., 2020; Tuckson et al., 2017). Examples of telehealth are emails, telephones,videos, et cetera.
- **Telemedicine** is the treating of patients through real time interactions between patient and physician from distant locations (Kichloo et al., 2020)

The purpose of digital/ virtual healthcare was to impact access to healthcare, improve on the inefficacies that are present in healthcare systems (Ronquillo et al., 2020). Digital health has been designed to improve the quality of care that's provided to patients (Ronquillo et al., 2020). This system of healthcare was created to reduce healthcare costs, and create prognoses that tailor to higher success rates in patient care (Ronquillo et al., 2020). Through virtual healthcare and digital health, apps and websites were created to make medical information more accessible. Call hotlines and databases were formed to gain valuable information to improve the quality of care for patients and to improve the standard care provided. Digital health has been declared the way forward for medicine.

THE EFFECTS OF DIVERSITY EQUITY ON VIRTUAL CARE

"Without diversity and health equity, there is no quality. Quality, cost, equity, diversity, and community health are inextricably linked."- Bhatt & Bathija (2018)

The basis of equity is providing fair treatment based on the need of the person. Hence achieving equity in medicine and within communities means that each individual no matter their race, their ethnicity, social status, gender, education, or age will receive the most fair treatment for their need(s). It means that there will not be any hindrance to their access to care, their treatment would be conditioned on the physician's ability to provide efficiency and efficacy (Hardcastle & Ogbogu, 2020). As well as the resources that should be available to the community are not contingent upon who the individual is, but based upon their need. Once implemented correctly, virtual care should provide the level of equity that the healthcare industry aspires to attain, and increase diversity in both patients and medical staff. Lastly, it will improve the medical outcome rates for minorities and reduce the gap in health disparities between people of color and non hispanic, white individuals.

Virtual health services are aimed towards expanding coverage and improving access to healthcare services. Millions of Americans suffer in their health, because they are impacted by one or more of the following barriers (Bhatt & Bathija, 2018). They either have no or limited insurance coverage, the facilities that are within their community lack the basic services necessary to diagnose or control their condition (Bhatt & Bathija, 2018). Barriers such as high medical costs and limited cultural competency also impede an individual's ability to access the care that they need (Bhatt & Bathija, 2018). Virtual care can improve access to basic health services. In order for virtual healthcare to provide equitable care the access to health services has to be culturally competent to yield optimal treatment outcomes (Hilty et al., 2020).

In 2020 telehealth became the way forward as the world was halted by Coronavirus (Covid-19). Ceasing operations for many medical facilities, businesses and schools, comprehensively hindering the access to healthcare. Telehealth methods became the way forward to reduce the opportunity of viral transmission from patient to patient and patient to physician (Hoffman, 2020; Koonin et al., 2020). By implementing these measures it helped to reduce the impact that Covid-19 had already had on the United States (Hoffman et al., 2020; Koonin et al., 2020; Schwann et al., 2020). Thousands of people were dying daily while information on the virus was still being revealed as time progressed. In rural populations where health facilities are limited, virtual health can be a viable solution to accommodate the basic healthcare needs (Haun, et al., 2015; Lynch & Franklin, 2019; Sieck, et al., 2018; Grossman et al., 2019). Virtual care can aid in servicing the 57. 6 million people that live in rural communities and the 272.91 million Americans that live in urban communities (Statistica Research department, 2021).

Equity and diversity are the foundational pillars that virtual care has been designed on. Their goals for healthcare availability, treatment methods, along with improved quality of care and patient outcomes are the reason that so much focus has been placed on its existence. By utilizing virtual care methodologies people are getting the help they need while obtaining the care they deserve (Kaliya-Perumal et al., 2020). Minorities and non minorities have had so many differentiates in the ways they have been treated and the ways they are viewed that their outlook on medical care has become warped (Sorkin et al., 2010). The application of equity and diversity tools in virtual care indicates that the complaints of patients and the community have been heard and medical facilities are placing more of an emphasis on the care of everyone versus those who are of a certain race, ethnicity, social and education class, gender, sex, and/ or age. According to McGrail et al., (2017) virtual visits have proven to become a cost effective tool that proves and provides for itself over time. Equity and diversity in virtual care encourage healthcare providers to focus on ways that they can implement and use virtual care to the benefit of themselves and their patients (Shaw, et al., 2021). The transformation of healthcare through virtual care will be transmitted throughout the healthcare industry. The rapid inception and adjustment to virtual care in 2020 is only an indication of the magnitude to which virtual care will impact healthcare as the population continues to change and adapt to those changes (Hoffman, 2020; Koonin et al., 2020).

Equity and diversity will also continue to engage and educate providers current, present, and old of the manner in which cultural competency is important in their line of work (Hilty et al., 2020). Many of the huge gaps in health disparities are due to the ignorance of providers on the culture of their patients. Equity and diversity will force them to research and understand the extent of gender, racial and ethnic diseases that are not common or that adhere to a different course of action for minorities (people of color). Equity and diversity will allow for substantial amounts of data to be analyzed and used to assist in the making of vaccinations, new treatment methodologies, and drug development (Raghupathi & Raghupathi, 2014; Wouters et al., 2021). Equity and diversity have a directly proportional relationship with each other. In order to truly obtain equity in healthcare diversity must be present to ensure that each individual that turns on that computer or enters that medical facility is receiving justified and the best optimal treatment that can be offered. Only then, once the diverse nature of people is understood, can the patient treatment outcomes and preventable ailments truly be reversed and projected in the direction that healthcare needs it to go.

CHALLENGES FACED IN VIRTUAL HEALTHCARE

Measuring equity in healthcare would mean acknowledging that all patients- current and prospective - would be afforded full access to healthcare, distribution of resources, expenditures, and utilization (Waters, 2000). Virtual care was designed to mitigate the inefficiencies experienced in face-to face medicine. Nevertheless, virtual care even though different has not been devoid of challenges and negative implications. In 2020 the use of virtual care methodologies has increased by over 150% (Koonin et al., 2020). The increase in usage was as a result of the rapid onset of Coronavirus (also known as Covid-19 virus) (Hoffman, 2020; Koonin et al., 2020). Patients had to adapt to new medical care treatment methods and providers had to become more open to new ways of caring for current and new patients. Unfortunately, many of the challenges faced in virtual care are not new but with the increased need for virtual care methods, it led to increased challenges for both providers and patients (Hoffman, 2020). One of the challenges faced by patients is the fact that many minorities do not have access to the materials needed to access virtual care. Having access to internet or broadband bandwidth are necessary for the utilization of virtual care access (Hoffman, 2020). This limitation obstructs a patient's ability to receive the appropriate treatment.

The intended design for virtual care was to aid in making patients' lives easier as it relates to managing their health. In areas where racial and ethnic minorities account for 22% of the rural population and 43% of the urban population it is imperative to take into account the impact that negative outcomes have had on the communities, the economy, medicine, society, and ultimately the patient (U. S. Department of Agriculture, 2020). Warshaw (2017) revealed that many of the disparities faced in rural areas occur because of economic, social, racial, ethnic, geographic, and health workforce factors. Therefore, resulting in limited access to healthcare treatment, difficulty diagnosing and treating the patient, and exposure to more critical problems and hardship for the community (Warshaw, 2017). Diagnostic errors or misdiagnoses are obstacles that oppose the success of virtual care for minorities and non minorities. For physicians, there are many reasons why they are unable to properly diagnose their patients. One being the fact that the patient may not give all of the necessary information. The patient's description of their illness or symptoms may not be clear enough for the physician to make a valid conclusion, or may require additional questioning. Lastly, diagnosis errors can occur because the physician is unable to perform a physical exam to truly understand the nature of the patient's ailment Glauser, 2020).

The rapid implementation of digital/ virtual health initiatives has opened patients and physicians to a world of possibilities. Patients are now comfortably afforded the opportunity to divulge vulnerable information with their physician, while having the ability to get treated in a place where they are most comfortable. Virtual care is doing something that healthcare has needed for decades. It has become the vehicle needed to improve the provider-patient relationship, reduce healthcare costs, and so much more. In a time where an ongoing pandemic has been running rampant around the world, equity and diversity in healthcare have been gravely impacted by the sudden shift in virtual healthcare usage.

By using virtual care, many healthcare insurance policies had to adapt to the increased use of virtual care access in the year 2020 (Schwamm et al., 2020). This method of treatment was necessary as many healthcare providers and facilities had to protect themselves and their staff from a disease that left many people dead or severely ill. Therefore, patients and medical facilities have found themselves compromised by the crippling costs of virtual healthcare (Hoffman, 2020; Schwamm et al., 2020). Due to the complex and unique nature of every patient, facilities are forced to charge uniquely to ensure that providers and their facilities receive enough funds to maintain their treatment measures and to be compensated for their work (Blandford et al., 2020; Hoffman et al., 2020). People were left paying a bill that consumed a substantial portion of their salary or pension, initially, because the cost associated with the installation of hardware and training for virtual health techniques and tools forced healthcare professionals and facilities to increase charges (Hoffman, 2020).

In an article written by (2018) they found that through their Medi-Cal initiative that they were not able to supply enough work hours or manpower needed to meet the financial capacity to fully manage their telehealth program (Purpose Hope, 2020). This impacted their ability to fully merge their telehealth goals into their daily operations (Purpose Hope, 2020). Regrettably, the inability to merge telehealth into the regular daily practice of healthcare facilities is not new. With equity being the ultimate goal on both an individual and national scale in healthcare, having an insufficiency in resources impairs the facility's capacity to reach as many members as possible. The Medi-cal telehealth initiative encountered reimbursement obstacles that added to the facility's inability to properly fund the program and accommodate the demands for patients and the facility. Ultimately they were awarded a grant to facilitate their operation demands and maintain their current coordinators (Purpose Hope, 2020).

CONCLUSION

In conclusion, Virtual care has always been seen as a major mitigating tool and the potential answer to many outstanding issues that the healthcare industry has been facing for years. It has been seen as the answer to access issues, especially for rural and urban communities (Hardcastle &, 2020). Virtual care became an answer to improving the quality of care provided to patients (Herzer & Pronovost, 2021), and an answer for improving patient education while reversing negative treatment outcomes. Virtual care is slowly but surely creating a standard of care that introduces and improves equity while demanding a change in professional diversity and cultural competency. Equity can and will be obtained in virtual and traditional healthcare. Unfortunately, the new advantages that virtual care proposes are not always the solutions needed for real-time medicine. The end all be all expectations that have been placed on virtual care are not what providers and patients are experiencing. The challenges associated with virtual care such as limited access to reliable technology, diagnosis errors, and associated cost seem to present the same hindrances to care for minorities are facing with receiving traditional healthcare. If virtual healthcare is to truly become a solution for the masses - including minorities - pressure should be continually applied to the backbone issue of diversification in healthcare providers to eliminate personal biases and ultimately establish a more inclusive, empathetic medical model that is equally beneficial for *all* patients. These challenges must be addressed to make virtual care equitable for all

REFERENCES

Alegría, M., Alvarez, K., Ishikawa, R. Z., DiMarzio, K., & McPeck, S. (2016). Removing Obstacles To Eliminating Racial And Ethnic Disparities In Behavioral Health Care. *Health Affairs (Project Hope)*, *35*(6), 991–999. doi:10.1377/hlthaff.2016.0029 PMID:27269014

Ateriya, N., Saraf, A., Meshram, V., & Setia, P. (2018). Telemedicine and virtual CONSULTATION: The Indian perspective. *The National Medical Journal of India*, *31*(4), 215. doi:10.4103/0970-258X.258220 PMID:31134926

Baker, R. B. (2014). The American Medical Association and race. *AMA Journal of Ethics*, *16*(6), 479–488. doi:10.1001/virtualmentor.2014.16.6.mhst1-1406 PMID:24955603

Bhatt, J., & Bathija, P. (2018). Ensuring access to quality health care in vulnerable communities. *Academic Medicine*, *93*(9), 1271–1275. doi:10.1097/ACM.0000000000002254 PMID:29697433

Blandford, A., Wesson, J., Amalberti, R., AlHazme, R., & Allwihan, R. (2020). Opportunities and challenges for telehealth within, and beyond, a pandemic. *The Lancet. Global Health*, *8*(11), 1–2. doi:10.1016/S2214-109X(20)30362-4 PMID:32791119

Braveman, P., & Gruskin, S. (2003). Defining equity in health. *Journal of Epidemiology and Community Health*, *57*(4), 254–258. doi:10.1136/jech.57.4.254 PMID:12646539

Carver, N., Gupta, V., & Hipskind, J. E. (2021 Jan). Medical Error. In *StatPearls*. StatPearls Publishing. Available from https://www.ncbi.nlm.nih.gov/books/NBK430763/

Centers for Disease Control and Prevention. (2019, September 6). *Racial and ethnic disparities continue in pregnancy-related deaths.* Retrieved May 01, 2021, from https://www.cdc.gov/media/releases/2019/p0905-racial-ethnic-disparities-pregnancy-deaths.html

Chauhan, A., Walton, M., Manias, E., Walpola, R. L., Seale, H., Latanik, M., Leone, D., Mears, S., & Harrison, R. (2020). The safety of health care for ethnic minority patients: A systematic review. *International Journal for Equity in Health*, *19*(1), 118. doi:10.118612939-020-01223-2 PMID:32641040

Chryssanthou, A., Varlamis, I., & Latsiou, C. (2009). Security and trust in virtual healthcare communities. *Proceedings of the 2nd International Conference on PErvsive Technologies Related to Assistive Environments - PETRA '09*, 1-8. 10.1145/1579114.1579186

Egede, L. E. (2006). Race, ethnicity, culture, and disparities in health care. *Journal of General Internal Medicine*, *21*(6), 667–669. doi:10.1111/j.1525-1497.2006.0512.x PMID:16808759

Evans, M. K., Rosenbaum, L., Malina, D., Morrissey, S., & Rubin, E. J. (2020). Diagnosing and treating systemic racism. *The New England Journal of Medicine*, *383*(3), 274–276. doi:10.1056/NEJMe2021693 PMID:32521155

Gaskin, D. J., Dinwiddie, G. Y., Chan, K. S., & McCleary, R. (2012). Residential segregation and disparities in health care services utilization. *Medical Care Research and Review: MCRR*, *69*(2), 158–175. doi:10.1177/1077558711420263 PMID:21976416

Gee, G. C., & Ford, C. L. (2011). Structural Racism and Health Inequities: Old Issues, New Directions. *Du Bois Review*, *8*(1), 115–132. doi:10.1017/S1742058X11000130 PMID:25632292

Glauser, W. (2020). Virtual care is here to stay, but major challenges remain. *Canadian Medical Association Journal*, *192*(30), E868–E869. Advance online publication. doi:10.1503/cmaj.1095884 PMID:32719027

Gogia, S. B., Maeder, A., Mars, M., Hartvigsen, G., Basu, A., & Abbott, P. (2016). Unintended Consequences of TeleHealth and their Possible Solutions. Contribution of the IMIA Working Group on Telehealth. *Yearbook of Medical Informatics*, *25*(1), 41–46. doi:10.15265/IY-2016-012 PMID:27830229

Grossman, L. V., Creber, R. M. M., Benda, N. C., Wright, D., Vawdrey, D. K., & Ancker, J. S. (2019). Interventions to increase patient portal use in vulnerable populations: A systematic review. *Journal of the American Medical Informatics Association: JAMIA*, *26*(8-9), 855–870. doi:10.1093/jamia/ocz023 PMID:30958532

Hall, W. J., Chapman, M. V., Lee, K. M., Merino, Y. M., Thomas, T. W., Payne, B. K., Eng, E., Day, S. H., & Coyne-Beasley, T. (2015). Implicit Racial/Ethnic Bias Among Health Care Professionals and Its Influence on Health Care Outcomes: A Systematic Review. *American Journal of Public Health*, *105*(12), e60–e76. doi:10.2105/AJPH.2015.302903 PMID:26469668

Hardcastle, L., & Ogbogu, U. (2020). Virtual care: Enhancing access or harming care? *Healthcare Management Forum*, *33*(6), 288–292. doi:10.1177/0840470420938818 PMID:32686506

Haun, J. N., Patel, N. R., French, D. D., Campbell, R. R., Bradham, D. D., & Lapcevic, W. A. (2015). Association between health literacy and medical care costs in an integrated healthcare system: A regional population based study. *BMC Health Services Research*, *15*(1), 249. Advance online publication. doi:10.118612913-015-0887-z PMID:26113118

Herzer, K. R., & Pronovost, P. J. (2021). Ensuring quality in the era of virtual care. *Journal of the American Medical Association*, *325*(5), 429. doi:10.1001/jama.2020.24955 PMID:33528544

Hilty, D. M., Gentry, M. T., McKean, A. J., Cowan, K. E., Lim, R. F., & Lu, F. G. (2020). Telehealth for rural diverse populations: Telebehavioral and cultural competencies, clinical outcomes and administrative approaches. *mHealth*, *6*, 20. doi:10.21037/mhealth.2019.10.04 PMID:32270012

Hoffman, D. A. (2020). Increasing access to care: Telehealth during covid-19. *Journal of Law and the Biosciences*, *7*(1), lsaa043. Advance online publication. doi:10.1093/jlb/lsaa043 PMID:32843985

Hoffman, K. M., Trawalter, S., Axt, J. R., & Oliver, M. N. (2016). Racial bias in pain assessment and treatment recommendations, and false beliefs about biological differences between blacks and whites. *Proceedings of the National Academy of Sciences of the United States of America*, *113*(16), 4296–4301. doi:10.1073/pnas.1516047113 PMID:27044069

Jackson, C. S., & Gracia, J. N. (2014). Addressing health and health-care disparities: the role of a diverse workforce and the social determinants of health. *Public Health Reports*, *129*(Suppl 2), 57–61. doi:10.1177/00333549141291S211

Jamieson, T., Wallace, R., Armstrong, K., Agarwal, P., Griffin, B., Wong, I., & Bahtia, S. (2015). Virtual Care: A Framework for a Patient-Centric System. *Women's College Hospital Institute for Health Systems Solutions and Virtual Care*. https://www.womenscollegehospital.ca/assets/pdf/wihv/WIHV_VirtualHealthSymposium.pdf

Kaliya-Perumal, A. K., Omar, U. F., & Kharlukhi, J. (2020). Healthcare virtualization amid COVID-19 pandemic: An emerging new normal. *Medical Education Online*, *25*(1), 1780058. doi:10.1080/10872981.2020.1780058 PMID:32529948

Kelly, R. L. (2015). *2015 Kelly Report: Health Disparities in America*. Congressional Black Caucus Health Braintrust.

Kichloo, A., Albosta, M., Dettloff, K., Wani, F., El-Amir, Z., Singh, J., Aljadah, M., Chakinala, R. C., Kanugula, A. K., Solanki, S., & Chugh, S. (2020). Telemedicine, the current COVID-19 pandemic and the future: A narrative review and perspectives moving forward in the USA. *Family Medicine and Community Health*, *8*(3), e000530. doi:10.1136/fmch-2020-000530 PMID:32816942

Koonin, L. M., Hoots, B., Tsang, C. A., Leroy, Z., Farris, K., Jolly, B., Antall, P., McCabe, B., Zelis, C. B. R., Tong, I., & Harris, A. M. (2020). Trends in the use of telehealth during the emergence of the Covid-19 Pandemic— United States, January–March 2020. *MMWR. Morbidity and Mortality Weekly Report*, *69*(43), 1595–1599. doi:10.15585/mmwr.mm6943a3 PMID:33119561

Lee, B. X., & Young, J. L. (2018). Clinicians' need for an ecological approach to violence reduction. *AMA Journal of Ethics*, *20*(1), 91–98. doi:10.1001/journalofethics.2018.20.1.msoc1-1801 PMID:29360032

Lustig, T. A. (2012). The role of telehealth in an evolving health care environment: Workshop summary. In *The Role of Telehealth in an Evolving Health Care Environment: Workshop Summary* (pp. 11-16). Washington, DC: National Academies Press

Lynch, M. A.-M., & Franklin, G. V. (2019). Health Literacy: An Intervention to Improve Health Outcomes. *Strategies to Reduce Hospital Mortality in Lower and Middle Income Countries (LMICs) and Resource-Limited Settings*. doi:10.5772/intechopen.86269

McGrail, K. M., Ahuja, M. A., & Leaver, C. A. (2017). Virtual Visits and Patient-Centered Care: Results of a Patient Survey and Observational Study. *Journal of Medical Internet Research, 19*(5), e177. doi:10.2196/jmir.7374 PMID:28550006

Nair, L., & Adetayo, O. A. (2019). Cultural Competence and Ethnic Diversity in Healthcare. *Plastic and Reconstructive Surgery. Global Open*, *7*(5), e2219. doi:10.1097/GOX.0000000000002219 PMID:31333951

Nong, P., Raj, M., Creary, M., Kardia, S. L., & Platt, J. E. (2020). Patient-Reported experiences of discrimination in the US health care system. *JAMA Network Open*, *3*(12), 1–11. doi:10.1001/jamanetworkopen.2020.29650 PMID:33320264

Nuriddin, A., Mooney, G., & White, A. I. (2020). Reckoning with histories of MEDICAL racism and violence in the USA. *Lancet*, *396*(10256), 949–951. doi:10.1016/S0140-6736(20)32032-8 PMID:33010829

Ossebaard, H. C., & Van Gemert-Pijnen, L. (2016). EHealth and quality in health CARE: Implementation time. *International Journal for Quality in Health Care*, *28*(3), 415–419. doi:10.1093/intqhc/mzw032 PMID:27029590

Park, Y. T. (2016). Emerging New Era of Mobile Health Technologies. *Healthcare Informatics Research*, *22*(4), 253–254. doi:10.4258/hir.2016.22.4.253 PMID:27895955

Peek, M. E., Odoms-Young, A., Quinn, M. T., Gorawara-Bhat, R., Wilson, S. C., & Chin, M. H. (2010). Racism in healthcare: Its relationship to shared decision-making and health disparities: a response to Bradby. *Social Science & Medicine, 71*(1), 13–17. doi:10.1016/j.socscimed.2010.03.018

Penner, L. A., Blair, I. V., Albrecht, T. L., & Dovidio, J. F. (2014). *Reducing racial health Care Disparities: A social psychological analysis*. Retrieved May 01, 2021, from https://journals.sagepub.com/doi/10.1177/2372732214548430

Penner, L. A., Dovidio, J. F., Gonzalez, R., Albrecht, T. L., Chapman, R., Foster, T., Harper, F. W., Hagiwara, N., Hamel, L. M., Shields, A. F., Gadgeel, S., Simon, M. S., Griggs, J. J., & Eggly, S. (2016). The Effects of Oncologist Implicit Racial Bias in Racially Discordant Oncology Interactions. *Journal of Clinical Oncology*, *34*(24), 2874–2880. doi:10.1200/JCO.2015.66.3658 PMID:27325865

Project Hope. (2018). Foundation funding for telehealth. *Health Affairs*, *37*(12), 2096–2097. doi:10.1377/hlthaff.2018.05150 PMID:30508391

Rivenbark, J. G., & Ichou, M. (2020). Discrimination in healthcare as a barrier to Care: Experiences of socially disadvantaged populations in France from a nationally representative survey. *BMC Public Health*, *20*(1), 31. Advance online publication. doi:10.118612889-019-8124-z PMID:31918699

Ronquillo, Y., Meyers, A., & Korvek, S. J. (2020). Digital Health. In *StatPearls*. StatPearls Publishing. Available from: https://www.ncbi.nlm.nih.gov/books/NBK470260/

Schwamm, L. H., Estrada, J., Erskine, A., & Licurse, A. (2020). Virtual care: New models of caring for our patients and workforce. *The Lancet. Digital Health*, *2*(6), 1–4. doi:10.1016/S2589-7500(20)30104-7 PMID:32382724

Shaw, J., Brewer, L. C., & Veinot, T. (2021). Recommendations for Health Equity and Virtual Care Arising From the COVID-19 Pandemic: Narrative Review. *JMIR Formative Research*, *5*(4), e23233. doi:10.2196/23233 PMID:33739931

Sieck, C. J., Hefner, J. L., & McAlearney, A. S. (2018). Improving the patient experience through patient portals: Insights from experienced portal users. *Patient Experience Journal*, *l*(5), 47–54. doi:10.35680/2372-0247.1269

Sorkin, D. H., Ngo-Metzger, Q., & De Alba, I. (2010). Racial/Ethnic Discrimination in Health Care: Impact on Perceived Quality of Care. *Journal of General Internal Medicine*, *25*(5), 390–396. doi:10.100711606-010-1257-5 PMID:20146022

Statista Research Department. (2021, January 20). *Size of urban and rural population of the U.S. 2020*. Retrieved May 01, 2021, from https://www.statista.com/statistics/985183/size-urban-rural-population-us/

Stepanikova, I., & Oates, G. R. (2017). Perceived Discrimination and Privilege in Health Care: The Role of Socioeconomic Status and Race. *American Journal of Preventive Medicine, 52*(1S1), S86–S94. doi:10.1016/j.amepre.2016.09.024

Taylor, J. (2019). *Racism, Inequality, and Health Care for African Americans*. The Century Foundation.

Thompson, T., & Brailer, D. (2004). *Health IT Strategic Framework*. US Department of Health and Human Services.

Tuckson, R. V., Edmunds, M., & Hodgkins, M. L. (2017). Telehealth. *The New England Journal of Medicine*, *377*(16), 1585–1592. doi:10.1056/NEJMsr1503323 PMID:29045204

U. S. Department of Agriculture. (2020, October 13). *Racial and ethnic minorities made up about 22 percent of the rural population in 2018, compared to 43 percent in urban areas*. Retrieved April 28, 2021, from https://www.ers.usda.gov/data-products/chart-gallery/gallery/chart-detail/?chartId=99538

Washington, H. A., Baker, R. B., Olakanmi, O., Savitt, T. L., Jacobs, E. A., Hoover, E., & Wynia, M. K. (2009). Segregation, civil rights, and health disparities: The legacy of African American physicians and organized medicine, 1910-1968. *Journal of the National Medical Association*, *101*(6), 513–527. doi:10.1016/S0027-9684(15)30936-6 PMID:19585919

Woolf, S. H., & Aron, L. (2013). U.S. health in international perspective, shorter lives, poorer health. In *U.S. health in international perspective shorter lives, poorer health* (Vol. 4). National Academies Press.

Wouters, O. J., Shadlen, K. C., Salcher-Konrad, M., Pollard, A. J., Larson, H. J., Teerawattananon, Y., & Jit, M. (2021). Challenges in Ensuring global access To COVID-19 vaccines: Production, affordability, allocation, and deployment. *Lancet*, *397*(10278), 1023–1034. doi:10.1016/S0140-6736(21)00306-8 PMID:33587887

Wykes, T. (2019). Racing towards a digital paradise or a digital hell? *Journal of Mental Health (Abingdon, England)*, *28*(1), 1–3. doi:10.1080/09638237.2019.158 1360 PMID:30888284

Yen, P. Y., McAlearney, A. S., Sieck, C. J., Hefner, J. L., & Huerta, T. R. (2017). Health Information Technology (HIT) Adaptation: Refocusing on the Journey to Successful HIT Implementation. *JMIR Medical Informatics*, *5*(3), e28. doi:10.2196/medinform.7476 PMID:28882812

Section 3

Creative Solutions: Reimagining the Future

Chapter 11
Closed Captioned:
Designed Constraints Impact on Millennial Preservice Teachers' Creativity

Timothy Hinchman
Midwestern State University, USA

ABSTRACT

This chapter investigates the constraints impact on the creativity of millennial preservice teachers when captioning a New Yorker Magazine cartoon. According to research, millennials have a decreased capacity for unique ideas, synthesis, articulation, and open-mindedness, which limits their ability to function in a diverse world. The study examined the topic using the constraint-based model of novelty (C-BMN; creativity problem, constraints, variability, and problem spaces) as a framework, as well as the new rubric, which assessed data on the impact of constraints on creativity among 90 millennial PST. Inferential statistics were used to compare differences in creativity, and the results indicated a statistically significant difference in total caption creativity between the experimental and control groups. The study corroborated previous research, concluding that constraints foster creativity. The findings of this study have implications for stakeholders interested in incorporating constraint-based strategies to increase cognitive diversity.

INTRODUCTION

The desire to assist educator preparation programs to prepare pre-service teachers (PST) to foster creative thinking, an important component of cognitive diversity, is pertinent to the current study. In this study, a millennial refers to any individual

DOI: 10.4018/978-1-7998-8028-8.ch011

born between 1982 and 2004 (Weinbaum et al., 2016). For a quarter of a century, instructional practices emphasizing high-stakes standardized testing material diminished American university students' abilities to manifest creative problem-solving abilities (Kim, 2016; Martz et al., 2017). Teaching only the testable content reduces domain skills necessary to generate new, unique, and appropriate solution pathways (Kim 2016), while creative problem solving is an economic desired commodity (Corgnet et al., 2016). Results of research focusing on factors relating to creativity indicate a continuing reduction in the number of unique ideas, frustration tolerance, synthesis, articulation, and open-mindedness (Kim, 2016). Millennials exposed to rote instructional practices limited their ability to creatively problem-solve (Martin 2016; Olszewski 2016; Tallent & Barnes 2015) manifesting in avoiding complex tasks (Erlam et al., 2018; Kuehn & Lingwall, 2015; Mercier et al., 2015; Roskes, 2015; Schiffrin & Liss, 2017). Research indicated that these instructional practices diminished opportunities for innovation and stunted academic development (Copp, 2018). As the literature revealed, this challenged educator preparation programs (EPP) to develop tools and resources to assist millennial PST think creatively, adversely affecting their ability to consider and respond to novel, uncertain, and complex situations. (Erenli, 2016; Hosek & Titsworth, 2016).

COGNITIVE DIVERSITY AND CREATIVITY

Cognitive diversity refers to perceived distinctions in one's variability, knowledge, abilities, values, and beliefs (Van der Vegt & Janssen, 2003) and is precondition to idea fluency and creativity (Amabile, 1996; Henriksen et al., 2015). There is a positive relationship between a cluster's cognitive diversity and their capacity to construct new ideas and products. (Wang et al., 2016). The creative process has been compartmentalized in the literature review as the "four P's of creativity": product, process, person, and place (Runco, 2004). A creative individual is one who is able to: (1) solve a problem in an unusual way, (2) devise a novel or unusual solution, or (3) significantly reshape the problem space (Amabile, 1996). To encourage creative actions and thoughts, it is necessary to place a premium on the location of those actions, also known as the field (Csikszentmihalyi et al., 2018).

In the educational setting, creativity is a result from the reciprocal action between the classroom (field) and the student's capacity to construct and produce creative task solutions (Amabile, 1996; Feldman et al., 1994; Henriksen et al., 2015; Csikszentmihalyi, 2018). Consensual interpretation of creativity requires that nonpartisan and experienced appraisers agree on the creative magnitude of a task resolution (Amabile, 1996; Feldman et al., 1994; Henriksen et al., 2015). Thus, classroom teachers can assess a student's creativity through the lens of the field. This classroom discipline is critical in the development of novel products and solution pathways, and it concludes the process with persistent product evaluation (Csikszentmihalyi, 2018). From start to finish, the classroom teacher acts as a field expert, monitoring creative production through time-sensitive encouragement and discouragement (Csikszentmihalyi, 2018). Teachers motivate their students to approach problems in novel and creative ways (Beghetto & Kaufman, 2014).

Creativity requires a balance of intrinsic and extrinsic motivation in all organizational systems, including the educational setting (Amabile, 1997). Synergy between intrinsic and extrinsic motivation catalyzes the development of creative task solutions through process validation between the creative individual and the gatekeeper of the existing field (Amabile, 1997; Amabile et al. 2005; Feldman et al., 1994). The field agent both encourages and facilitates a creative individual's alternative solution pathways (Amabile et al., 2005). When a field views work as novel and unique, that a positive emotional response is experienced (Amabile et al., 2005). The field agent's values and responses directly affect novel idea fluency and creative products generated (Wang et al., 2016). A transformational leader imposes the necessary constraints and supports that assists the group in developing novel and unique task solutions (Gute et al., 2008; Wang et al., 2016).

MILLENNIALS, STANDARDIZED EDUCATION, AND COGNITIVE DIVERSITY

Millennials were subjected to highly structured curriculum as a consequence of high stakes testing (Jakee & Keller, 2017; Morgan, 2016; Polson, 2018; Sanchez & Patel, 2017; Zohar & Agmon, 2018). Strict and prescribed curricula suffocate authentic learning and alienate students from natural world experiences (Noddings, 2013). Instructional practices that focus exclusively on these high-stakes tests do not improve students' overall learning (Copp, 2018) and cease treating them as diverse human beings, resulting in a stunted academic development (Grinell & Rabin, 2013). Classroom teachers disregard best practices, those involving a high degree of variability, instead focusing only on the knowledge required to pass tests, exploiting the students in the process (Grinell & Rabin, 2013; Hinchman, 2019). Parents and students are viewed as non-participants in the educational process (Knoester & Au, 2017; Morgan, 2016; Polson, 2018). Millennials who have been exposed to instructional practices geared toward passing standardized tests lack skills in process-based thinking (Martin, 2016; Olivant, 2015). This curriculum specialization provides students with a limited set of skills that have little real-world application (Martin, 2016).

Teachers whose primary focus is on high-stakes testing content have few opportunities for professional development in order to develop creativity-enhancing skills that can be used to teach for and teach creatively (Ashadi & Rice, 2016). By focusing exclusively on the knowledge and skills necessary to pass high-stakes testing, these instructional practices reinforce curriculum standardization (Sanchez & Patel, 2017). These instructional practices have demotivated teachers to teach subjects or content other than those covered by high-stakes standardized tests. (Gannon-Tagher & Robinson, 2016; Gunnulfsen & Roe, 2018; Morgan, 2016; Olivant, 2015; Sanchez & Patel, 2017; Zohar & Alboher Agmon, 2018). Teachers become compliant in a standardized education environment, eschewing the necessary teaching abilities to foster their students' creativity in favor of testable content (Hinchman, 2019; Sanchez & Patel, 2017). Kim (2016) argues that unless this trend is reversed, the United States will lack the cognitive diversity necessary to address future problems and challenges.

Millennial PST demonstrated the same pedagogical strategies emphasizing test-teaching (Sanchez & Patel, 2017), while lowering their personal aspirations for creativity and problem-solving ability (Jakee & Keller, 2017; Morgan, 2016; Polson, 2018; Sanchez & Patel, 2017; Zohar & Agmon, 2018). Millennial PST are not provided opportunities with creative classroom moments within their EPP (Schreiber, 2018). As a result, millennial PST are not exposed to environments that foster and support their individual creativity (Schreiber, 2018); rather, they emulate practices that promote linear, mechanical thinking (Kim, 2016). Kim (2016) identified this phenomenon as a public education crisis negatively impacting the creative problem-solving process as reative thinking and problem-solving significantly decreased in the 18-year period between 1990 and 2008. Findings indicated a significant decrease in the creative strength category for publicly educated K-3 millennial students from 1998 to 2008 (Kim, 2016). Millennials educated during these years demonstrated diminished abilities to express emotions and divergent viewpoints, engage in deep discourse, use humor in everyday situations, engage in imaginative thinking, pursue hobbies, and synthesize information, all of which inhibited their creative problem-solving process (Kim, 2016).

Millennial teachers constitute 42% of teachers in the educational workforce (OCED, 2021). Current EPPs do not provide enough creative opportunities to prepare millennial PST to teach creatively and teach for creativity (Schreiber, 2018). Millennial PST pattern their instructional delivery focusing only on the knowledge and skills necessary to pass high-stakes testing (Sanchez & Patel, 2017). This focus diminishes professional development opportunities for millennial teachers to cultivate creativity-building skills that will ultimately be used to teach for creativity in their classrooms (Ashadi & Rice, 2016). Due to an educational system that emphasizes rote memorization and a reliance on standardized testing, millennial PST lack critical knowledge (Krahl, 2018). These instructional practices have an effect on millennials' abilities to synthesize and apply their knowledge in new situations (McAllum, 2016). These fears have altered their worldview and created unique challenges for higher education and EPPs to address (Erenli, 2016; Hosek &Titsworth, 2016). Exposure to test-teaching instruction had a significant impact on their cognitive diversity, as it limited their exposure to diverse content, disconnected them from the natural world, and ignored their differences and interests (Noddings, 2013).

Federal mandates implemented curricular constraints for millennial PST through content standards and high accountability assessments and evaluation (Ashadi & Rice, 2016; Fusarelli & Fusarelli, 2018; Holloway & Brass, 2017; Martin, 2016; Ro, 2018; Wilcox & Lawson, 2017). High-stakes testing and accountability measures are any assessment/procedure that is utilized in the following: (a) inhibit student promotion, (b) withhold graduation based student's test scores, (c) ranking schools, including a failing designation, (d) associated with teacher/administration employment status, and (e) school funding (Johnson et al., 2008). The content standards era and high accountability assessments have demonstratively altered curriculum planning and pedagogical (Ashadi & Rice, 2016; Fusarelli & Fusarelli, 2018; Holloway & Brass, 2017; Jakee & Keller, 2017; Martin, 2016; Olivant, 2015; Wilcox & Lawson, 2017).

High accountability assessments reduce in-depth exploration in the many different forms of literature and creative writing that reinforce the imaginative skills in an effort to focus on the finite skills and strategies that increase student's proficiency on these high-stakes tests (Martin, 2016). Creative and expressive writing time and space were limited due to the alignment with standardized high-stakes assessment content (Frawley, 2014). The millennial student was not exposed to effective teaching practices due to standardized teaching practices (Erskine, 2014). The "drill and kill" review practices reduced classroom opportunities for millennial students to acquire new creative skills (Kim, 2011). This shift has demonstrated a disruption in established methods of training and educational practices (Ayudhya & Smithson, 2016; Weinbaum et al. 2016).

Wilson (2014) also noted that millennial generation students constitute the most diverse generation in the history of the United States; for them, instructions usually need to be more varied. Millennial PST see themselves as accepting diversity, hesitant toward evaluations, excited about their teaching abilities, but lack skills to both offer feedback and adequate critiques (Clark & Byrnes, 2015). As such, millennial PST pose a challenge to traditional EPPs due to their unique educational experiences, as mentioned earlier (Hosek & Titsworth, 2016).

ROLE OF EPPS

Buckner and Strawser (2016) advise that to meet millennial PST' needs, university EPPs and higher education calls for (a) consistent and concise purpose, (b) modifying instructional strategies to compensate for deficiencies, and (c) continued research that benefits their educational experiences. Millennial PST pose new challenges to traditional EPPs due to their unique educational experiences (Hosek & Titsworth, 2016). An EPP needs tools and research that can address their PST needs and deficits (Erenli, 2016). These tools are essential in developing strategies to prepare the PST to deliver effective instructional practices (Erenli, 2016). The EPPs can promote instructional practices enhancing creativity and provide learning environments where student creativity is both encouraged and supported (Abramo & Reynolds, 2015; Alismail & McGuire, 2015; Ellis & Childs, 2019; Harris & Ammermann, 2016; Henriksen et al. 2016).

Faculty at EPPs can initiate best practices in creativity that improve their PST skill to teach creatively through carefully constructed strategies (Pang, 2015). The experiences provided by EPP practices for PST to have successful teaching careers and provide an adequate opportunity for creativity enhancement (Abramo & Reynolds, 2015; Alismail & McGuire, 2015; Ellis & Childs, 2019; Harris & Ammermann, 2016; Henriksen et al., 2016). Although there is a growing emphasis on preparing better and more teachers, and despite the almost universal presence of student teaching across EPPs, many questions regarding how student teaching experiences contribute to the development of preservice teachers remain unanswered (Anderson & Stillman, 2013).

According to Hallman (2017), it is widely recognized in EPPs that the process for PST to become effective teachers is complex. They have the ultimate responsibility for preparing PST to enter the workforce. The modern challenge for classroom teachers is structuring lessons that stimulate creativity and provide competencies to do so (Tan, 2015). Focusing on creativity enhancement is essential in educational pedagogy (Tan, 2015); unfortunately, not all EPPs emphasize creativity (Newton & Beverton, 2012). Research indicates that millennial PST are unable to identify instructional strategies to increase student creativity (Newton & Beverton, 2012). EPPs need both space and time to improve creativity (Henriksen et al., 2016) through developing pedagogy that encourages curiosity, flexibility, fluency, and self-confidence to increase a teacher candidate's self-motivation (Tan, 2015). The purpose of this study was to fill a knowledge gap regarding the effect of designed constraints on the creativity of millennial PSTs. The C-BMN framework proposed by Stokes (2009) was used to examine how designed constraints can positively influence individuals' creativity, resulting in novel, unique, and appropriate task resolutions.

CONSTRAINT-BASED MODEL OF NOVELTY

The primary concept of C-BMN is that constraints generate a positive influence on an individual's creativity. Prior studies suggested that constraints initiate the necessity for creative solutions (Caniels & Rietzschel, 2015; Haught, 2015; Haught-Tromp, 2017; Stokes, 2009). Notably, novel, effective, and whole resolutions are the direct result of a problem's constraints (Caniels & Rietzschel, 2015; Hatchuel & Chen, 2017; Haught 2015; Haught-Tromp 2017; Torrents-Martin et al., 2015). Stokes (2006) noted that individuals that use strategic constraints increase their ability to generate novel resolutions. Stokes C-BMN has the following main concepts: (a) creativity problem, (b) constraints, (c) variability, and (d) problem spaces.

Creativity Problem

Any physical or mental constructs that are novel, unique, and contextually appropriate to the larger domain are considered creative (Amabile, 1996; Henriksen et al., 2015). Stokes (2006) recognizes that solving a problem creatively involves some level of stratagem and structure. The creativity problem starts as an ill-structured problem, and its solution depends on the designed use of these constraints (Stokes, 2006). These tactical constraints structure the problem space in a two-fold process by limiting traditional solution pathways and encouraging creative ones (Stokes, 2006).

Constraints

This term refers to any boundary, hindrance, or instruction that limits the number of potential solutions to any prescribed task (Hatchuel & Chen, 2017; Haught-Tromp, 2017; Henriksen et al., 2015; Stokes, 2009; Torrents-Martin et al., 2015). Stokes (2006) establishes creativity constraints as an obstruction that guides a person towards innovation. A constraint influences individual creativity by requiring redirecting cognitive pathways to resolve any proposed/given task (Hatchuel & Chen, 2017; Haught-Tromp, 2017; Stokes, 2006). Constraints assist novices in expanding their creativity by simplifying the process through guided direction (Stokes 2006). Constraints come in pairs where one type restricts normal solution pathways, and the other facilitates new ones (Stokes, 2006).

Variability

Variability determines the number of appropriate solution pathways and the dimensions within the problem space (Stokes 2006). The expertise needed to construct products that are considered novel, unique, and appropriate relies on a variability spectrum that includes constraints (Stokes 2006, 2009). This concept is associated with one's cognitive diversity (Van der Vegt & Janssen, 2003).

Problem Spaces

This term refers to the identification of a problem and its relevant solution pathways (Stokes, 2008). A problem space that is ill-structured is vital in the construction of creative products and solution pathways (Stokes, 2008). Strategic use of paired constraints assists the creative individual to resolve the problem space (Stokes, 2006). These paired constraints structure the problem space through the establishment of boundaries and limits (Stokes, 2006). A dual-purpose is served by the paired constraints within the problem space, limits the common pathways, and promotes unique solutions to the problem space (Stokes, 2006). Strategic manipulations of constraints that limit and promote allow the individual to resolve the problem space by increasing response variability (Stokes, 2006).

Solution pathways from complex problem spaces provide more variability than simple ones (Stokes, 2006). Constraints that involve predictable and routine responses are considered low variability (Stokes, 2006). Those responses that encourage innovation and the unique are high variability (Stokes, 2006). When low variability behavior is reinforced, and rewarded individuals are more likely to use those strategies to resolve a problem space (Skinner, 1990). The early rewarding of high variable behavior in students increased domain-specific knowledge, skills, and abilities (Stokes, 2006). Constraint pairs within the problem space limit common and familiar responses and promote novel and unique solution pathways (Stokes, 2006).

APPLICATIONS OF C-BMN

Problem limits, boundaries, and constraints create the need for creative solutions (Caniels & Rietzschel, 2015; Haught, 2015; Haught-Tromp, 2017; Stokes 2006). Novel, unique, and appropriate solutions can only emerge from problems that have predefined constraints (Caniels & Rietzschel, 2015; Hatchuel & Chen, 2017; Haught 2015; Haught-Tromp 2017; Torrents-Martin et al. 2015). Constraints, boundaries, and restrictions encourage individuals to formulate creative solution pathways and ideas through their manipulation (Caniels & Rietzschel, 2015; Hatchuel & Chen, 2017; Haught, 2015; Haught-Tromp, 2017; Torrents-Martin et al., 2015).

Constraints facilitate narrowing of solution options, which allows a more in-depth exploration into a problem space (Haught-Tromp, 2017). The in-depth exploration encourages individuals to hyper-focus on plausible task solutions by reducing those pathways to those that will suitably resolve the designated task (Haught-Tromp 2017).

This hyper-focus provides and architecture to the problem space where the product can be creatively constructed within the traditional confines (Haught-Tromp, 2017). A constraint is a permanent concept that exists in all tasks and problem spaces, and the antithetical to a constraint is the ability to have total and complete freedom (Torrents-Martin et al., 2015; Stokes, 2006).

Constraint manipulation can be used to strengthen a teacher's capacity for teaching for and about creativity (Stokes, 2010). Stokes (2010) identifies five distinct types of educational constraints that affect individual creativity: cognitive constraints, talent constraints, curricular constraints, domain constraints, and variability constraints. The first two types of constraint, cognitive and talent, are the result of neurological and biological evolution (Stokes, 2010). Cognitive and talent constraints have an effect on how new information is processed and how new domain skills are acquired (Stokes, 2010). Due to the fact that these constraints involve biological, genetic, and individual variation, they are beyond the scope of this study. Federally mandated content standards established high accountability assessments for use in evaluation, are examples of curriculum constraints (Stokes, 2010). Domain constraints are defined as distinct subject areas (math, science, and ELAR) and the performance criteria that apply to them (Stokes, 2010). The ability to generate a novel, unique, and unexpected response is considered a constraint on variability that is directly related to the student's domain expertise and content mastery (Stokes, 2010).

CONSTRAINT TYPES

Constraint pairs that influence the process and product creativity phases are input and output constraints (Rosso, 2014). Constraints influencing the initial process stage of an individual's creativity are defined as the input type of constraints (Rosso, 2014); those constraints influencing the construction/manufacturing/production creativity stages are defined as the output constraints type (Rosso, 2014). Inputs, or process constraints, influence a creative individual's task resolution pathway and methodology approach to a designated problem space (Rosso, 2014). Inputs establish boundaries of the problem space where the creative individual attempt to resolve the task within the established confines (Rosso, 2014). This constraint type influences how a creative individual manages the constraints of the problem space and hyper-focuses on novel and unique solution pathways without being overwhelmed by an infinite number of possibilities (Stokes, 2013).

A constraint that reduces the number of appropriate possibilities or alternatives a creative individual is able to construct is defined as an output constraint (Rosso, 2014). Output constraints can be considered goal oriented, where individuals are limited by a select number of possible construct outcomes (Stokes, 2013). They restrict and narrow appropriate and acceptable task outcomes by outlining the resolution's needs. The field and larger domain influences an individual's ability to construct a creative product (Csikszentmihalyi et al., 2018). The field can act as an output constraint through gatekeeping, and the definition of the type of responses/products are considered to be appropriate and acceptable (Csikszentmihalyi et al., 2018).

RESEARCH QUESTION AND HYPOTHESIS

The researcher addressed the following overarching question: Do designed constraints affect the caption creativity of Midwestern millennial PST? The research sub-question aligns with the study problem, purpose, and conceptual model. The following sub-research question and hypothesis was evaluated.

RQ*:* Will there be a statistical difference between the means of the rubric scores of the input-constraint group, output-constraint group and the control group in terms of caption creativity for millennial PST at a Midwestern U.S. public university EPP?

H_0: No difference exists in caption creativity between the output constraint, input constraint, and control groups.

$$\mu_1 = \mu_2 = \mu_3$$

H_a: There is a difference between the input constraint group, output constraint group, and control group concerning caption creativity.

$$\mu_1 \neq \mu_2 \neq \mu_3$$

METHOD

Research Design

A nonequivalent posttest only quasi-experimental evaluation design using a comparison group (NPQCG) research design was utilized for this inquiry due to the natural setting and grouping of students at the Midwestern U.S. EPP (Creswell & Guetterman, 2015; Leedy et al., 2019). A NPQCG separates the respondents into nonequivalent groups that undergo different treatments and are compared with a control group (Leedy et al., 2019).

As this design indicates, the researcher employed a convenience sample, and therefore attention was provided to minimize the potential threat to internal validity. The NPQCG compared the caption creativity of two experimental groups (ICG and OCG) with that of a control group (CCG). This NPQCG (Leedy et al., 2019) had 30 participants in each group to ensure this large sample size (Krishnan, 2018). The ICG and OCG each consisted of 30 participating PST where constraint instructions were enforced in their completion of *The New Yorker Magazine* cartoon caption. The CCG comprised of a different group of 30 participating PST from the same professional courses, who were constraint-free in their captioning of *The New Yorker Magazine* cartoon caption.

Sample

A nonprobability convenience sampling technique was utilized in selecting millennial PST participants. This sampling technique involves nonprobability sampling, where the researcher elects to use willing and available participants (Creswell & Guetterman, 2015). According to the Midwestern EPP's data, over seventy percent of the PST (75%) of the 215 enrolled are millennials. The minimum sample size needed for this study was 90 as calculated based on the following assumptions: $\alpha = .05$; statistical power = .90; $k = 3$, three study variables, and a medium effect size (Leedy et al., 2019).

Research Protocols/Instrumentation

As indicated, the researcher used a NPQCG (Leedy et al., 2019) by comparing the product creativity of three groups; two experimental groups (ICG and OCG) and a control group (CCG) to measure caption creativity. The dependent variable of caption creativity as measured by NEW rubric (Henriksen et al., 2015) as influenced by the independent variables (input constraint, output constraint, and constraint free). The input constraint group (ICG), output constraint group (OCG), and constraint-free control group (CCG) were randomly assigned under the experimenter's control (Leedy et al., 2019). Previous exposure and training using constraints increases creative production (Medeiros et al., 2014); with this in mind a pretest may have a possible effect on the experimental treatment (Leedy et al., 2019).

The ICG (N = 30) participants were instructed to use an input constraint by including one concrete noun and a concrete verb in their novel, unique, and appropriate caption. The concrete nouns were on the left side of the cartoon and the verbs on the right. These words were selected from Category norms for verbal items word list (Van Overshelde et al., 2004). The OCG (N = 30) participants were required to provide their novel, unique and appropriate caption to the cartoon in precisely seven words. The output number seven was chosen due to its specific syntactic form that has demonstrated increase association and appropriateness (Froese, 1977). The CCG (N = 30) were instructed to produce their novel, unique and appropriate caption without any constraints. No further directions were provided with the CCG.

The researcher collected the captions generated by the participants and anonymously provided them to three judges for assessment. The judges utilized the NEW rubric to assess the captions on novelty, effectiveness, wholeness, and total creativity. The NEW rubric, as developed by Henriksen et al. (2015), was adapted for this study to measure participants' caption creativity. The rubric categories and scoring remained identical to the original, while the descriptions were modified to address caption creativity, as the original rubric was utilized in assessing entire projects. The researcher received permission and approval from the authors.

The instrument includes statements using a Likert-like scale ranging 1 = *minimal*, and 5 = *exceptional* to measure creativity along three subdimensions: novelty, effectiveness, and wholeness. Subdimension scores were summed to create a total creativity score. A product's novelty (caption) is assessed on its uniqueness and interest to other individuals (Henriksen et al., 2015). Effectiveness is assessed on the degree of clarity and comprehensiveness (Henriksen et al., 2015). Wholeness refers to a product's contextual connection with the assigned task that has some individualized style (Henriksen et al., 2015). This evaluative instrument provides the rater with a product description range that can be readily assessed at each point level (Henriksen et al., 2015). The instrument provides the rater flexibility in assessing the creative products and provides a framework, which outlines characteristics to which each product can be assessed thoroughly (Henriksen et al., 2015).

Participants were asked to provide answers to eight additional demographic questions upon completing the caption. These data included questions on gender, ethnicity, and employment status. Participants were also asked to identify their K-12 educational experience and certification area. The data were summarily used to support analysis and assist in research conclusions (Table 1).

Content Validity and Inter-Rater Reliability

Creative products were assessed and scored between one and five along the dimensions of novel, effective, and whole (Henriksen et al., 2015). The researcher calculated Lawshe's content validity ratio (CVR) with the three independent university professors to judge and determine NEW rubric's (Henriksen et al., 2015) alignment with the concepts measured (Ayre & Scally, 2014). The NEW rubric CVR value was calculated as a 1.0, a CVR or .99 and above are acceptable for construct and measurement alignment (Ayre & Sally, 2014). Inter-rater reliability was assessed using Fleiss kappa (Fleiss et al. 2013) and Krippendorf's alpha (Krippendorff, 2004).

Cohen (1960) indicates that Kappa values that are higher than .61 have substantial agreement between the raters. Krippendorff writes "[I]t is customary to require $\alpha \geq .800$. Where tentative conclusions are still acceptable, $\alpha \geq .667$ is the lowest conceivable limit" (2004, 24).

Table 1. Study sample demographics (N = 90)

Characteristic	*n*	%
Gender		
Male	30	33.3
Female	59	65.6
Other	1	1.1
Educational Background		
Public School	80	88.9
Charter School	3	3.3
Private Parochial	4	4.4
Private Nonparochial	1	1.1
Other	2	2.2
Certification Seeking Type		
EC-6	22	24.4
EC-6 SPED	22	24.4
4-8 Math	3	3.3
4-8 Science	2	2.2
4-8 ELAR	1	1.1
4-8 Social Studies	4	4.4
4-8 Generalist	2	2.2
7-12 Math	4	4.4
7-12 Science	2	2.2
7-12 Social Studies	4	4.4
All Kinesiology	16	17.8
All Level Art	3	3.3
All Level Music	5	5.6
Ethnicity		
American Indian/Alaska Native	2	2.2
Asian	1	1.1
African American	7	7.8
White	66	73.3
Other	14	15.6

Fleiss κ statistics were calculated to examine inter-rater reliability between three university professors' judgment on 90 millennial PST' caption novelty. Substantial agreement occurred between the three professors' judgements, κ = .728 (95% CI, .652 to .804), $p < .0005$. Fleiss statistics calculated to examine inter-rater reliability for effectiveness indicated substantial agreement between the three professors' judgements, κ = .691 (95% CI, .614 to .768), $p < .0005$. Fleiss statistics calculated to examine inter-rater reliability for caption wholeness indicated substantial agreement between the three professors' judgements, κ = .738 (95% CI, .660 to .815), $p < .0005$. Finally, Fleiss statistics calculated to examine inter-rater reliability for total caption creativity indicated substantial agreement between the three professors' judgements, κ = .629 (95% CI, .580 to .678), $p < .0005$.

This researcher also ran Krippendorff α. As shown in Table 2, there was substantial agreement between the three university professors' judgments, α = .822 for novelty. There was also substantial agreement between the three university professors' judgments, α = .795 for effectiveness and wholeness, α = .826. Finally, Krippendorff's α was run to determine if there was agreement on the caption's total creativity score. There was substantial agreement between the three university professors' judgments, α = .824.

Table 2. Inter-rater reliability

Variable	Fleiss κ	Krippendorf's α
Novelty	.728	.822
Effectiveness	.691	.795
Wholeness	.738	.826
Caption Creativity	.629	.824

Test of Statistical Assumptions

When data conformed to ANOVA statistical assumptions for normality and using Shapiro Wilk and homoscedasticity using Levene's test, an ANOVA was employed (Field, 2018). When an assumption for use of ANOVA was not met, Kruskal Wallis test was employed (Dinno, 2015).

RESULTS

Based on the conceptual framework utilized for this quantitative study, the following hypothesis was considered for investigation: (H_a): There is a difference between the input constraint group, output constraint group, and control group concerning caption creativity.

To address the hypothesis, a non-parametric Kruskal-Wallis *H* statistic was calculated for caption novelty, caption effectiveness, caption wholeness, and ANOVA for total caption creativity. Levene's test showed that variances for caption novelty scores were not equal, $F(2,87) = 3.15, p = .048$, and the results of the Shapiro-Wilk test indicated a violation of normality ($W = .863, p = .001$), indicating a Kruskal-Wallis *H* statistic was used to compare means. A Kruskal-Wallis *H* test indicated a statistically significant difference in caption novelty between at least one pair of constraint groups, $\chi^2(2) = 8.422, p = .015$. Therefore, Null Hypothesis for caption novelty was rejected; there was a significant difference in mean caption novelty scores between the input constraint group, output constraint group, and control group. The effect size ($\eta_{p2} = .102$) demonstrated a medium effect size (Cohen 1988).

Levene's test showed that variances for caption effectiveness scores were equal, $F\ (2,87) = 0.013, p = .987$, and the results of the Shapiro-Wilk test indicated a violation of normality ($W = .896, p = .007$), indicating use of Kruskal-Wallis *H* statistics to compare means. Kruskal-Wallis *H* indicated no statistically significant difference in caption effectiveness scores between at least one pair of constraint groups, $\chi^2(2) = 5.613, p = .06$, however the results were very close to significance at $p<.05$. Therefore, the researcher failed to reject Null Hypotheses 1 for caption effectiveness; there was no significant difference in mean caption effectiveness scores between the input constraint group, output constraint group, and control group. The effect size ($\eta_{p2} = .063$) demonstrated a medium effect size (Cohen, 1988).

Levene's test showed that variances for caption wholeness scores were equal, $F\ (2,87) = 0.066, p = .936$, and the results of the Shapiro-Wilk test indicated a violation of normality ($W = .896, p = .007$), indicating use of Kruskal-Wallis *H* statistics to compare means. Kruskal-Wallis *H* indicated statistically significant difference in caption wholeness scores between at least one pair of constraint groups, $\chi^2(2) = 6.423, p = .04$. Therefore, Null Hypothesis for caption wholeness was rejected; there was a significant difference in mean caption wholeness scores between the input constraint group, output constraint group, and control group. The effect size ($\eta_{p2} = .066$) demonstrated a medium effect size (Cohen, 1988).

Levene's test showed that variances for caption total creativity scores were equal, $F(2,87) = 0.153, p = .858$, and the results of the Shapiro-Wilk test did not indicate a violation of normality ($W = .948, p = .145$), indicating use of ANOVA statistics to compare means. ANOVA indicated a statistically significant difference in caption total creativity scores between at least one pair of constraint groups, $F\ (2,87) = 4.254, p = .017$. Therefore, Null Hypothesis for caption total creativity was rejected; there was a significant difference in mean caption total creativity scores between the input constraint group, output constraint group, and control group.

Caption Creativity Problem Space

Data analyses conducted revealed a significant difference between constraints and the caption creativity of a Midwestern U.S. public university's EPP millennial PST. The results identified through a one-way ANOVA indicated a statistically significant difference exists (.017) in total caption creativity between the three constraint groups. Constraints are obstructions that guide individuals towards creativity and innovation (Stokes, 2006). Constraints are necessary in the problem space that restrict normal pathways enabling new ones (Stokes, 2006). The strategic use of constraints assists novices by simplifying the problem space (Stokes, 2006). This study's results extended the existing scholarship in the field of constraints and creativity (Stokes, 2006). Constraint usage was demonstrated as beneficial to creative problem-solving and product construction (Caniels & Rietzschel, 2015; Eckert et al. 2012; Haught, 2015; Haught-Tromp, 2017; Medeiros et al., 2014; Torrents-Martin et al., 2015) applying the concepts outlined in Stokes (2009) C-BMN. Hatchuel and Chen (2017) concluded that deliberate exercises that redefine the problem space enhance creativity in individuals. Constraint-based practices increase an individual's creativity level and can increase solution variability in a problem space (Haught, 2015; Haught-Tromp, 2017).

Interpretations of Results

Perhaps most importantly, study results confirmed earlier research (Caniels & Rietzschel, 2015; Haught, 2015; Haught-Tromp, 2017) that Stokes (2009) C-BMN can increase response creativity. Additionally, this study will inform future research focusing on designed constraints. The study examined Stokes (2009) C-BMN on millennial PST caption creativity at a Midwestern EPP, and future research may enhance the conclusions. The lack of literature examining constraint usage on millennial PST was noted, one suggestion for future research is to examine the impact of constraints on millennial PST creativity through different mediums related to Stokes' variables. A suggestion for future research is to examine mechanisms that can increase a millennial PST effectiveness on teacher-specific tasks.

Implications

The results extended Stokes (2009) C-BMN to a crucial new population, Millennial PST, which represent future teachers. While considerable research literature existed on the use of constraints (Caniels & Rietzschel, 2015; Haught, 2015; Haught-Tromp, 2017), the researcher was unable to locate any studies that test Stokes (2009) C-BMN on millennial PST. The study contributes to EPP literature by providing a pathway to develop mechanisms through Stokes (2009) C-BMN to increase millennial PST cognitive diversity, and their capacity to teach creatively, and teach for creativity.

The results provided further validation for the NEW rubric (Henriksen et al., 2015) to assess PST creativity. The researcher was not successful in finding previous research that used the rubric and performed any statistical analyses on the NEW rubric (Henriksen et al., 2015). In other words, previous studies suggested that creativity can be assessed through a three-dimensional rubric but did not perform a content validity ratio to determine the rubric's alignment with the measured concepts.

Finally, the study contributes to the limited research on millennial PST cognitive diversity. The majority of literature focused on proposals relating to constraint usage/manipulation without directly targeting millennial PST. While study results confirmed Stokes (2009) C-BMN to predict creative responses, the research design did not include the effect differences in the problem space in terms of timing, or complexity of time constraints, which were raised by Medeiros et al. (2014). In addition, no effort was made to determine whether differences in intrinsic motivation between groups explained differences between the input-constraint group and the control group (Rosso, 2014).

By demonstrating the C-BMN's utility, school administrators and decision makers can confidently develop pedagogical tools and practices based on the C-BMN concepts of novelty, wholeness, and effectiveness. Second, and more precisely, EPP school policymakers and decision makers can develop pedagogical tools for PSTs and their future students that promote creativity through the use of problem limits, boundaries, and constraints (Haught-Tromp, 2017; Stokes 2013, 2014). Innovative, one-of-a-kind, and effective solutions emerge as a result of the creative pathways and processes triggered by designed constraints (Caniels & Rietzschel, 2015; Hatchuel & Chen, 2017; Haught, 2015; Haught-Tromp, 2017; Torrents-Martin et al., 2015). Thirdly, school administrators can strengthen teachers' capacity to teach creativity by implementing virtual C-BMN-based professional development programs.

According to the Interstate Teacher Assessment and Support Consortium (InTASC), a teacher's disposition is critical to effective teaching (Council of Chief State School Officers, 2013). Dispositions are defined as "habits of professional action and moral commitments that underlie the performances" (Council of Chief State School Officers, 2013, p. 6). Historically, EPPs have required dispositions as a condition of certification. InTASC anticipates that new teachers will not only engage students in creative problem-solving but will also understand how the mechanisms that support their creativity work (Council of Chief State School Officers, 2013). Integrating the findings enables EPP to develop a framework for improving their teaching disposition.

EPPs face unique and complex challenges as a result of the millennial generation (Anderson & Stillman, 2013; Hallman, 2017; Hosek & Titsworth, 2016). The findings of this study may aid in the development of a virtual model that can be used consistently in professional education courses through their online Learning Management System (LMS). By incorporating strategies throughout their program coursework, engineering design models have demonstrated increased cognitive diversity when confronted with constraints (Hatchuel & Chen, 2017). Sinclair (2018) argues that higher education should place a greater emphasis on creativity in both the physical and digital realms. EPP policymakers can implement designed constraint-based strategies in each professional course to encourage teacher candidates to maximize their creativity by utilizing the assignment's problem space boundaries.

Finally, through professional development and faculty forums, EPP policymakers can enhance their faculty's capacity to teach creatively. Anagnostopoulos, Levine, Roselle, and Lombardi (2017) advocate for a multi-organizational overhaul aimed at providing tools to improve teacher education. University professors' pedagogical practices have an impact on their students' classroom performance (Major & Mulvihill, 2018). Jain & Jain (2019) argue that in order for students to be engaged, they must internalize the process's objectives and the critical role of university faculty. The findings of this study can be used to develop EPP-wide professional development opportunities that can positively impact their candidates' ability to teach creatively through repeated practice.

Suggestions for Future Research

Stokes (2009) C-BMN has the potential to inform researchers' and educators' efforts to explore constraints to promote cognitive diversity. Study results suggest that policymakers develop pedagogical tools to promote creativity directly in terms of assignments that promote creative thinking, and indirectly by adding constraints to existing pedagogical tools (Bereczki & Kárpáti, 2018; Kettler et al., 2018; Maley & Kiss, 2017). Given that millennials are digital natives (Silvestru et al., 2019), it will be necessary to investigate constraint manipulation in virtual environments in order to ascertain precisely which constraint type (input and output) affects the millennial PST cognitive diversity.

Future researchers can increase the sample size to enhance generalizability and obtain more robust results. When the sample size is larger, the researcher's ability to generalize the results is increased (Field, 2018). Additionally, including participants from a broader geographic range would improve the study's ability to generalize its findings. Additional research is needed to establish the validity of the NEW rubric (Henriksen et al., 2015) as a tool for assessing the creativity of diverse populations and, if successful, to validate the instrument across a range of populations and platforms. While educational institutions value creativity, commercial concerns require rapid innovation to maintain a competitive edge in a global marketplace. NEW rubric adaptations are reasonable extensions for commercial product development, scientists, and marketers.

CONCLUSION

Framed within Stokes' (2009) C-BMN, this study sought to compare the impact of designed constraints on A Midwestern U.S. public university millennial PST caption creativity. After a literature review, the researcher discovered that constraints are a necessary process in creative problem-solving and production. This led the researcher to conduct a quantitative analysis comparing input and output constraints using a NPQCG design. This design involved millennial PST captioning a *New Yorker Magazine* cartoon using different constraint instructions to answer the C-BMN based research questions related to the study's purpose and problem. The statistical analysis of the one-way ANOVA and Kruskal-Wallis H test provided the researcher with the capability to compare the different treatment groups' mean creativity scores.

To address the research question, a non-parametric Kruskal-Wallis H statistic was calculated for caption novelty, caption effectiveness, caption wholeness, and ANOVA for total caption creativity. Levene's test showed that variances for caption novelty scores were not equal, $F(2,87) = 3.15$, $p = .048$, and not normally distributed ($W = .863$, $p = .001$), indicating a Kruskal-Wallis H statistic was used to compare means. Results indicated a significant difference between the means of the rubric scores of the input-constraint group, output-constraint group, and the control group in terms of caption creativity for millennial teacher-candidates at a Midwestern U.S. public university. The results indicate a statistically significant difference exists in total caption creativity (.017) because it is less than the .05 level.

The results may provide EPP policymakers and administrators with the ability to increase PST' dispositions through course augmentations and faculty professional development. The results may provide school administration and faculty a framework to develop constraint-based workshops that will assist in teaching for creativity and ultimately increase student engagement. The millennial generation has posed some challenges to EPP; however, according to Stokes (2006), "It's never too late to pose a new creativity problem" (p. 135).

REFERENCES

Abramo, J. M., & Reynolds, A. (2015). Pedagogical creativity as a framework for music teacher education. *Journal of Music Teacher Education*, *25*(1), 37–51. doi:10.1177/1057083714543744

Al-Balushi, S. M., & Al-Abdali, N. S. (2014). Using a moodle-based professional development program to train science teachers to teach for creativity and its effectiveness on their teaching practices. *Journal of Science Education and Technology*, *24*(4), 461–475. doi:10.100710956-014-9530-8

Alismail, H. A., & McGuire, P. (2015). 21st Century standards and curriculum: Current research and practice. *Journal of Education and Practice*, *6*(6), 150–154. http://tarjomefa.com/wp-content/uploads/2017/06/6879-English-TarjomeFa.pdf

Amabile, T. M. (1996). *Creativity in context: Update to the social psychology of creativity*. Westview Press.

Anagnostopoulos, D., Levine, T., Roselle, R., & Lombardi, A. (2017). Learning to redesign teacher education: A conceptual framework to support program change. *Teaching Education*, *29*(1), 61–80. doi:10.1080/10476210.2017.1349744

Anderson, L. M., & Stillman, J. A. (2013). Student teaching's contribution to preservice teacher development: A review of research focused on the preparation of teachers for urban and high-needs contexts. *Review of Educational Research*, *83*(1), 241–259. doi:10.3102/0034654312468619

Ashadi, A., & Rice, S. (2016). High stakes testing and teacher access to professional opportunities: Lessons from Indonesia. *Journal of Education Policy*, *31*(6), 727–741. doi:10.1080/02680939.2016.1193901

Atay, A., & Ashlock, M. Z. (2018). *Millennial culture and communication pedagogies: Narratives from the classroom and higher education*. Lexington Books.

Ayre, C., & Scally, A. J. (2014). Critical values for Lawshe's content validity ratio: Revisiting the original methods of calculation. *Measurement & Evaluation in Counseling & Development*, *47*(1), 79–86. doi:10.1177/0748175613513808

Ayudhya, U., & Smithson, J. (2016). Entitled or misunderstood? Towards the repositioning of the sense of entitlement concept in the generational difference debate. *Community Work & Family*, *19*(2), 213–226. doi:10.1080/13668803.2016.1134116

Bereczki, E. O., & Kárpáti, A. (2018). Teachers' beliefs about creativity and its nurture: A systematic review of the recent research literature. *Educational Research Review*, *23*, 25–56. doi:10.1016/j.edurev.2017.10.003

Buckner, M. M., & Strawser, M. G. (2016). Millennials and the paralysis of choice: Reigniting the purpose of higher education. *Communication Education*, *65*(3), 361–363. doi:10.1080/03634523.2016.1177845

Caniels, M. J., & Rietzschel, E. F. (2015). Organizing creativity: Creativity and innovation under constraints. *Creativity and Innovation Management*, *24*(2), 184–196. doi:10.1111/caim.12123

Clark, S. K., & Byrnes, D. (2015). What Millennial preservice teachers want to learn in their training. *Journal of Early Childhood Teacher Education*, *36*(4), 379–395. doi:10.1080/10901027.2015.1100148

Cohen, J. (1960). A coefficient of agreement for nominal scales. *Educational and Psychological Measurement*, *20*(1), 37–46. doi:10.1177/001316446002000104

Cohen, J. (1988). *Statistical power analysis for the behavioral sciences* (2nd ed.). Erlbaum.

Copp, D. T. (2018). Teaching to the test: A mixed methods study of instructional change from large-scale testing in Canadian schools. *Assessment in Education: Principles, Policy & Practice*, *25*(5), 468–487. doi:10.1080/0969594X.2016.1244042

Corgnet, B., Espin, A. M., & Hernan-Gonzalez, R. (2016). Creativity and cognitive skills among Millennials: Thinking too much and creating too little. *Frontiers in Psychology*, *7*(1626), 1–9. doi:10.3389/fpsyg.2016.01626 PMID:27826268

Council of Chief State School Officers. (2013). *Interstate teacher assessment and support consortium InTASC Model Core Teaching Standards and Learning Progressions for Teachers 1.0: A resource for ongoing teacher development*. Author.

Creswell, J. W., & Guetterman, T. C. (2015). *Educational research: Planning, conducting, and evaluating quantitative and qualitative research*. Prentiss Hall.

Csikszentmihalyi, M., Montijo, M. N., & Mouton, A. R. (2018). Flow theory: Optimizing elite performance in the creative realm. In S. I. Pfeiffer, E. Shaunessy-Dedrick, & M. Foley-Nicpon (Eds.), *APA Handbooks in Psychology. APA Handbook of Giftedness and Talent* (pp. 215–229). American Psychological Association., doi:10.1037/0000038-014

DeLeon, J. (2018). *Leading the Millennial Generation: An investigation of leadership practices that encourage and support education's youngest teachers* (PhD diss.). Dallas Baptist University.

Dinno, A. (2015). Nonparametric pairwise multiple comparisons in independent groups using Dunn's test. *The Stata Journal*, *15*(1), 292–300. doi:10.1177/1536867X1501500117

Dunn, O. J. (1964). Multiple comparisons using rank sums. *Technometrics*, *6*(3), 241–252. doi:10.1080/00401706.1964.10490181

Eckert, C. M., Stacey, M., Wyatt, D., & Garthwaite, P. (2012). Change as little as possible: Creativity in design by modification. *Journal of Engineering Design*, *23*(4), 337–360. doi:10.1080/09544828.2011.639299

Ellis, V., & Childs, A. (2019). Innovation in teacher education: Collective creativity in the development of a teacher education internship. *Teaching and Teacher Education*, *77*, 277–286. doi:10.1016/j.tate.2018.10.020

Erenli, K. (2016). Generation I (mmersion)–how to meet learner expectations of tomorrow. *International Journal of Advanced Corporate Learning*, *9*(1), 19–25. doi:10.3991/ijac.v9i1.5241

Erlam, G., Smythe, L., & Clair, V. W. (2018). Action research and millennials: Improving pedagogical approaches to encourage critical thinking. *Nurse Education Today*, *61*, 140–145. doi:10.1016/j.nedt.2017.11.023 PMID:29197689

Erskine, J. L. (2014). It changes how teachers teach: How testing is corrupting our classrooms and student learning. *Multicultural Education*, *21*(2), 38–40. https://eric.ed.gov/?id=EJ1045812

Field, A. P. (2018). *Discovering statistics using IBM SPSS Statistics*. Sage.

Fleiss, J. L., Levin, B., & Paik, M. C. (2013). *Statistical methods for rates and proportions*. John Wiley and Sons.

Frawley, E. F. (2014). No time for the airy fairy: Teacher perspectives on creative writing in high stakes environments. *English in Australia, 49*(1), 17-26. https://search.informit.com.au/documentSummary;dn=381501270146252;res=IELHSS

Fusarelli, L. D., & Fusarelli, B. C. (2018). Instructional supervision in an era of high-stakes accountability. In S. J. Zepeda & J. A. Ponticell (Eds.), *The Wiley Handbook of Educational Supervision* (pp. 131–155). Wiley-Blackwell. doi:10.1002/9781119128304.ch6

Gunnulfsen, A. E., & Roe, A. (2018). Investigating teachers' and school principals' enactments of national testing policies: A Norwegian study. *Journal of Educational Administration*, *56*(3), 332–349. doi:10.1108/JEA-04-2017-0035

Hallman, H. L. (2017). Millennial teachers and multiculturalism: Considerations for teaching in uncertain times. *Journal for Multicultural Education, 11*(3), 194–205. doi:10.1108/JME-10-2016-0055

Harris, A., & Ammermann, M. (2016). The changing face of creativity in Australian education. *Teaching Education, 27*(1), 103–113. doi:10.1080/10476210.2015.1077379

Hatchuel, A., & Chen, M. K. (2017). Creativity under strong constraints: The hidden influence of design models. *European Review (Chichester, England), 25*(2), 194–207. doi:10.1017/S1062798716000557

Haught, C. (2015). The role of constraints in creative sentence production. *Creativity Research Journal, 27*(2), 160–166. doi:10.1080/10400419.2015.1030308

Haught-Tromp, C. (2017). The green eggs and ham hypothesis: How constraints facilitate creativity. *Psychology of Aesthetics, Creativity, and the Arts, 11*(1), 10–17. doi:10.1037/aca0000061

Henriksen, D., Mishra, P., & Fisser, P. (2016). Infusing creativity and technology in 21st Century education: A systemic view for change. *Journal of Educational Technology & Society, 19*(3), 27–37. https://www.jstor.org/stable/jeductechsoci.19.3.27

Henriksen, D., Mishra, P., & Mehta, R. (2015). Novel, effective, whole: Toward a NEW framework for evaluations of creative products. *Journal of Technology and Teacher Education, 23*(3), 455–478. https://www.learntechlib.org/p/151574/

Hinchman, T. (2019). Expanding the Curriculum With Creativity. In *Critical Literacy Initiatives for Civic Engagement* (pp. 48–68). IGI Global. doi:10.4018/978-1-5225-8082-9.ch003

Hodges, G. C. (2005). Creativity in education. *English in Education, 39*(3), 47–61. doi:10.1111/j.1754-8845.2005.tb00624.x

Holloway, J., & Brass, J. (2017). Making accountable teachers: The terrors and pleasures of performativity. *Journal of Education Policy, 33*(3), 361–382. doi:10.1080/02680939.2017.1372636

Hosek, A. M., & Titsworth, S. (2016). Scripting knowledge and experiences for millennial students. *Communication Education, 65*(3), 357–359. doi:10.1080/03634523.2016.1177844

Jain, N., & Jain, T. K. (2019). Sustainable development through student engagement: Innovations and initiatives in higher education. SSRN *Electronic Journal*, 1-8. doi:10.2139/ssrn.3314509

Jakee, K., & Keller, E. (2017). The price of high-stakes educational testing: Estimating the aggregate costs of Florida's FCAT exam. *Journal of Education Finance*, *43*(2), 123–151. https://muse.jhu.edu/article/688009/summary

Johnson, D. D., Johnson, B., Farenga, S. J., & Ness, D. (2008). *Stop high-stakes testing: An appeal to America's conscience*. Rowman and Littlefield.

Kajzer-Mitchell, I., & Walinga, J. (2017). The creative imperative: The role of creativity, creative problem-solving and insight as key drivers for sustainability. *Journal of Cleaner Production*, *140*, 1872–1884. doi:10.1016/j.jclepro.2016.09.162

Kennedy, M. M. (1999). The role of preservice teacher education. In L. Darling-Hammond & G. Sykes (Eds.), *Teaching as the Learning Profession: Handbook of Teaching and Policy* (pp. 54–86). Jossey Bass.

Kettler, T., Lamb, K., Willerson, A., & Mullett, D. (2018). Teachers' perceptions of creativity in the classroom. *Creativity Research Journal*, *30*(2), 164–171. doi:10.1080/10400419.2018.1446503

Kim, K. H. (2011). The creativity crisis: The decrease in creative thinking scores on the Torrance Tests of Creative Thinking. *Creativity Research Journal*, *23*(4), 285–295. doi:10.1080/10400419.2011.627805

Kim, K. H. (2016). *The creativity challenge: How we can recapture American innovation*. Prometheus Books.

Krahl, D. (2018). Resisting the influence of neoliberalism on Millennial faculty through critical communication pedagogy and critical mentoring. In A. Sutherland & M. G. Strawser (Eds.), *Millennial Culture and Communication Pedagogies: Narratives from the Classroom and Higher Education* (pp. 3–19). Lexington Books.

Krippendorff, K. (2004). *Content analysis: An introduction to its methodology*. Sage.

Krishnan, P. (2018). A review of the non-equivalent control group post-test-only design. *Nurse Researcher*, *27*(1), 114–124. https://www.ncbi.nlm.nih.gov/pubmed/30226337 PMID:30226337

Kruskal, W. H., & Wallis, W. A. (1952). Use of ranks in one-criterion variance analysis. *Journal of the American Statistical Association*, *47*(260), 583–621. doi:10.1080/01621459.1952.10483441

Kuehn, S., & Lingwall, A. (2015). In their own words: A thematic analysis of students' comments about their writing skills in mass communication programs. *Journal of Public Relations Education*, *1*(1), 1–12. https://aejmc.us/jpre/wp-content/uploads/sites/25/2015/08/JPRE-11-All-Three-Articles.pdf#page=21

Leedy, P. D., Ormrod, J. E., & Johnson, L. R. (2019). *Practical research: Planning and design*. Pearson Education.

Major, T., & Mulvihill, T. M. (2018). Problem-based learning pedagogies in teacher education: The case of Botswana. *The Interdisciplinary Journal of Problem-Based Learning*, *12*(1), 1–11. doi:10.7771/1541-5015.1543

Maley, A., & Kiss, T. (2017). Pre-conditions for classroom creativity. *Creativity and English Language Teaching*, 205-217. doi:10.1057/978-1-137-46729-4_10

Martin, P. C. (2016). Test-based education for students with disabilities and English language learners: The impact of assessment pressures on educational planning. *Teachers College Record*, *118*(14), 1–24. https://eric.ed.gov/?id=EJ1110765

Martz, B., Hughes, J., & Braun, F. (2017). Creativity and problem-solving: Closing the skills gap. *Journal of Computer Information Systems*, *57*(1), 39–48. doi:10.1080/08874417.2016.1181492

McAllum, K. (2016). Managing imposter syndrome among the trophy kids: Creating teaching practices that develop independence in Millennial students. *Communication Education*, *65*(3), 363–365. doi:10.1080/03634523.2016.1177848

Medeiros, K. E., Partlow, P. J., & Mumford, M. D. (2014). Not too much, not too little: The influence of constraints on creative problem-solving. *Psychology of Aesthetics, Creativity, and the Arts*, *8*(2), 198-210. https://psycnet.apa.org2014-12651-001

Mercier, E., Rattray, J., & Lavery, J. (2015). Twitter in the collaborative classroom: Micro-blogging for in-class collaborative discussions. *International Journal of Social Media and Interactive Learning Environments*, *3*(2), 83–99. doi:10.1504/IJSMILE.2015.070764

Morgan, H. (2016). Relying on high-stakes standardized tests to evaluate schools and teachers: A bad idea. *The Clearing House: A Journal of Educational Strategies, Issues and Ideas*, *89*(2), 67–72. doi:10.1080/00098655.2016.1156628

Newton, L., & Beverton, S. (2012). Pre-service teachers' conceptions of creativity in elementary school English. *Thinking Skills and Creativity*, *7*(3), 165–176. doi:10.1016/j.tsc.2012.02.002

Noddings, N. (2013). Standardized curriculum and loss of creativity. *Theory into Practice*, *52*(3), 210–215. doi:10.1080/00405841.2013.804315

OECD. (2021). *Teachers by age (indicator)*. doi:10.1787/93af1f9d-en

Olivant, K. F. (2015). I am not a format: Teachers' experiences with fostering creativity in the era of accountability. *Journal of Research in Childhood Education, 29*(1), 115–129. doi:10.1080/02568543.2014.978920

Olszewski, P. (2016). Teaching Millennials how to study under the 21st Century sky. *Pyrex Journal of Educational Research and Reviews, 2*(1), 1-9. https://www.pyrexjournals.org/pjerr

Pang, W. (2015). Promoting creativity in the classroom: A generative view. *Psychology of Aesthetics, Creativity, and the Arts, 9*(2), 122–127. doi:10.1037/aca0000009

Polson, C. (2018). TAKS-ing students? Texas exit exam effects on human capital formation. *Economics of Education Review, 62*, 129–150. doi:10.1016/j.econedurev.2017.09.009

Ro, J. (2018). Learning to teach in the era of test-based accountability: A review of research. *Professional Development in Education, 45*(1), 87–101. doi:10.1080/19415257.2018.1514525

Roskes, M. (2015). Constraints that help or hinder creative performance: A motivational approach. *Creativity and Innovation Management, 24*(2), 197–206. doi:10.1111/caim.12086

Rosso, B. D. (2014). Creativity and constraints: Exploring the role of constraints in the creative processes of research and development teams. *Organization Studies, 35*(4), 551–585. doi:10.1177/0170840613517600

Sanchez, J. G., & Patel, L. (2017). Growing up tested: Teachers' lived experiences of testing as students. *Peabody Journal of Education, 92*(4), 440–444. doi:10.1080/0161956X.2017.1349444

Schiffrin, H. H., & Liss, M. (2017). The effects of helicopter parenting on academic motivation. *Journal of Child and Family Studies, 26*(5), 1472–1480. doi:10.100710826-017-0658-z

Schreiber, J. B. (2018). Creativity and pre-service teacher education: What you see is what you get. *Global Education Review, 5*(1), 5–19. https://ger.mercy.edu/index.php/ger/article/view/421

Silvestru, C. I., Mehedinţu, G., Silvestru, R. C., & Icociu, V. C. (2019). Millennials and the Need for Adapting e-Learning Methods for Business Performance–a Case Study. In *The International Scientific Conference eLearning and Software for Education* (*Vol. 3*, pp. 225-231). "Carol I" National Defence University.

Skinner, B. F. (1990). *The behavior of organisms: An experimental analysis*. B. F. Skinner Foundation.

Stokes, P. D. (2006). *Creativity From constraints: The psychology of breakthrough*. Springer Publishing Company.

Stokes, P. D. (2008). Creativity from constraints: What can we learn from Motherwell? from Modrian? from Klee? *The Journal of Creative Behavior*, *42*(4), 223–236. doi:10.1002/j.2162-6057.2008.tb01297.x

Stokes, P. D. (2009). Using constraints to create novelty: A case study. *Psychology of Aesthetics, Creativity, and the Arts*, *3*(3), 174–180. doi:10.1037/a0014970

Stokes, P. D. (2013). The effects of constraints in a mathematics classroom. *Journal of Mathematics Education at Teachers College*, *4*(2), 25–31. https://jmetc.columbia.edu/

Stokes, P. D. (2014). Thinking inside the box: Creativity, constraints, and the colossal portraits of Chuck Close. *The Journal of Creative Behavior*, *48*(4), 276–289. doi:10.1002/jocb.52

Suarez, R. D. A. (2021). *Collaboration for Healthcare Virtual Teams: Supporting Creativity Through Diversity and Knowledge Management* (Doctoral dissertation). Capella University.

Tallent, R. J., & Barnes, J. J. (2015). Think bubbles and Socrates: Teaching critical thinking to Millennials in public relations classes. *Universal Journal of Educational Research*, *3*(7), 435–441. doi:10.13189/ujer.2015.030702

Tan, O. S. (2015). Flourishing creativity: Education in an age of wonder. *Asia Pacific Education Review*, *16*(2), 161–166. doi:10.100712564-015-9377-6

Torrents-Martin, C., Ric, A., & Hristovski, R. (2015). Creativity and emergence of specific dance movements using instructional constraints. *Psychology of Aesthetics, Creativity, and the Arts*, *9*(1), 65–74. doi:10.1037/a0038706

Van der Vegt, G. S., & Janssen, O. (2003). Joint impact of interdependence and group diversity on innovation. *Journal of Management*, *29*(5), 729–751. doi:10.1016/S0149-2063_03_00033-3

Van Overschelde, J. P., Rawson, K. A., & Dunlosky, J. (2004). Category norms: An updated and expanded version of the Battig and Montague (1969) norms. *Journal of Memory and Language*, *50*(3), 289–335. doi:10.1016/j.jml.2003.10.003

Wang, X. H. F., Kim, T. Y., & Lee, D. R. (2016). Cognitive diversity and team creativity: Effects of team intrinsic motivation and transformational leadership. *Journal of Business Research, 69*(9), 3231–3239. doi:10.1016/j.jbusres.2016.02.026

Weinbaum, C., Girven, R., & Oberholtzer, J. (2016). *The millennial generation: Implications for the intelligence and policy communities*. RAND, National Defense Research Institute. doi:10.7249/RR1306

Wilcox, K. C., & Lawson, H. A. (2017). Teachers' agency, efficacy, engagement, and emotional resilience during policy innovation implementation. *Journal of Educational Change, 19*(2), 181–204. doi:10.100710833-017-9313-0

Wilson, M. E. (2004). Teaching, learning, and Millennial students. *New Directions for Student Services, 2004*(106), 59–71. doi:10.1002s.125

Zohar, A., & Agmon, V. A. (2018). Raising test scores vs. teaching higher order thinking (HOT): Senior science teachers' views on how several concurrent policies affect classroom practices. *Research in Science & Technological Education, 36*(2), 243–260. doi:10.1080/02635143.2017.1395332

Chapter 12
Promoting Equity in Elementary School Mathematics:
Vignettes From Virtual Learning Environments

Drew Polly
https://orcid.org/0000-0003-2370-4409
University of North Carolina at Charlotte, USA

Julie Bacak
University of North Carolina at Charlotte, USA

Madelyn Colonnese
University of North Carolina at Charlotte, USA

Kaitlyn O. Holshouser
University of North Carolina at Charlotte, USA

Wendy Lewis
University of North Carolina at Charlotte, USA

ABSTRACT

This chapter describes the frameworks of mathematics tasks, equity-based mathematics teaching practices, technological pedagogical and content knowledge (TPACK), and how they relate to teaching for equity in virtual learning environments (VLE). The frameworks are used to describe vignettes from VLEs with elementary school learners and elementary school teacher candidates (future teachers). Implications of these vignettes include a need for an empirically based trajectory related to equity-based teaching practices as well as a need to consider which technologies in a VLE can be used to effectively promote equity-based mathematics teaching practices.

DOI: 10.4018/978-1-7998-8028-8.ch012

INTRODUCTION

The spring of 2020 caused teachers around the world to consider how to effectively teach students from preschool to college through virtual settings. In this chapter we define virtual learning environments (VLEs) as live online synchronous classes. While students from preschool through university-levels participated in VLEs, students also completed academic work on their own time in asynchronous settings, or a combination of synchronous and asynchronous which scholars have labeled as bichronous (Martin et al., 2020). The United Nations (2020) reported that 98.6 of children across the world experienced a shift in their mode of learning. In some instances, national, state, district and school-based leaders provided ample resources and support to teachers, but a majority of teachers had to go into "emergency teaching mode" and determine how to most effectively support their students without adequate support (Basilaia & Kvavadze, 2020; Doucet et al., 2020; Martin et al., 2021; United Nations, 2020). Regardless of students' and teachers' technological skills and knowledge, their beliefs about virtual learning, and their motivation to participate in online or virtual school, the pandemic forced teachers into that situation.

Prior to the pandemic there had been an increase in the educational opportunities in VLEs for learners from pre-Kindergarten through high school (Martin et al., in press). The origin of these VLEs was to increase access and opportunities to diverse groups learners in underserved communities who may not have access to specific courses, classes, or for individuals who may be unable to physically attend schools (Barbour, 2011; Barbour & Reeves, 2009). Despite the increase in K-12 VLEs most K-12 educators did not teach virtually until the COVID-19 pandemic (United Nations, 2020).

The COVID-19 pandemic came at a time when in the United States there was an intentional focus on equity-based pedagogies, culturally sustaining pedagogy, and including instructional pedagogies that support students who share demographics from those who have been traditionally marginalized, lacked access to effective teaching, and/or been labeled by policy makers and others as at-risk or low-performing (Garcia & Weiss, 2020). Scholars who are experts on equity-based pedagogies advocate that educational leaders and teachers aim to reduce barriers to students' access to high-quality teaching, resources, and educational support (Gutierrez, 2009; Paris, 2012; Paris & Alim, 2017). In this chapter we posit that equity-based pedagogies are effective strategies for teaching diverse classrooms of learners.

In this chapter we describe the theoretical frameworks of mathematical tasks, equity-based mathematics teaching, and Technological Pedagogical Content Knowledge (TPACK). We then provide vignettes from both elementary schools and elementary teacher education programs about how equity-based mathematics practices can be established and maintained in VLEs in order to effectively teach diverse classrooms of learners.

THEORETICAL FRAMEWORKS

Mathematical Tasks

Mathematics education research advances that effective teaching is situated in the use of cognitively-demanding mathematical tasks (NCTM, 2014; Polly & Hannafin, 2011; McGee et al., 2013). Recommendations call for teachers to provide opportunities for students to explore tasks that are cognitively-demanding. Scholars have organized tasks into four categories based on their level of cognitive demand (NCTM, 2014; Smith & Stein, 1998; Stein, et al., 2007). Table 3 shows the different types of mathematical tasks. While there is a role for each type of task in mathematics classrooms, recommendations from mathematics education leaders cite the need for students to engage in productive struggle and develop dispositions related to problem solving and perseverance by exploring and solving tasks with high cognitive demands (NCTM, 2014).

Multiple studies indicate that teachers tend to primarily pose tasks with low cognitive demands, even when tasks are written or intended to have high cognitive demand (Polly, 2010, 2011, 2016; Henningsen & Stein, 1997; Stein, Grover, & Henningsen, 1996). Further, research indicates that when trying to integrate technology into mathematics classrooms, teachers struggle maintaining the cognitive demand of technology-rich tasks (Polly & Hannafin, 2011; Polly, 2016). In some cases teachers have reported that the pressure of high-stakes assessments, their knowledge of students' abilities, and students' behaviors during tasks influence whether they maintain or decrease the cognitive demand of tasks that they use in their classroom (Munter & Wilhelm, 2021; Stein et al., 2007; Wilhelm et al., 2017).

Equity-Based Mathematics Teaching

Leading scholars in equity-based teaching advocate for educational leaders to innovate practices to increase access and opportunities for diverse classrooms learners whose identity aligns with subgroups who have been marginalized by either educational systems or society in general. Gutierrez (2009) proposed a framework for equity-based teaching that includes four categories. The categories and examples are detailed in Table 2.

Table 1. Types of Mathematical Tasks

Low Cognitive Demand	High Cognitive Demand
Memorization What is 27 + 25?	**Procedures with Mathematical Connections** There are 27 children on the playground. Then 25 more children come. How many are there now? Show your answer using a picture and an equation. Explain how your equation matches the numbers in your picture.
Procedures without Mathematical Connections There are 27 children on the playground. Then 25 more children come. How many are there now?	**Doing Mathematics** There are 27 children on the playground. More children come. If there are between 50 and 55 children on the playground how many children came? Find all of the possible combinations. For each combination show your work with a picture and an equation.

(Adapted from Smith & Stein, 1998)

Table 2. Dimensions to Address Equity in Mathematics

Dimension	Description
Access	Access to rigorous and current curriculum and resources Access to high-quality mathematics teachers Access to mathematical tools (e.g., manipulatives, technology)
Achievement	Engagement in mathematics Scores on assessments Preparation for STEM-based fields
Identity	Incorporation of personal and cultural backgrounds in mathematics Opportunity for students to select and use their own strategies Opportunity for students to use their own language
Power	Able to discuss and share their thinking and strategies Ownership of their own understanding and have opportunities to make sense Use mathematics to investigate problems relevant to their community

(from Gutiérrez, 2009)

This framework is critical when considering teachers' use of mathematical tasks in VLEs with diverse groups of learners. While it does not explicitly prescribe a specific type of instruction. The Identity and Power dimensions focus on providing students with agency and choice, which align well with task-focused lessons where a teacher poses a task for students to explore and then supports their exploration. The framework for equity-based mathematics teaching does not align with the traditional model of teachers giving a mini lesson, demonstrating how to solve problems and then giving students sufficient opportunities to practice. Therefore, mathematics lessons in VLEs that do not provide students with agency, choice, or the opportunity to personally choose strategies and make sense of the mathematics cannot be considered equitable.

Further, scholars in mathematics education advocate that all students have access to rigorous content and tasks (NCTM, 2014; Smith & Stein, 1998). When considering Gutierrez' (2009) framework with the framework of cognitively-demanding mathematics tasks (Smith & Stein, 1998) in Table 1, there is a need to ensure that all students have access to rigorous tasks as described in the equity framework, which includes cognitively-demanding tasks. Equity-based mathematics teaching is not whole classes of students working on basic tasks with low cognitive-demand and then providing access to only those who prove to be capable, but rather providing access to tasks with high cognitive-demand and providing scaffolding through collaboration, questioning, and other pedagogies to support students' mathematics understanding. Equity cannot happen if all students do not have access to tasks with high cognitive-demand.

Technological Pedagogical Content Knowledge (TPACK)

The third framework taken into account in this chapter is Technological Pedagogical Content Knowledge (TPACK) which was originally advanced by various scholars (e.g., Mishra & Koehler, 2006; Niess, 2005). Scholars advocate that the effective use of technology only occurs after teachers are able to apply in action the knowledge and skills related to pedagogies associated with teaching with technology in the context of specific subject areas (Mishra & Koehler, 2006; Mouza, 2009; Polly, 2011; Polly et al., 2021). Table 3 describes the various aspects of the TPACK framework.

In the context of this chapter, it is impossible to focus on the ideas of promoting equity-based mathematics teaching through the use of mathematics tasks that have high cognitive-demand in VLEs without considering teachers' TPACK. The interplay of TPACK and teachers' use of mathematics pedagogies in face-to-face classrooms has been documented in earlier studies (Polly, 2011; Urbina & Polly, 2017), but there is a need to consider TPACK in the context of teaching in VLEs.

There is promise in the intersection of the frameworks of equity-based mathematics instruction (Gutierrez, 2009), cognitively-demanding mathematical tasks (NCTM, 2014), and TPACK (Mishra & Koehler, 2006) there is a need to synthesize and provide tangible examples about how these frameworks can work together to advocate for equity-based virtual teaching. In the following section we provide vignettes from elementary schools and elementary school teacher education programs in order to provide tangible examples about how virtual mathematics instruction leverages tasks and instructional moves to promote equity-based mathematics instruction.

Table 3. Aspects of TPACK

Aspect	Description	What It Looks Like
Technological Knowledge (TK)	Knowledge of technologies used by teachers or students without a specific reference to pedagogies or content.	Modifying the settings in a generative learning program used in a classroom.
Pedagogical Knowledge (PK)	Knowledge of general pedagogies without specific reference to technologies or content.	Asking open-ended questions to begin a discussion.
Content Knowledge (CK)	Knowledge of content without specific reference to technologies or pedagogies	Explaining strategies to solve a multi-digit multiplication task.
Technological Pedagogical Knowledge (TPK)	Knowledge of technologies and pedagogies without specific reference to content.	Modeling for students how to use an iPad application.
Technological Content Knowledge (TCK)	Knowledge of content-specific technologies without reference to pedagogies.	Using an interactive whiteboard to annotate text features in a story.
Pedagogical Content Knowledge (PCK)	Knowledge of content-specific pedagogies without reference to technology.	Asking open-ended questions about the mathematical representations students used to solve a task
Technological Pedagogical Content Knowledge (TPACK)	Knowledge of instances that include content-specific technologies and pedagogies.	Using an interactive whiteboard to notate, record, and organize mathematical representations students used to solve a task.

VIGNETTES AND EXAMPLES OF VIRTUAL MATHEMATICS INSTRUCTION

Examples From VLEs With Elementary School Students

Promoting Equity in a VLE With Task Exploration and Discussion

In a third grade classroom students engage in a task in a Virtual Learning Environment (VLE). The teacher, Mrs. Berty provides the following task to students during a live Zoom mathematics class held on the platform Zoom.

Part 1: There are 8 pencils in a box of pencils. How many pencils are in the following number of boxes? 2 boxes? 3 boxes? 4 boxes? 5 boxes?

Part 2: Fiona has 6 boxes of pencils. If there are 10 crayons in a box how many boxes of crayons are needed to have more crayons than pencils?

Mrs. Berty launched the task by having students read it and then talk over Zoom to retell the task in their own words and determine what the various numbers represented. After students collaboratively discussed the task they explored the problem individually while logged on virtually. Mrs. Berty asked students every few minutes to explain what they were doing and to allow students to share with and hear from their classmates. After the task has been explored for a few minutes Mrs. Berty brings students back together.

In a face to face classroom with 75 minutes for math Mrs. Berty would have a brief discussion where students share their work and she helps them to see connections between strategies before differentiated activities. The discussion was led by students who shared their work, and responded to Mrs. Berty's frequent questions, "How did you know how to get started" and "How are these approaches similar or different from each other?" By posing mathematics tasks with high cognitive-demand (Table 1) and allowing students to explore them using a variety of strategies, Mrs. Berty was addressing the Access, Power, and Identity aspects of the framework for equity-based mathematics teaching practices (Table 2). Mrs. Berty's TPACK was evident during students' exploration and discussion of tasks with her use of a digital whiteboard to help students show their work. Typically she would either have students show written paper-based work on their camera while they talked or she would quickly sketch on the whiteboard the representation while students were sharing.

One of the complications that Mrs. Berty talked about while moving from face-to-face to a VLE environment is the lack of time for a mathematics discussion. In the VLE, in an effort to not overburden elementary students with too much screen time students logged off and instead used Flipgrid to make a video or Jamboard to post a written explanation about their strategy. Mrs. Berty used both Flipgrid and Jamboard as tools to allow students to share their thinking.

Mrs. Berty commented during a conversation, "I need students to be able to share their thinking and their strategies so Aligned with Gutierrez' (2009) framework Mrs. Berty used the Zoom sessions as well as tools Flipgrid and Jamboard to give students Power and help develop their mathematical Identity. Her tasks were cognitively-demanding so all students had access to rigorous mathematics as well. Mrs. Berty provided evidence of her TPACK by using general technologies such as Flipgrid to have students record and share their thinking with their classmates and Mrs. Berty herself. Mrs. Berty also used the Zoom VLE to engage students in sharing their thinking with their classmates orally and using the chat window. These uses of non-mathematics specific technology align to the TPK dimension of the TPACK framework.

In summary, the third grade students participated in lessons with mathematics tasks with high cognitive demand and the Access, Identity, and Power dimensions of Gutierrez' (2009) framework for equity based mathematics tasks. Further, in her

VLE, Mrs. Berty provided evidence of TPACK with her use of the chat window and a virtual whiteboard to provide pictorial representations of student work.

Promoting Students' Identity and Power in Mathematics Small Groups

Another example of equity-based mathematics teaching in a VLE was done by UNC Charlotte elementary students during their senior year clinical placement. During the semester prior to their student teaching experience, teacher candidates complete a mathematics small group teaching assignment over the course of the semester. Teacher candidates are instructed to teach small groups settings and pose mathematics tasks with a high level of cognitive demand (Table 1). While this assignment was designed to be done in person with students, candidates during the Fall 2020 and Spring 2021 semesters taught these small groups in a VLE using either Zoom, Google Meet, or Microsoft Teams.

Candidates concurrently participated in a book study with Lisa Delpit's (2012) *Multiplication is for White People; Raising Expectations for Other People's Children* and took turns leading chapter discussions of the book. Candidates had previously been introduced to both the frameworks on mathematical tasks (Table 1) and equity-based mathematics teaching (Table 2) in a previous semester during a mathematics pedagogy course. In each small group lesson candidates intended to incorporate mathematical tasks with a high cognitive-demand in ways that aligned to the framework for equity-based mathematics teaching. Candidates worked with students whose data suggested that they were not meeting grade-level expectations, who often are students who do not have access to rigorous mathematics experiences nor do they have a positive identity of themselves as mathematics learners. In the case of this experience, teacher candidates gave them access to rigorous tasks and allowed them to have agency and choice which developed their identity and power aspects of the framework.

This was a challenge during the pandemic as many teacher candidates were teaching their small math groups virtually during synchronous sessions. Teacher candidates were expected to reflect on their experiences throughout the semester. In their reflections many students acknowledged Gutiérrez's (2009) dimensions of equity in mathematics. For example, one teacher candidate reported:

When meeting with my students each week during small groups, I encourage them to tell me one fact about them each week, so I can better get to know them. Learning virtually makes it harder to build relationships but asking each student a question and touching base with them sets up a positive learning environment where the students feel comfortable to share. I have learned from them their favorite school subject (none was math) their favorite books, their favorite thing about being back in school, and this week I asked them about their favorite Halloween candy...By learning things, they are interested in, I am able to make the problems more enjoyable for them. They also told me they enjoy using manipulatives when doing math, so I provided them with virtual math tools to show their work.

This teacher's reflection exemplifies their attention to equity, especially in the midst of a global pandemic. By taking time during each lesson to relate to students on a personal level, this teacher candidate appeals to students' identities. Additionally, the teacher attends to the dimension of access, specifically access to mathematical tools by discussing some of the virtual math tools they found to support students in lieu of the hands-on materials they would have been able to use if instruction was taking place in person.

This was also evident in another teacher candidate's reflection when they stated:

Over the course of the month, I have seen them develop their knowledge of the area model and standard algorithm. Partial products have been the area that have grown the most in. At first, students were having a lot of trouble figuring out what numbers go in what specific box for the area model, but some students prefer this visual over the standard algorithm and partial product. Using the virtual whiteboard where both students can see everyone elses work and my work and compare them to each other, really helps them to see where they may be missing a step in their problem and is the most effective and similar way to simulate having an actual whiteboard.

This teacher candidate's use of a collaborative whiteboard where students could see the work of their classmates aligns with Guiterrez's (2009) dimension of power in the framework of equity-based mathematics practices. In this vignette candidates utilized equity-based mathematics practices and provided access to cognitively-demanding mathematics tasks to elementary school students.

Promoting Equity Through an Open-Ended Mathematics Task

Another way that equity-based mathematics practices can be supported with elementary school learners is with mathematics routines. In this vignette we describe the implementation of Three Act Tasks, originally created by Dan Meyer (2011) and adapted for the elementary grades by Graham Fletcher (2016). Three Act Tasks are problem solving structures which encourage the teacher and gifted math student to approach mathematics as a team and make meaning of the mathematics together (Fletcher, 2016). This team-oriented approach lends itself well to the infusion of problem posing because it positions the students to be co-creators of the learning.

These relatable scenarios accompany online visual aids and videos to help students develop internal representations that can support problems posing and cognitively demanding thinking Meyer, 2011).

Three Act Tasks promote cognitive demand virtually through the gradual release of visual aids with an element of suspense, which opens up inquiry by the teacher asking the students what they notice and wonder. The students are given an ill-structured problem or image to make sense of which is accessible via youtube (Meyer, 2011). Students then engage collaboratively to solve the Ill-structured problems in Act 2 with flexibille and utilize creative thinking due to the unknowns (Meyer, 2011; Fletcher, 2016). Act 3 involves a reveal of the solution or the creation of a sequel where gifted students can showcase their mathematical creativity (Meyer, 2011).

In a VLE of fourth grade students they explored the Krispy Kreme 3 Acts Task (https://gfletchy.com/krispy-kreme-me/). Mrs. Levins used the VLE to show the picture of the task (Figure 1). She then allowed students to use the tool Jamboard to have students put their initial thoughts about what they noticed and wondered about the task. After students had recorded their ideas on Jamboard Mrs. Levins provided an opportunity for students to orally share their thoughts with their classmates.

Figure 1. Krispy Kreme Picture from the 3 Acts Task
Used with Permission from https://gfletchy.com/krispy-kreme-me/

Mrs. Levins asked students to share their thinking about the question, " How do we know how many rows of donuts are in the picture? Can someone share their estimates and give an explanation about their reasoning?"

Alyssa, one student, commented, "It looks like the box has some depth so I am guessing that there are many donuts we cannot see in the picture. What does everyone else think?"

Jamir, another student, said, "I think there are 3 stacks of donuts in the

Mrs. Levins then spent time showing the email to students in Act 2 of the task which gave the dimensions of the box and provided more information. Students in the VLE then used the chat window to provide their thoughts and ideas about the size of 1 donut and how many donuts would fit in each row and column of the donut box. Mrs. Levins used the chat window to have students quickly provide their estimates of the number of donuts, which were revised estimates from the first Act of the task.

Students were then given the third part of the task which had the specific dimensions of the donut box. Mrs. Levins put the students in breakout groups to have them collaborate to determine the total number of donuts in the box. Students used the digital whiteboard to work together and create shared pictorial representations of the task.

In this vignette, Mrs. Levins used an open-ended 3 Act Task to provide students with Access to rigorous activities with a high cognitive demand. Further, Mrs. Levins used the tool Jamboard, along with digital whiteboards, breakout rooms, and the chat feature of the VLE to address the Identity and Power aspects of Gutierrez' (2009) framework of equity-based mathematics teaching practices. Similar to previous cases, Mrs. Levins' TPACK was reflected in her use of general, non-mathematics specific technologies.

Examples of Equity in VLEs With Elementary Education Teacher Candidates

The following vignettes provide examples of how equity-based mathematics practices, mathematical tasks, and TPACK are evident in VLEs which are pedagogy courses for elementary education teacher candidates.

Promoting Equity-based Practices through Mathematical Discussions in a VLE

I would ask Student B to share the array they used to solve first. Even though their final answer is incorrect, they have a strong visual representation of the equal groups in the problem that I think many other students would connect with. I would ask Student D to share next. They created equal groups using the base ten blocks.

By sharing these two strategies back-to-back, I think students would be able to make connections between the two different visual representations of the problem. Also, Student D's model connects nicely to the decomposing strategy that Student A used. Student A's strategy is the most efficient, but also the most abstract, so I would have them share last. Sequencing the strategies this way would hopefully help all students discover the different ways they can arrange the equal groups in the problem to find the total.

The response shared above represents a teacher candidate's reasoning when deciding how to sequence student work samples for a mathematical discussion where elementary school students would hypothetically share their work. Cognitively-demanding mathematical tasks provide opportunities for students to discuss and share their thinking and strategies and aligns with equity-based teaching practices (Gutiérrez, 2009). When teachers thoughtfully sequence students' responses for a discussion it helps facilitate connections between students' ideas and connections to key mathematical understanding (NCTM, 2014).

In a synchronous VLE course meeting teacher candidates were given a brief overview of research-based strategies for facilitating a mathematics discussion, and talked about how discussions can be used as a way to promote equity by strengthening students' identity as mathematics learners as well as give them power and agency to personally explore and makes sense of both mathematics concepts and tasks.

Following the synchronous VLE meeting, sample student responses to a given mathematical task were shared with teacher candidates on an interactive Google slide deck. Teacher candidates completed this sequencing assignment asynchronously outside of the VLE. The interactive slide deck allowed candidates to rearrange student responses to the task in a way that highlights the strengths in different student strategies, addresses developing ideas by facilitating connections between ideas, and contributes to student understanding of key mathematical ideas. Completing this sequencing task asynchronously gave teacher candidates time to analyze student responses and consider multiple sequences for sharing ideas. With practice, the ability to sequence students' strategies in meaningful ways will help prepare teacher candidates to facilitate focused and productive discussion of mathematical tasks during VLE sessions.

In the subsequent VLE lesson candidates worked with the course instructors to debrief and discuss their reasoning for their proposed sequence. Candidates also discussed with their peers how the selection process promoted equity for specific students. The teacher educators used a combination of synchronous and asynchronous formats in the VLE to give teacher candidates opportunities to analyze and work with actual student work from a mathematics task that had a high cognitive demand. Teacher candidates spent time thinking through the processes of how to promote equity through their sequence of student work with special attention to the Identity and Power aspects of Gutierrez' (2009) framework. Further, teacher candidates developed their TPACK as they worked with the interactive Google slide deck and spent time annotating the slides and making digital notes about the student work samples.

Promoting Mathematical Reasoning in a VLE

Do you agree or disagree with Ally's idea?" the teacher education candidate Kristy asked the small group of teacher education candidates. In response to this question, one candidate unmuted to share that she disagreed with Ally's idea, while others responded in the chat they agreed with Ally's idea. Kristy then followed up by asking the candidates to explain why they agreed or disagreed. As each candidate shared their ideas, Kristy showed the candidate's slide with their thinking.

The scenario shared here is from a VLE for an elementary mathematics pedagogy course. The scenario is an example as to how teacher candidates might engage with their peers during a virtual synchronous teaching rehearsal. The virtual synchronous teaching rehearsal is one way to provide teachers with an opportunity to design a mathematics task and try out methods for implementing the task with coaching support. To prepare for the synchronous teaching rehearsal, teacher candidates first created the task for the lesson.

To help encourage teacher candidates to create a task with a high level of cognitive demand, the teacher candidates used the word problem structures reflected in the North Carolina Mathematics Standards and included sentence frames such as this one: "Student A thinks ___. Do you agree or disagree and why?". An example of a task was, "Abby bought 5 books. Kim bought 8 books. Kim says that she purchased 13 more books than Abby. Do you agree or disagree with Kim? Why?" This activity is cognitively demanding since the correct answer is 3 and teacher candidates are expected to provide reasoning about why the example given is incorrect.

After developing the task, the teacher candidates then created virtual spaces with embedded virtual manipulatives for their peers to show why they agreed or disagreed. The variety of virtual manipulatives provided an "opportunity for students to select and use their own strategies'' (Gutiérrez, 2009). To implement their task, the teacher candidates were placed in a breakout room to teach their lesson to a small group of their peers. First, their peers solved the task before the teacher candidates led a discussion based on their peers' thinking. Teacher candidates, when leading the discussion, asked questions such as "How do you know?" and "Why did you select that strategy?". The course instructor provided feedback during the candidate's instruction and after. One area candidates were encouraged to continue working on was using the virtual tools to annotate and record the ideas shared to revise and enhance the group's mathematical thinking.

In this vignette the teacher educator provided a space for teacher candidates to engage as learners in mathematics tasks with high cognitive demand with one teacher candidate playing the role as the classroom teacher. The teacher candidate pretending to be a classroom teacher practiced providing opportunities to develop mathematical identity and power by providing time to collaborate on the task, make sense of the mathematics concepts in their own way, and share their thinking with others. TPACK in this case involved both the general technological tools of a VLE as well as mathematics specific virtual mathematics manipulatives.

DISCUSSION

In the examples above we provide ways in which equity-based mathematics teaching practices can be included in a VLE with a focus on mathematics tasks with a high cognitive demand, allowing learners to explore and discuss tasks in ways that promote individual agency, sensemaking, and autonomy. Lastly, TPACK is at play as elementary school teachers and elementary school teacher candidates engaged in teaching and learning in a VLE. In this section we focus on three topics that warrant further discussion: 1) a continuum of equity-based mathematics teaching practices, and 2) the role of TPACK and technology in promoting equity-based teaching.

A Continuum of Equity-Based Mathematics Teaching Practices

Gutierrez (2009) provides a succinct yet thorough perspective on equity-based mathematics teaching practices in her framework (Table 2). The framework has a high level of utility for those teaching diverse classrooms of learners, including researchers, professional development facilitators, teacher educators, and classroom teachers due to the simple language of the dimensions as well as the concrete examples provided. Still, though, with any list or structure provided there is the unintended consequence of individuals treating the framework as a dichotomous construct, which could lead to teachers being promoters of or barriers to Access, Achievement, Identity, and Power.

Rather than seeing these four dimensions of haves and have nots, we encourage scholars, educational leaders, and educators to think about these dimensions as separate continuums in which teachers' work in these dimensions may be tenuous and shift based on the unit, lesson, task, or even the question posed. For example, a teacher who provides students with a rigorous task that is engaging to students would be addressing the Access and Achievement dimensions. Yet if the teacher decides that they should provide students with an example strategy for them to use to solve it, the teacher is decreasing the likelihood that students will develop Identity and Power since students are likely to gravitate towards using the teachers' strategy (Polly & Hannafin, 2011).

In another example a teacher could provide students with an easy mathematical task that is not engaging for the entire class which does not meet the recommendations for Access and Achievement. If some of the students are able to finish that task they then move onto a more rigorous task that is more engaging, but only some of the students have that opportunity, then the teacher would be providing opportunities for some but not all of their students. As we think about ways to promote equity-based teaching practices in mathematics there is a need to think of these dimensions and constructs as continua instead of dichotomous categories.

Implications for research regarding this idea of continuum related to equity-based practices includes two broad areas. First, there is a need for researchers to gather research data on various instructional practices and establish an empirical trajectory related to the various dimensions of Gutierrez' framework. The dimensions, as stated previously, may include more of a focus on the various descriptive statements in Table 2 rather than the broad framework. For example, when considering the Power dimension of Gutierrez' (2009) framework there may be a research-based trajectory formed related to each of the three parts of that dimension (Table 2). In the mathematics education field scholars have already studied and published about topics included in the framework, such as the exploration of tasks, discussing mathematics concepts, and personal sensemaking of concepts. Those findings hopefully can inform the development of an empirically-based trajectory.

Second, a research agenda is needed to examine ways to effectively support teachers' implementation of equity-based mathematics teaching practices. This line of scholarship could include both teacher education programs and professional development efforts. There is a need to contribute to the equity literature in and outside the field of mathematics education and share findings from carefully designed studies about how to support teachers' use of equity-based practices. Teachers' enacted practices and their use of specific pedagogies is directly related to their knowledge, beliefs, and backgrounds (Polly & Hannafin, 2010, 2011; Stein et al., 2007), which means these teacher education and professional development efforts must concurrently address knowledge, beliefs, and backgrounds.

The Role of TPACK and Equity-based Teaching

These vignettes provided tangible examples of how equity-based practices were advanced in both elementary school classrooms as well as teacher education courses for elementary education teacher candidates in a virtual learning environment (VLE).

Based on these vignettes, the interplay between TPACK and equity-based teaching practices requires more attention. The creators of the TPACK framework Mishra and Koehler (2006) posit that the framework is agnostic related to learning theory and beliefs about teaching, but that contexts should be considered in all circumstances. In the case of this chapter, vignettes were biased towards constructivist, student-centered pedagogies that align well with Gutierrez' (2009) framework for equity-based mathematics teaching practices. Further, teachers' and teacher educators' use of technology in these vignettes included providing visual displays through the VLE, communication via the chat window and the audio components of the VLE, and providing opportunities for students and teacher candidates to engage in written reflections or writing through the use of Jamboard, the chat window, or Google slide decks. These technologies were content agnostic and their use aligned closely to the Technological Pedagogical Knowledge (TPK) and Technological Knowledge (TK) aspects of the TPACK framework. Those aspects of TPACK were more than sufficient to implement equity-based mathematics teaching practices in the VLEs that were described.

That begs the question- is there a need for teachers' to possess and be able to apply aspects of the TPACK framework that are related to specific concepts such as TCK or TPACK in order to promote equity-based teaching. An example of teachers' use of TCK or TPACK that was not included in these vignettes would have focused on content-specific technologies, such as website or web-based tools or simulation that are focused specifically on concepts such as geometry, fractions, or other mathematics concepts. While TCK and TPACK are beneficial to effective teaching and the effective use of equity-based mathematics teaching practices, a possible take away from this chapter is that teachers and teacher educators were able to use general technologies that are commonly used in order to enact equity-based practices. While previous studies have found that professional development focused on integrating technology effectively into teaching should include a combination of general and content-specific technologies (Polly, 2011; Polly & Hannafin, 2011) these vignettes provide examples about how general technologies can be used in conjunction with equity-based practices. As a result, both professional development facilitators and teacher educators who are preparing teachers for both VLEs and face-to-face teaching, should determine the current level of technology use of their learners, and thoughtfully consider whether general or content-specific technologies should be integrated into the learning experience, and how to best support teachers use of equity-based practices (Byker et al., 2018).

CONCLUSION

This chapter provided five vignettes from elementary school virtual learning environments (VLEs) related to teaching mathematics tasks with a high cognitive demand (NCTM, 2014; Smith & Stein, 1998), equity-based mathematics instructional practices (Gutierrez, 2009), and Technological Pedagogical Content Knowledge (Mishra & Koehler, 2006). The vignettes provided tangible examples about how teachers leveraged equity-based teaching practices and tasks with high cognitive demand in VLEs for elementary school students or elementary school teacher candidates.

Of note, vignettes point to a need for an empirically-based continuum or trajectory for equity-based teaching practices as well as further research on how to support teachers' and teacher candidates' use of equity-based instructional practices. Further, teachers and teacher educators used general, non-mathematics specific technologies in the VLEs, which suggests that the work to promote teachers' use of equity-based practices in VLEs may need to focus more on the use of equity-based practices and not on hard to learn technological tools.

REFERENCES

Barbour, M. K. (2011). *The promise and the reality: Exploring virtual schooling in rural jurisdictions. Keynote paper from the 2010 Annual SPERA Conference.* University of the Sunshine Coast. Retrieved from https://core.ac.uk/download/pdf/231035933.pdf

Barbour, M. K., & Reeves, T. C. (2009). The reality of virtual schools: A review of the literature. *Computers & Education*, *52*(2), 402–416. doi:10.1016/j.compedu.2008.09.009

Basilaia, G., & Kvavadze, D. (2020). Transition to online education in schools during a SARS-CoV-2 coronavirus (COVID-19) pandemic in Georgia. *Pedagogical Research*, *5*(4), 10. doi:10.29333/pr/7937

Byker, E. J., Putman, S. M., Handler, L., & Polly, D. (2018). Examining elementary education teachers and pre-service teachers' self-efficacy related to Technological Pedagogical and Content Knowledge (TPACK). In C. Hodges (Ed.), *Self-Efficacy in Instructional Technology Contexts* (pp. 119–140). Springer. doi:10.1007/978-3-319-99858-9_8

Delpit, L. (2012). Multiplication is for white people. In *Raising expectations for other people's children*. The New Press.

Doucet, A., Netolicky, D., Timmers, K., & Tuscano, F. J. (2020). *Thinking about pedagogy in an unfolding pandemic* (An Independent Report on Approaches to Distance Learning during COVID-19 School Closure). Work of Education International and UNESCO. https://issuu.com/educationinternational/docs/2020_research_covid-19_eng

Fletcher, G. (2016). *Modeling with mathematics through three-act tasks*. Math Tasks to Talk About. Retrieved from: https://www.nctm.org/Publications/TeachingChildren-Mathematics/Blog/Modeling-withMathematics-through-Three-Act-Tasks/

Garcia, E., & Weiss, E. (2020). *COVID-19 and student performance, equity, and U.S. education policy*. Economic Policy Institute. Retrieved from: https://www.epi.org/publication/the-consequences-of-the-covid-19-pandemic-for-education-performance-and-equity-in-the-united-states-what-can-we-learn-from-pre-pandemic-research-to-inform-relief-recovery-and-rebuilding/

Gutiérrez, R. (2009). Framing equity: Helping students "play the game" and "change the game.". *Teaching for Excellence and Equity in Mathematics, 1*(1), 4–8.

Martin, C., Harbour, K., & Polly, D. (2021). Transitioning the elementary mathematics classroom to virtual learning: Exploring the perspectives and experiences of teachers. In A. Slapac, P. Balcerzak, & K. O'Brien (Eds.), *Handbook of Research on the Global Empowerment of Educators and Student Learning through Action Research* (pp. 343–365). IGI Global. doi:10.4018/978-1-7998-6922-1.ch015

Martin, F., Bacak, J., Dymes, L., & Polly, D. (in press). A systematic review of research on K12 online teaching and learning: A comparison of research from two decades 2000 to 2019. *Journal of Research on Technology in Education*.

Martin, F., Polly, D., & Ritzhaupt, A. (2020). Bichronous online learning: Blending asynchronous and synchronous online learning. *Educause Review*. Retrieved from: https://er.educause.edu/articles/2020/9/bichronous-online-learning-blending-asynchronous-and-synchronous-online-learning

Meyer, D. (2011). *The three acts of a mathematical story*. Dy/Dan. Retrieved from: http://blog.mrmeyer.com/2011 /the-three-acts-of-a-mathematical-story/

Mishra, P., & Koehler, M. (2006). Technological pedagogical content knowledge: A framework for teacher knowledge. *Teachers College Record, 108*(6), 1017–1054. doi:10.1111/j.1467-9620.2006.00684.x

Mouza, C. (2009). Does research-based professional development make a difference? A longitudinal investigation of teacher learning in technology integration. *Teachers College Record, 111*, 1195–1241.

Munter, C., & Wilhelm, A. G. (2021). Mathematics teachers' knowledge, networks, practice, and change of instructional visions. *Journal of Teacher Education, 72*(3), 342–354. doi:10.1177/0022487120949836

National Council of Teachers of Mathematics. (2014). *Principles to actions: Ensuring mathematics success for all.* National Council of Teachers of Mathematics.

Niess, M. L. (2005). Preparing teachers to teach science and mathematics with technology: Developing a technology pedagogical content knowledge. *Teaching and Teacher Education, 21*(5), 509–523. doi:10.1016/j.tate.2005.03.006

Paris, D. (2012). Culturally sustaining pedagogies: A needed change in stance, terminology, and practice. *Educational Researcher, 41*(3), 93–97. doi:10.3102/0013189X12441244

Paris, D., & Alim, H. S. (2017). *Culturally sustaining pedagogies: Teaching and learning for justice in a changing world.* Teachers College Press.

Polly, D. (2011). Developing teachers' technological, pedagogical, and content knowledge (TPACK) through mathematics professional development. *The International Journal for Technology in Mathematics Education, 18*(2), 83–96.

Polly, D. (2016). Examining Elementary School Teachers' Enactment of Mathematical Tasks and Questions. *Research in the Schools, 23*(2), 61–71.

Polly, D., Byker, E. J., & Colonnese, M. W. (2021). Future directions for K-12 technology enhanced learning environments. *TechTrends, 65*(3), 240–242. doi:10.100711528-021-00602-y PMID:33817701

Polly, D., & Hannafin, M. J. (2010). Reexamining technology's role in learner-centered professional development. *Educational Technology Research and Development, 58*(5), 557–571. doi:10.100711423-009-9146-5

Polly, D., & Hannafin, M. J. (2011). Examining how learner-centered professional development influences teachers' espoused and enacted practices. *The Journal of Educational Research, 104*(2), 120–130. doi:10.1080/00220671003636737

Smith, M. S., & Stein, M. K. (1998). Selecting mathematical tasks: From research to practice. *Mathematics Teaching in the Middle School, 3*, 344–350. doi:10.5951/MTMS.3.5.0344

Stein, M. K., Remillard, J. T., & Smith, M. S. (2007). How curriculum influences student learning. In F. Lester (Ed.), *Second handbook of research on mathematics teaching and learning* (pp. 319–370). Information Age Publishing.

United Nations. (2020). *Policy brief: Education during COVID-19 and beyond.* United Nations. https://www.un.org/development/desa/dspd/wp-content/uploads/sites/22/2020/08/sg_policy_brief_covid-19_and_education_august_2020.pdf

Urbina, A., & Polly, D. (2017). Examining elementary school teachers' integration of technology and enactment of TPACK in mathematics. *International Journal of Information and Learning Technology, 34*(5), 439–451. doi:10.1108/IJILT-06-2017-0054

Wilhelm, A. G., Munter, C., & Jackson, K. (2017). Examining relations between teachers' explanations of sources of students' difficulty in mathematics and students' opportunities to learn. *The Elementary School Journal, 117*(3), 345–370. doi:10.1086/690113

Chapter 13
Challenges and Victories of Virtual Environments With Students With Disabilities

Dennis Cavitt
Midwestern State University, USA

Danielle Kovach
Tulsa Trail Elementary, USA

ABSTRACT

The COVID-19 pandemic that began in 2020 altered the educational environment for everyone. Moving from an in-person environment to a virtual one can be highly problematic for individuals (specifically, students with disabilities) whose teachers most often use direct instruction strategies that require face-to-face interaction. This chapter will help the reader understand the variations of the characteristics of the different disability categories served under the Individuals with Disabilities Education Act (IDEA). This chapter will provide the readers with instructional strategies that effectively mitigate difficulties with instruction in a virtual environment. Additionally, this chapter will examine the challenges observed in students with disabilities that make accessing the virtual educational setting problematic. These areas include race, culture, disability, and low socioeconomic factors. Finally, this chapter will provide the reader with tips from teachers and students as they have struggled to navigate this new educational environment.

DOI: 10.4018/978-1-7998-8028-8.ch013

INTRODUCTION AND BACKGROUND

The Institute of Education Science in *The Condition of Education 2020* (Husser et al., 2020) indicated the number of students age 3-21 who received special education services under IDEA was 7.1 million which was 14 percent of all public-school students. That means that 14 out of every 100 students in school have been identified with a significant disability that adversely affects their educational performance.

Ensuring that a child receives the full benefit of a free, appropriate public education (FAPE) according to their IEP is an ongoing process. FAPE was especially significant during the COVID 19 pandemic as school districts halted their face-to-face programs and moved to a fully virtual learning environment. This move has created some very significant challenges to providing FAPE. Three significant court cases have shaped FAPE policy in the United States (Yell, 2019). Precedence of these cases should shape the way districts implement FAPE during virtual learning. These include *Rowley v. Hendrick Hudson Board of Education*, *Doe v. Board of Education of Tullahoma City Schools*, and *Endrew F. v. Douglas County School District RE-1.*

This chapter aims to identify the significant challenges that students with special needs encountered during virtual learning and highlight the victories of virtual environments during the COVID19 pandemic through a review of the legal systems, historical challenges experience by the disability community, and specific instructional issues exposed by the pandemic. Based on the disability categories according to IDEA, particular challenges students experienced during virtual instruction in each classification will be discussed, explaining how teachers have overcome these challenges. Identifying various evidence-based practices (EBP) and stories from the field will provide specific suggestions and recommendations from teachers, practitioners, service providers, and students.

Learning Objectives

By the end of this Chapter, the reader should be able to…

1. Explain the history and the legislation of special education and why it is important in virtual environments.
2. Articulate the legal issues associated with students identified with a disability.
3. Explain the key provisions of the Individuals with Disabilities Education Act
4. Explain potential legal landmines, including
    a. Zero Reject
    b. Free Appropriate Public Education
    c. Least Restrictive Environment
    d. Appropriate Evaluation

5. Explain how demographic information (Race, Culture, Language, and Socioeconomic) has an effect on the learning process for students with disabilities in virtual environments.
6. Identify interventions that will align with a student's IEP goals and objectives and how to implement them in virtual environments.

LEGAL ISSUES

Disability Discrimination

Students with disabilities have a long history of being excluded from school (Yell, 2019). In 1893, the Massachusetts Supreme Court ruled that children who were "weak in mind" and would not receive benefit from instruction, were a problem to other children, made unusual noises and were unable to take "ordinary, decent, physical care of himself" could be removed from public school.

In 1934, the *Cuyhoga County Court of Appeals in Ohio* ruled that the compulsory attendance for children ages 6 through 18 provided the State Education Agency the authority to exclude certain students (Winzer, 1993). In 1958, The Illinois Supreme Court in *Department of Public Welfare v. Haas,* ruled that the state's existing compulsory attendance laws did not require the state to provide free public education for the "feeble minded" or children identified "mental deficient". In 1969, the North Carolina legislature made it a crime for parents to continue trying to force the attendance of a child with a disability after they were excluded (Weber, 2008).

The 1950's and 1960's was filled with individuals and groups seeking changes to society in order to provide minorities with equal opportunities. The landmark case of *Brown v. Board of Education* of (1954, henceforth referred to as *Brown*), was a significant victory for the civil rights movement. The case not only removed barriers for the education of minorities, but it eventually affected many areas of educational law and services for individuals with disabilities (Yell, 2019). Advocates for students with disabilities believed that these students have the same rights as students without disabilities. Additionally, the advocates argued that many students with disabilities were being denied an education. The advocates used the *Brown* decision as the stimulus for ensuring that students with disabilities are provided educational rights. Advocates and individuals with disabilities challenged the disparities experienced by students with disabilities in the public education system. This movement also included treatment of individuals in institutions.

In the Fall of 1965, Robert Kennedy and several of his senate colleagues, visited many of the nation's institutions for the mentally retarded and their comments about these visits created an uproar. Blat and Kaplan wrote:

These disclosures shocked millions of Americans and Infuriated scores of public office holders and professional persons responsible for the care and treatment of the mentally retarded. A segment of the general public was numbed because it is difficult for "uninvolved" people to believe that in our country, today, human beings are being treated less humanely, with less care, and under more deplorable conditions than animals. (1974, p. v).

During December 1965, Blat and Kaplan, visited 5 state run institutions in 4 eastern states where they "were able to visit the darkest corridors and vestibules that humanity provides for its journey to purgatory (1974, p.vi)." Kaplan, a photojournalist, took many pictures during their visit. Blat and Kaplan put their notes, photos, and comments into a book, *Christmas in Purgatory.* The first edition of the book was distributed to legislators, professionals, professors, and other leaders who were decision makers in the field (1974). This book, the court cases mentioned previously, and the ground swell of parent and stake holder groups, began a movement to chip away at the systems that discriminated against individuals with disabilities that could not be ignored by decision makers. The following lawsuits and subsequent legislation produced significant changes to the field of disabilities that came out of this movement.

Beginning with the Elementary and Secondary Education Act of 1965 (ESEA- the federal law that guides all education programs), Congress passed several federal laws that would provide national guidance and funding to rectify the unconstitutional issues for disadvantaged children. The Education of the Handicapped Act (EHA) of 1970 replaced title VI of the ESEA. The purpose of the EHA was to streamline the previous federal grant programs focused on individuals with disabilities. The EHA provided funds for state programs in order to initiate, expand, or improve programs for students with disabilities. Additionally, the EHA provided funding for Teacher Preparation programs in order to train teachers for students with disabilities. However, it became clear that state programs were not meeting the needs of students with disabilities.

In 1973, Congress passed the Rehabilitation Act. A short section of the Act, Section 504, was the first federal civil rights law that protected the rights of individuals with disabilities. The first sentence of Section 504 reads:

No otherwise qualified individual with a disability in the United States… shall, solely by reason of his or her disability, be excluded from the participation in, be denied the benefits of, or be subjected to discrimination under any program or activity receiving Federal financial assistance. (U. S. Department of Education, 2010)

Section 504 reflects the language of other federal civil rights laws that prohibit discrimination on the basis of race and sex, however the focus is on individuals identified with a handicap (Yell, 2019). Section 504 defined a handicapped individual as a person who has a physical or mental impairment that substantially limits one or more of that person's major life activities, or a person who has a record of such impairment, or who is regarded as having such impairment.

Students with disabilities were experiencing limited educational opportunities prior to the passage of Education of All Handicapped Children Act (EAHCA). According to researchers, (Katsiyannis, Yell & Bradley, 2001; Yell, Drasgow, Bradley, & Justesen, 2004) students with disabilities were limited in two significant ways. First, many students were excluded from public schools. Congressional findings at the time identified more than 1.75 million students with disabilities were not receiving educational services. Second, more than 3 million students with disabilities in public schools were not receiving an appropriate education. The EAHCA promised to provide students with disabilities educational rights and federal funding to the states. The funding was intended to supplement not supplant state and local monies. Congress promised full funding of EAHCA. Full funding was defined as supplemental funds to cover 40 percent of the "excess costs" of serving students with disabilities. Unfortunately, the federal share was only 13 percent in Fiscal Year (FY) 2020, leaving states and school districts to pay the balance and increasing the burden on local taxpayers (Council for Exceptional Children, 2020). Advocates such as the Council for Exceptional Children (CEC), Consortium for Citizens with Disabilities (CCD), Committee for Education Funding, Friends of IES, and many others have been encouraging congress to fulfill their promise they made in 1975 to provide full funding for special education.

The amendments to EAHCA renamed the law the Individuals with Disabilities Education Act (IDEA) in 1990. The IDEA removed the term handicap and inserted disability throughout the law. Additionally, the language in the law began to use "people first" language and added autism and traumatic brain injury disability conditions. Finally, the law clarified types of related services, assistive technology and rehabilitation services.

The amendments to IDEA in 1997 were the most significant since the laws initial passage in 1975. Congress stated that the goal of the amendments was to improve the effectiveness of specially designed education by requiring measurable improvements in the students with disabilities educational achievement.

The most current reauthorization of the IDEA was passed nearly unanimously by both the house and senate on December 3, 2004. The name of the law was changed to the Individuals with Disabilities Education Improvement Act (IDEIA). The legislators' goal in passing the IDEIA was to align it to the Elementary and

Secondary Education Act (ESEA). This alignment would increase accountability for districts to increase academic results for students with disabilities (Yell, 2019).

MAJOR PRINCIPLES OF SPECIAL EDUCATION LEGISLATION

Researchers (Heward, 2017; Yell, 2019) identified six major principles of IDEA that the states must ensure are fully applied to all students identified with a disability. These are:

Zero-Reject

All public schools must provide educational services to all children identified with disabilities. No student from age 3-21 can be rejected from receiving a public education, in fact, there is a provision in the IDEA the requires school districts to identify, locate, and evaluate all children with disabilities who need specially designed instruction and related services regardless of the severity of their disability. This provision is called Child Find (U.S. Department of Education, 2010). The zero-reject principle in the IDEA came from the decisions of PARC and Mills court cases.

Nondiscriminatory Identification and Evaluation

"An evaluation that does not discriminate on the basis of language, culture, and students background must be provided for each individual identified for special education" (Vaughn, S. R., Bos, C. S. & Schumm, J. S. 2018, p. 7). The evaluation is the foundation of the individual education program (IEP) team's determination of a disability. It is important to remember that eligibility for specially designed instruction that is provided in special education programs is a formula, A+B=C. A stands for the determination of a condition of a disability. A is determined by the full individual evaluation that is not discriminatory. B stands for the determination of educational need for specially designed instruction. B is also determined by the full individual evaluation and confirmed by the IEP team. In this formula, you must have both A and B in order to be eligible for services through special education, the C. Without A or B, an individual can not reach C. The nondiscriminatory identification and evaluation principle were founded in the court decision of *Diana v. State Board of Education* (henceforth, Diana). This case originated when the state of California engaged in the evaluation practice of using a single English only test in the identification of native Spanish speaking students with the disability of ID. This class action lawsuit was brought on behalf of 9 students whose native language was Spanish and were placed in an Educably Mentally Retarded (EMR) class, based solely on a single IQ test. The case was settled by consent decree and the holding found the California State Board of Education violated the Equal Protection Clause of the 14th amendment of the U.S. Constitution. The holding of Diana was integrated into the nondiscriminatory identification and evaluation principle in special education legislation. (Yell, 2019).

Free Appropriate Public Education

The elements of FAPE are critical for the protection of individuals with disabilities from discrimination. The most litigated area of education is special education, and the most litigated area in special education is FAPE. FAPE is defined as: Free: A free education, an education that is at no cost to a student with a disability. Appropriate: An appropriate education must be individualized based on the unique needs of the individual. Additionally, the education must be reasonably calculated to ensure the individual makes progress in mastering the individualized education program (IEP) goals. Public Education: The education must be tied to the state educational standards in schools that are funded by federal and state educational funds (Section 504 Regulations, 34 C.F.R. §§ 100.6-100.10) (U. S. department of Education 2010). Over the history of disability legislation, 2 court case decisions stand out that refined the elements of FAPE. The first case was *Rowley v. Hendrick Hudson School district* (henceforth, Rowley). In Rowley, the parents of a student with a disability filed a due process request seeking relief that the school district maximize their child's education. The court ruled that Congress never intended for schools to maximize the educational progress of a student with disabilities, but to make available an appropriate educational program. According to Osborne and Russo (2021) the Rowley decision defined

"an appropriate education is one that is formulated in accordance with all of the IDEA's procedures and is 'sufficient to confer some educational benefit' on children with disabilities. The IDEA requires an educational program reasonably calculated to enable children to make progress that is appropriate in light of their circumstances" (p. 20).

In 2017 the Supreme Court issued its ruling in the *Endrew F. v. Douglas County School District* (*Endrew*). The court was unanimous in its decision. The Court set a new, more rigorous standard that was "markedly more demanding than the 'merely more than de minimis' test applied by the tenth circuit." (*Endrew*, 2017, p.14).

Least Restrictive Environment (LRE)

The IDEA requires to the maximum extent appropriate, children with disabilities, including children in public or private institutions or other care facilities, are educated with children who are not disabled, and that special classes, separate schooling, or other removal of children with disabilities from the regular educational environment occurs only when the nature or severity of the disability is such that education in regular classes with the use of supplementary aids and services cannot be achieved satisfactorily. (IDEA, 20 U.S.C.§ 1412(a)(5)(A)).

In 1989, the Fifth Circuit Court of Appeals rendered a decision in the *Daniel R.R. v. State Board of Education* (*Daniel*) that has become the seminal decision regarding the LRE mandate (Yell, 2019). According to Yell (2019), the court developed a two-part test for districts to determine compliance with the LRE requirements.

1. Can education in the general education classroom with supplementary aids and services be achieved satisfactorily?
2. If a student is placed in a more restrictive setting, is the student integrated to the maximum extent appropriate? (p. 11).

Due Process Safeguards

According to the IDEA, schools must make sure that due process safeguards protect children's rights with disabilities and their parents. These include: The districts must provide notice and obtain parental consent for initial assessment and all re-evaluations. Additionally, they must obtain parental consent for the initial placement of the student. School districts must maintain the confidentiality of all records on a child with a disability. Additionally, school districts must make the records available to the parents. Parents have the right to pursue an independent educational evaluation at public expense if the parents and district disagree on the identification, evaluation, placement, or provision of FAPE and related services for the child. The parents may request a due process hearing. States are also required to offer the opportunity to resolve the conflict through mediation before holding the hearing (Getty & Summey, 2004; Heward, 2017; Osborne & Russo, 2021; Yell, 2019). It is important to remember, "due process hearings are a last resort to resolve conflicts or problems between school districts and parents" (Getty & Summey, 2004, p. 40).

Parent and Student Participation and Shared Decision-Making

Parent participation in the development of their child's IEP and full participation in the IEP meeting is the most basic requirement of the IDEA (Yell, 2019). Districts must follow specific procedures to ensure that parents are fully involved. The procedures include

- Notifying parents of the meeting early enough to ensure participation.
- Scheduling the meeting at a mutually agreeable time and place.
- Including the following content in the notice:
    - The purpose, time, and location of the meeting and who will be in attendance.
    - For a transition IEP, an invitation to the student and the name of the additional agencies invited.
- Using alternative methods if neither parent can attend (e.g., individual or conference telephone calls).
- Giving the parents a copy of the IEP if they request it (Yell, 2019, p. 229).

According to Yell (2019), the U.S. Court of Appeals in their decision in *Amanda J. v. Clark County School District* "held that interference with parental participation in the IEP development undermines the very essence of the IDEA" (p. 228).

The Pandemic, the Laws and Services for Students With Disabilities

The reader may be asking themselves, "Why is the history of disability law important to me?" It is all about context. It is human nature for individuals to narrow their focus when they experience a crisis. This is also true for human systems such as the education system. Turner and Klein (2021, June 16), in a National Public Radio (NPR) report on the Morning Edition, interviewed several families who had students with significant disabilities regarding their experiences during the pandemic. Turner and Klein shared a quote from one of the attorneys for the parents they interviewed during the pandemic. "School systems are bureaucracies. They are not set up to be flexible" (para. 21). The closing of face-to-face instruction and moving to virtual instruction created significant issues for all areas of education. Turner and Klein stated that many of the special services, instructional and related services were "feebly" reproduced online. The parents told Turner and Klein that their children were regressing with their academic, social, and physical skills without direct, face-to-face services. The parents of these children strongly felt their children were denied a FAPE. Consequently, they filed due process seeking compensatory education to obtain the additional services needed to help regain the skills lost during virtual learning and gain the skills they intended to achieve during the school year.

RACE, CULTURE AND LANGUAGE ISSUES

Over and Underrepresentation/Demographic Data

Minority over and underrepresentation in special education is problematic. The 2018 racial and ethnic makeup of the U.S. population of students ages 5-17: White 51%, Hispanic 25%, Black 14%, Asian 5%, Pacific Islander <1%, American Indian 1%, and two or more races 4%. The disability categories of ED, ID, and SLD have one or more racial groups overrepresented. In ED 24% of black students served in special education are identified in this category. In ID 26% of black students served in special education are identified in this category. In SLD 31% of Hispanic students and 19% of Black students served in special education are identified in this category. Analysis of this information suggests that the education system has not been diligent in ensuring appropriate racial representation. Howard (2019) examined the relationship between race, class, expulsions, and placement in special education. Howard found "African American, Latinx, and Native American students are significantly higher in the term 'judgement categories,' or disciplinary areas that are determined less on clinical judgement and more on individuals' subjective interpretations" (p. 21). The diagnostic categories of ID, SLD, and ED being suspiciously higher for students of color and individuals identified with Limited English proficiency, requires a closer look. According to Howard (2019) African American and Latinx males are more likely to be suspended or expelled from school. This information aligns with the information that African American males are identified under the disability category of ED more often than any other group (Barshay, 2018). Howard (2019) stated that these issues of over or under-representation is due to the lack of appropriate training in race and culture issues.

According to Husser et al. (2020), 28% of students with disabilities live in poverty. This is compared approximately 16% among children with disabilities that live in poverty. Additionally, approximately 20% of Children with disabilities are homeless. Husser et al. (2020) discussed that 6% of students in public school did not have the ability to access the internet. 43% of those said that it was too expensive. The majority of these families tend to be homeless. The issue of accessibility to education is a significant due to the switch to virtual learning from COVID 19.

Students With Disabilities and Socioeconomic Status

According to the American Federation of Teachers (2014), Children who live in poverty by 4 years old, are behind 18 months in cognitive development compared to their middle or upper socioeconomic age group. By the time they enter third grade, they tend to demonstrate a vocabulary that is one-third that of their peers who come from middle-income homes. These children tend to be more at risk for developmental delays and learning disabilities. These families are less likely to have the equipment and connectivity to access the education environment. According to UNICEF (2020) more than two-thirds of girls and boys ages 3 to 17 years lack internet access at home. Without internet access, students cannot access learning via virtual platforms. This digital divide contributes to and increased risk developmental delays and learning disabilities. High quality instruction also suffered because not all teachers had access to home internet and devices to teach their students virtually. Approximately 300,000 to 400,000 public school teachers (8%) lacked access to adequate connectivity, and 100,000 (3%) lacked devices, limiting the distance learning potential for entire classrooms of students (Chandra, et al., 2020).

LEGAL LANDMINES

Zero Reject

PARC and *Mills* rulings made excluding children from public school based on the student having a disability unconstitutional based 14th amendment of the U.S. Constitution. However, there continue to be issues where students are being excluded. This often includes districts creating procedures that are inadvertently discriminatory. These may include the procedures established in a Multi-Tiered Systems of Supports (MTSS) commonly known as the Response to Intervention (RTI) model. The procedures could be where a district create arbitrary timelines where children are not receiving tiered support. The issue of zero reject is complex, the landmine here is, if a district provides online instruction without providing equipment or connectivity to families without the means to obtain access to instruction, they are not providing educational opportunity? This seems very similar to school districts refusing to allow students to enroll in public school if they had a disability.

Free Appropriate Public Education (FAPE)

The elements of FAPE are critical for the protection of individuals with disabilities from discrimination. The most litigated area of education is special education, and the most litigated area in special education is FAPE. As mentioned previously, the IEP must be reasonably calculated to ensure the individual makes progress in mastering their goals. The teacher is the responsible professional that makes sure the student is receiving an appropriate education. For example, the court case of Doe v. Withers. In 1992, the parents of a student with a learning disability filed a complaint that their son's history teacher refused to follow the accommodation of oral testing even though the superintendent, the principal and the special education director all told Withers he must apply the accommodations as agreed upon for Doe. Wither's refusal to apply the accommodations resulted in Doe failing the class, making him ineligible for athletics. Mr. Withers was a state legislator for West Virginia and took leave to complete his legislative responsibilities. While Withers was on leave, A substitute was placed in his class. The substitute followed the accommodations during Withers absence and the student was successful. The parents sued Withers as well as the superintendent of Taylor County School Corporation, the principal of Grafton High School where the student attended, and Taylor County Board of Education. The plaintiffs argued that D.D. was not afforded the right to a Free and Appropriate Public Education promised to all students by IDEA. Doe Vs. Withers is a landmark case in special education law because it is the first special education case to be tried in front of a jury and the first to seek monetary damages. The jury decided that the school corporation, the superintendent and the principal were not responsible since there was documentation that Withers was instructed to follow the accommodations. The jury found that since Withers acted alone and refused to follow directions from his administration, he did not provide FAPE. He was ordered to personally pay $15,000 in damages. It is important to remember, that it is the teacher's responsibility to provide FAPE even in a virtual environment (Wright, P. 2021). Turner and Klein (2021, June 16) found that parents of students with disabilities strongly believe that "schools are legally required to do whatever it takes to get their special education students to where they would have been had there been no pandemic at all" (para. 26).

Least Restrictive Environment

As mentioned previously, LRE is a significant issue where children with disabilities are ensured a comparable educational environment to their nondisabled peers to the maximum extent appropriate. Districts have minimized the more obvious discriminatory issues associated with LRE, however, district personnel often do not realize that they are inadvertently engaging in LRE issues. For example, Students being served in a special life skills program where they are learning a job skill and they have the related service of transportation. The campus administrator makes a unilateral decision to have the students board their bus 30 minutes before all the other students on campus. What this does is reduce the instructional day of all the students in the class by 30 minutes. Which actually changes the placement of all the students without holding an IEP meeting. This ultimately denies the students of an appropriate education as determined by the LRE provision of IDEA. Teaching and learning life skills in a virtual educational environment can be challenging since most skills center around personal interactions. How does a teacher create strategies to interact with others face-to-face? LRE is more than providing a physically comparable education experience, the unique educational needs of the students must be taken into consideration in order for the student to engage in and make progress in their enrolled grade level general education curriculum.

Appropriate Evaluation

As mentioned previously, the evaluation is the foundation of the individual education program (IEP) team's determination of a disability. Additionally, it is important to understand that the evaluation must not discriminate on the basis of language, culture, and students background (Vaughn, S. R., Bos, C. S. & Schumm, J. S., 2018). The potential landmines include making sure that the evaluation is in the native language of the child. Most of the testing instruments are in English and there are a few in Spanish. The common practice is to use an evaluator trained in the use of the tests in the child's language. Districts often will contract these assessments with professionals that are trained. A potential problem is finding evaluators trained that speak uncommon languages. Districts may have to rely on training interpreters to help in the evaluation process. It is important that the interpreters are properly trained so the tests reliability and validity is not impacted.

The move to a virtual assessment format is fraught with potential problems in this area. The main issue is that the assessment tools have been designed to be administered in an individual face-to-face setting. The test development companies have created procedures and training that helps the educational professional make the move to a virtual assessment environment.

DISABILITY CATEGORIES AND CHALLENGES TO VIRTUAL LEARNING ENVIRONMENTS

High Incidence Disabilities

Students identified most often, can be listed under the label often referred to as high incidence disabilities. These disabilities include: Specific Learning Disability (SLD), Speech or Language Impairment (SLI), Other Health Impairment (OHI), and Autism (AU). According to the National Center for Education Statistics this grouping of disabilities makes up 78% of the students identified under IDEA. Additionally, the students identified with these disabilities most often are served 80 percent or more of the time in a general education classroom (Husser et al, 2020).

Challenges

According to Smith, Tyler, and Skow (2018), individuals with high incidence disabilities share several common problematic characteristics that negatively affect their ability to acquire the necessary life skills related to decision-making, social competency, and planning and carrying out a plan. These characteristics include holding negative attributions, poor social competence, non-strategic approach to learning/problem solving, poor generalization strategies, and poor information processing. Each of these characteristics can have a significant impact on a child's ability to behave responsibly. Individuals with disabilities who demonstrate one or more of these of these characteristics tend to depend on outside sources (parents, teachers, etc.) to make life decisions (Doll, Sands, Wehmeyer, & Palmer, 1996). This is referred to as external locus of control. Compare that to individuals that are organized and motivated individually. This is referred as internal locus of control. In a virtual setting, individuals with external locus of control will have much more difficulty staying on task and completing their work. Teachers often use proximity control to help the student stay on task and complete their work. For example a student who depends on an external locus of control (the teacher or a paraprofessional) to move from one item to the next will have difficulty completing their work if the teacher is not standing next to them and pointing to the next item and saying "now do the next one".

Low Incidence Disabilities

Students identified with the next several disability categories are often grouped as low incidence disabilities. These include emotional disturbance (ED), intellectual disability (ID), Orthopedic impairments (OI), Auditory impairments (AI), and visual impairments (VI). According to the National Center for Education Statistics this grouping of disabilities makes up 4% of the students identified under IDEA. The students identified with these disabilities most often are served in segregated learning environments (Husseret al, 2020).

Challenges

According to Smith, Tyler, and Skow (2018), this group of disability categories share many of the same common problematic characteristics as the High Incidence disabilities group. Additionally, the ED disability condition provides some unique challenges due to the behaviors often exhibited. Students identified with ED tend to engage in either limited or excessive behaviors. Often teachers need to work with behavior specialists or school psychologist in order to develop strategies for these students to stay engaged in the learning environment.

Students identified with ID often have difficulty with adaptive behaviors. Most state departments of education view adaptive behavior as the effectiveness and degrees to which the individual meets the standards of personal independence and social responsibilities.

Sparrow, Cicchetti and Saulnier (2016) adaptive behavior is a construct that is based on age. "for most individuals, the repertoire of adaptive behaviors becomes larger and more complex with increasing age (p. 10)." For example, young children the adaptive behaviors would focus on self-care (dressing, toileting, etc.) and relationships with peers while Older students and adults would be using community resources, working, managing money, etc. Working with student identified with ID in a virtual environment can be extremely difficult. Mainly it depends on the individual's ability to work independently. Even though the teacher's presence is there in the virtual environment, historically these individuals had either teachers or support personnel that were directly working in close proximity with them. Prompting and direct instruction is extremely difficult to conduct in a virtual environment.

Students with OI, AI, and VI often need specialized technology in order to access instruction and curriculum. Some of this technology can be very expensive and require specialized training. It is important to remember that the foundation of all services offered in special education requires an evaluation. Assistive technology (AT) is a related service that is a service or support that enables the child with a disability to make progress in the general education curriculum. In order to determine what AT a student needs, the IEP team needs to recommend and carry out a full Assistive Technology Evaluation (Husser et al, 2020). Once the assessment is completed, the IEP team recommends the AT and services for the child, including training for the teacher and support staff in implementing the AT.

Very Low Incidence Disabilities

Students identified with Traumatic Brain Injury (TBI), Multiple Disabilities (MD), Deaf-Blind (DB) disabilities offer teachers very unique challenges. Not only do these disabilities occur at very low incidence, individuals identified with these disabilities tend to have significantly severe and pervasive educational needs.

According to Centers for Disease Control and Prevention (2015) a TBI is an injury to the head that is caused by a blow or jolt to the head or a penetrating head wound that "disrupts the normal function of the brain" (p.2). Additionally, individuals with a TBI, often remember that they were able to engage the learning environment differently than what they are currently. This can create a lot of frustration and anxiety for the student.

According to the IDEA, the definition of MD:

34 Code of Federal Regulations § 300.8 Child with a disability.

300.8(c)(7)

(c) Definitions of disability terms. The terms used in this definition of a child with a disability are defined as follows:

(7) Multiple disabilities means concomitant impairments (such as intellectual disability-blindness or intellectual disability-orthopedic impairment), the combination of which causes such severe educational needs that they cannot be accommodated in special education programs solely for one of the impairments. Multiple disabilities do not include deaf-blindness (U.S. Department of Education, 2006).

According to the IDEA, the definition of DB:

34 Code of Federal Regulations § 300.8 Child with a disability.

300.8(c)(2)

(c) Definitions of disability terms. The terms used in this definition of a child with a disability are defined as follows:

(2) Deaf-blindness means concomitant hearing and visual impairments, the combination of which causes such severe communication and other developmental and educational needs that they cannot be accommodated in special education programs solely for children with deafness or children with blindness (U.S. Department of Education, 2006).

Challenges

In traditional settings with students identified with TBI, MD, and DB, teachers need to identify ways to modify materials and technology in order for students to access the curriculum and to demonstrate what they have learned. This becomes even more challenging in virtual environments. The integration of various assistive technologies (AT) can be extremely helpful in these cases. However, the virtual environment limits the teacher's options of using more than two modalities, visual or auditory. Many of these students need the integration of touch and movement in order to acquire the material. Additionally, many students with these highly complex disabilities have Occupational Therapies, Physical Therapies, or other related services that historically are provided in a face-to-face, hands on setting.

EVIDENCED-BASED/PEER-REVIEWED INTERVENTIONS

Prior to examining Evidenced-Based/Peer-Reviewed Interventions, it is important to remember that services and interventions are not based on disability specific needs, they are based the unique needs of each individual. With that being said, the following are Evidenced-Based/Peer-Reviewed Interventions that Teachers can select to meet the unique needs of their students.

High Leverage Practices for Students with Disabilities

High Leverage Practices is a set of 22 practices that are critical that every special education teacher master and apply to their teaching practice. These practices are used frequently and if appropriately applied, are shown to improve student outcomes (McLesky, J. et al., 2017 (The *High-Leverage Practices for Students with Disabilities* webpage provides teachers with all the information necessary to implement these

Serving students with disabilities is a collaborative effort from a team. This team includes the student's general education teacher, their special education teacher, related service providers, paraprofessionals, and their parents. In face-to-face environments, collaboration is an active, hands on process. This is especially true between the teacher and their instructional paraprofessional who may be working directly with the student with disabilities. The role of a paraprofessional is to provide additional hands on support with the student with a disability. The paraprofessional's job is to provide additional support of the student during the guided practice section of the lesson cycle. Moving to a virtual environment will require training the child's parent to serve the role of the paraprofessional. They will add the hands-on support as needed.

The HLP area of assessment is where the teacher identifies where the student is currently functioning and how much they have progressed in meeting their educational goals. In virtual learning environments there are many tools available such as online polls, virtual games and more that can be used to evaluate progress.

The HLP area of Social/Emotional/Behavioral are a set of practices that are used to create a supportive, organized and respectful learning environment. The specific practices in this area include: HLP 7: Establish a Consistent, Organized, and Respectful Learning Environment; HLP 8: Provide Positive and Constructive Feedback to Guide students' Learning and Behavior; HLP 9: Teach Social Behaviors; & HLP 10: Conduct Functional Behavioral Assessments to Develop Individual Student Behavior Support Plans. These practices can easily be adapted to a virtual learning environment.

The HLP area of Instruction is a set of practices that are use a teacher's content knowledge, pedagogical knowledge, and student data in order to design and implement effective instruction. These practices will help the teacher be more strategic and flexible in designing their instruction. The specific practices in this area include: HLP 11: Identify and prioritize long- and short-term learning goals, HLP 12 Systematically design instruction toward a specific learning goal, HLP 13: Adapt curriculum tasks and materials for specific learning goals, HLP 14: Teach cognitive and metacognitive strategies to support learning and independence, HLP 15: Provide scaffolded supports, HLP 16: Use explicit instruction, HLP 17: Use flexible grouping, HLP 18: Use strategies to promote active student engagement, HLP 19: Use assistive and instructional technologies, HLP 20: Provide intensive instruction, HLP 21: Teach students to maintain and generalize new learning across settings, & HLP 22: Provide positive and constructive feedback to guide students' learning and behavior. CEC and the CEEDAR Center have developed recorded webinars and videos aimed at preparing current and preservice teachers to implement evidenced-based practices in order to help children with disabilities succeed. These webinars and videos also include specific strategies for virtual learning environments.

Table 1. High Leverage Practices (HLP) for students with disabilities

HLPs are a set of 22 practices that address the "most critical practices that every K-12 special education teacher should master" (McLeskey et al., 2017 p. 15). These HLPs are divided among 4 areas: collaboration, assessment, social/emotional/behavioral practice and Instruction. Each area is essential for a special education teacher to be effective.
Collaboration According to McLeskey et al. (2017) special education teachers must collaborate with parents, instructional and related service professionals, and paraprofessionals in order to ensure a student with a disability receives an appropriate education. The area of collaboration includes the following HLPs:
Assessment The foundation for all services in special education must be supported by the foundation of a full individual evaluation prior to the implementation of the IEP.
Social/Emotional/Behavioral Practices Special education teachers create educational environments that are consistent, organized and respectful in order to support student success.
Instruction Special education teachers must be strategic and flexible using content knowledge, pedagogical knowledge, and student data in order to design and implement effective instruction.
The High Leverage Practices in Special Education book is available as a free download from: https://ceedar.education.ufl.edu/wp-content/uploads/2017/07/CEC-HLP-Web.pdf
High Leverage Practices resources and videos are available: https://highleveragepractices.org/welcome-our-new-series-high-leverage-practices

Table 2. Online tools

The following websites provide information that will help teachers identify evidenced based instructional strategies in both face-to-face and virtual environments:
National Center on Intensive Interventions (NCII): http://www.intensiveintervention.org/
What Works Clearing House: https://ies.ed.gov/ncee/wwc/
Evidenced Based Intervention Network: http://ebimissouri.edu/
The IRIS Center https://iris.peabody.vanderbilt.edu/

Table 3. Supporting students with disabilities in Virtual Learning Environments

The following resources help support students with disabilities complete tasks and assignments and adjust to new routines in the transition to distance learning.
Pennsylvania Training and technical Assistanc Network(PaTTAN): PaTTAN COVID19 Resources: https://www.pattan.net/Home/Mini-Slideshow-With-Image/COVID-19-Resources-1
Educating All Learners Alliance (EALA): https://educatingalllearners.org/ EALA Resource Library: https://educatingalllearners.org/resource-library/ EALA Tech Tool Library: https://educatingalllearners.org/tech-tool-library/
Can do Kids!: http://www.ucando.org/kids.html
The Special Student Services Blog: https://pickettsmill.typepad.com/pritchard/home.html
FunBrain: https://www.funbrain.com/pre-k-and-k-playground
AAA Math: http://aaamath.com/
Storyline Online: https://storylineonline.net/
Into the Book: https://reading.ecb.org/index.html
Starfall: https://www.starfall.com/h/
The Exploratorium: https://www.exploratorium.edu/explore
Do2Learn: https://do2learn.com/
Reading Rockets: https://do2learn.com/
Adapted Mind: https://www.adaptedmind.com/index.php
Arcademic Skill Builders: https://www.arcademics.com/
Learning Ally: https://learningally.org/Browse-Audiobooks

Table 4. Organizations/associations Providing technical support and professional development

The Council for Exceptional Children (CEC)-the largest international professional organization dedicated to improving the educational success of children and youth with disabilities and/or gifts and talents: https://exceptionalchildren.org/
The Advocacy Institute is a non-profit, tax-exempt organization dedicated to the development of products, projects and services that work to improve the lives of people with disabilities: https://www.advocacyinstitute.org/
The Center for Learner Equity is a non-profit organization to focus on working with states, charter authorizers, special education, and charter school advocates to raise awareness and address barriers that prevent students with disabilities from accessing a igh-quality public education: https://www.centerforlearnerequity.org/
Council of Parent Attorneys and Advocates (COPAA)- mission is to protect and enforce the legal and civil rights of students with disabilities and their families: https://www.copaa.org/
Learning Disabilities Association of America (LDA)- LDA's mission is to create opportunities for success for all individuals affected by learning disabilities through support, education, and advocacy: https://ldaamerica.org/
National Association of School Psychologists (NASP)-As the world's largest organization of school psychologists, NASP works to advance effective practices to improve students' learning, behavior, and mental health. Our vision is that all children and youth thrive in school, at home, and throughout life: https://www.nasponline.org/
National Center for Learning Disabilities (NCLD)-improves the lives of the 1 in 5 with learning and attention issues by empowering parents and young adults, transforming schools, and advocating for equal rights and opportunities: https://www.ncld.org/
National Down Syndrome Congress is dedicated to an improved world for individuals with Down syndrome. We are the leading national resource of support and information for anyone touched by or seeking to learn about Down syndrome: https://www.ndsccenter.org/
National Disability Rights Network (NDRN)- promotes the network's capacity, ensures that Protection and Advocacy Systems and Client Assistance Programs remain strong and effectively by providing training and technical assistance, and advocates for laws protecting the civil and human rights of all people with disabilities: https://www.ndrn.org/
Resource: The organizations listed above collaborated together and created a resource titled: **9 Recommendations for Inclusive Learning Recovery for Students with Disabilities**. The Resource may be downloaded from the following: https://www.ncld.org/wp-content/uploads/2021/03/9-Recommendations-for-Inclusive-Learning-Recovery-for-Students-with-Disabilities-Final-1.pdf

VICTORIES/STORIES FROM THE FIELD

Working Virtually With Paraprofessionals/Teaching Assistants

Paraeducators and Specialized Instructional Support Personnel (SISP) are team players in the classroom, whether virtual or in-person. Technology training for support staff is vital when students are virtual. To help students, both paraeducators and SISPs can work with students using breakout rooms and Google Classroom.

For students having difficulty in a virtual platform, small break rooms are helpful when delivering instruction and to help with student social-emotional learning.

Integrating Speech Virtually (Can Also Be Used for Life Skills, Mild ID)

Delivering therapies in a virtual world is a challenge, especially when implementing speech therapy. Teachers and Speech-Language Pathologists (SLP) can work collaboratively to engage students in a virtual environment creatively. While a global pandemic changed much of how classrooms typically run, it should not stop the learning but instead, motivate teachers to adapt and find another way to make things work. One example is when special education teachers and the SLP bring students to their Cafe Kids kitchen. Cafe Kids helps students learn through cooking with lessons focused on math, science, reading, language skills, and speech skills. The special education teachers and SLP decided the pandemic was not going to stop their students from cooking. Through the "Cafe Kids Virtual Food Network," students cook from home as they watch the teachers cook in the cafe at school. Every week, the three classes come together virtually to listen to a story. Weekly recipes are based on each story. Students join at home via Google Meet and cook along with their families. Students who cannot cook from home join the virtual class and interact with others and watch them cook. As one student described it, "I like cooking at home with mom every week because I never cooked with her before. It's cool to watch my teacher's cook because I'm learning so much!"

Teaching Technology With SLD

When technology goes from an accessory to a necessity, students need instruction on using different platforms first before diving into academics. The key to successful virtual learning is to work with students by demonstrating and showing them how to access digital media first. A resource that helps with virtual learning is to create digital videos through Screencast. Keeping a digital library of videos to access technology allows students and families to learn how to navigate a virtual world.

Google Classroom for ED

Virtual learning can be stressful, especially for students who struggle with emotional challenges. Providing online tools in a Google classroom, such as calming strategies, links to soft music, or a virtual fish tank, can help students when they are not in the school. One teacher also has her students take a mindful minute throughout the school day. Using picture cues, students choose the face that matches how they are feeling.

They act upon their feelings. For example, if they are doing ok, they can choose to keep going. If they are angry or upset, they can choose a calming strategy. This strategy helps the teacher stay in check with how the students feel while teaching them how to self-regulate when they are upset.

ADHD

Keeping a student's focus and attention when learning virtually is often a challenge for many educators. Teachers need to think creatively and out of the box when maintaining student's attention. Fun and engaging songs, sound bites, and frequent breaks often help students face long periods of time in front of a screen. Letting students have a fidget tool is also helpful.

Parent Participation

Virtual environments have changed the way we communicate and interact with parents. Traditionally, parents would need to take off work to attend IEP meetings or arrange for childcare to participate in parent-teacher conferences or back-to-school nights. Using virtual technology, parents can join in the meetings without leaving home or the office. Many teachers have seen an increase in parent participation. Additionally, communication tools such as REMIND, Google Classroom, Class DoJo, and Parent Square allow for accessible and open communication throughout the school day.

CONCLUSION

The historical movement from face-to-face to online learning environments in 2020 due to the COVID pandemic was unprecedented. When major shifts occur in professions such as education, returning to a "normal" environment will not necessarily occur. It seems to be the consensus among teachers, administrators and educational researchers, that online learning will be here to stay in one form or another. This significant shift provided major challenges for all educational professionals. This chapter addressed the challenges the special education field of professionals experienced and how these challenges were surmounted. Additionally, this chapter provided educational professionals evidenced based/peer reviewed instructional strategies/interventions that would help them provide opportunities for success with their students with special needs.

REFERENCES

American Federation of Teachers. (2014). *When poverty comes to school.* Retrieved from https://www.aft.org/periodical/psrp-reporter/winter-2014/when-poverty-comes-school#:~:text=According%20to%20the%20National%20Center,them%20to%20school%20every%20day

Barshay, J. (2018). *The debate over students with disabilities, suspensions and race.* The Hechinger Report Retrieved from https://hechingerreport.org/the-debate-over-students-with-learning-disabilities-suspensions-and-race/

Blat, B., & Kaplan, F. (1974). *Christmas in purgatory: A photographic essay on mental retardation.* Human Policy Press. https://www.canonsociaalwerk.eu/1966_Kerstmis/Xmas-Purgatory.pdf

Brown v. Board of Education, 347 U. S. Centers for Disease Control and Prevention (2015). *Report to Congress on Traumatic Brain Injury in the United States Epidemiology and Rehabilitation.* National Center for Injury Prevention and Control; Division of Unintentional Injury Prevention.

Chandra, S., Chang, A., Day, L., Fazlullah, A., Liu, J., McBride, L., & Weiss, D. (2020). *Closing the K–12 digital divide in the age of distance learning.* Common Sense and Boston Consulting Group. Retrieved on 5/2/2021 from: https://exceptionalchildren.org/sites/default/files/2020-07/AllBriefs_2020.pdf

Doll, E., Sands, D. J., Wehmeyer, M. L., & Palmer, S. (1996). Promoting the development and acquisition of self-determined behavior. In D. J. Sands & M. L. Wehmeyer (Eds.), *Self-determination across the lifespan: Independence and choice for people with disabilities* (pp. 63–88). Brookes.

Getty, L. A., & Summy, S. E. (2004). The course of due process. *Teaching Exceptional Children, 36*(3), 40–43. doi:10.1177/004005990403600305

Heward, W. L. (2017). *Exceptional children: An introduction to special education* (11th ed.). Pearson.

Howard, T. C. (2019). *Why race and culture matter in schools.: Closing the achievement gap in America's classrooms* (2nd ed.). Teachers College Press.

Hussar, B., Zhang, J., Hein, S., Wang, K., Roberts, A., Cui, J., Smith, M., Bullock Mann, F., Barmer, A., & Dilig, R. (2020). *The Condition of Education 2020* (NCES 2020-144). U.S. Department of Education. Washington, DC: National Center for Education Statistics. https://nces.ed.gov/pubsearch/pubsinfo. asp?pubid=2020144

Katsiyannis, A., Yell, M. L., & Bradley, R. (2001). Reflections on the 25th anniversary of the Individuals with Disabilities Education Act. *Remedial and Special Education*, *22*, 324–334.

McLesky, J., Barringer, M. D., Billingsley, B., Brownell, M., Jackson, D., Kennedy, M., Lewis, T., Maheady, L., Rodriguez, J., Scheeler, M. C., Winn, J., & Ziegler, D. (2017). *High leverage practices in special education.* Council for Exceptional Children & CEEDAR Center. https://ceedar.education.ufl.edu/wp-content/uploads/2017/07/CEC-HLP-Web.pdf

Osborne, A. G., & Russo, C. J. (2021). *Special education and the law: A guide for practitioners.* Corwin.

Turner, C., & Klein, R. (2021, June 16). *After months of special education turmoil, families say schools owe them.* NPR. https://www.npr.org/2021/06/16/994587239/after-months-of-special-education-turmoil-families-say-schools-owe-them

U. S. Department of Education. (2006). Assistance to states for the education of children with disabilities and preschool grants for children with disabilities; final regulations. *Federal Register,* 34. CFR parts 300 and 301.

United Nations Children's Fund and International Telecommunication Union. (2020). *How many children and young people have internet access at home? Estimating digital connectivity during the COVID-19 pandemic.* UNICEF.

U.S. Department of Education. (2010). Office for Civil Rights, Free Appropriate Public Education for Students With Disabilities: Requirements Under Section 504 of the Rehabilitation Act of 1973. Washington, DC: Author.

Vaughn, S. R., Bos, C. S., & Schumm, J. S. (2018). *Teaching Students Who are at Risk in the General Education Classroom.* Pearson.

Weber, M. C. (2008). *Special education law and litigation treatise* (3rd ed.). LRP Publications.

Winzer, M. A. (1993). *History of special education from isolation to integration.* Gallaudwt University Press.

Wright, P. (2015). *In the circuit court of Taylor County, West Virginia.* Retrieved on 4/2/2021 from: https://www.wrightslaw.com/law/caselaw/case_Doe_Withers_Complaint.html

Yell, M. (2019). *The law and special education.* Pearson.

Chapter 14
Envisioning the Rural Virtual College Classroom:
Some Proposals for Equity and Inclusivity

Pamela B. June
Ohio University, USA

ABSTRACT

This chapter addresses intersecting challenges faced by rural college students as online learning becomes more widespread. The chapter begins by discussing the ways in which geography—specifically access to campuses and broadband—can impact rural student inclusivity. It then discusses the multifaceted challenges of rural students, whose disproportionate rates of class disparity and mental health issues can amount to personal crises during the semester. Because the boundaries between personal and academic life are less clear when taking classes from home, students may find that online learning is more difficult than traditional face-to-face classes. Therefore, the chapter offers some suggestions for professors related to simplicity and transparency. It then describes creative, flexible, and empathetic ways of approaching course construction in synchronous online courses. Finally, it offers broader suggestions for decision makers in order to maximize equity and inclusivity moving forward, as synchronous online learning becomes more normalized.

INTRODUCTION

Early in 2020, a vast swath of people found themselves scrambling to balance work and home life, as many aspects of life moved online. Schools and workplaces moved into people's homes by shifting online. Meetings and classes, professional and social

DOI: 10.4018/978-1-7998-8028-8.ch014

lives—all went virtual. Practically every person's life changed in some manner, and each demographic met general and particular challenges. Of the countless groups who have been severely affected by the global pandemic, undergraduate students have faced a number of unique and intersecting challenges. At the brink of young adulthood, eager to enter into the so-called "real world," leaning into new academic experiences and new personal connections—college students have seen unprecedented changes. Consider the youngest of college students. These students typically find themselves working to understand the basics of navigating college—how to register for courses, how to meet with an advisor, how to apply for financial aid. But now, those students have had to learn how to do so in a virtual world.

These experiences are widespread among undergraduates in general, but more specifically, rural college students encounter additional challenges and opportunities. Data from the National Center for Education Statistics suggests that nearly a third of traditional college students are from rural areas (Weissman, 2020, p. 8), and just around five percent of four-year institutions are located in rural areas (McNamee et al., 2020, p. 13). Rural undergraduates are often place-bound and attend small, nonresidential, often open-enrollment campuses. Such students are frequently first generation, working class, dual enrollment, academically underprepared, or some combination of these factors. Anecdotally, these students are generally enthusiastic and eager learners. They choose small, local, or branch campuses for a number of reasons. First, such campuses tend to have lower tuition costs, and they lack expenses related to housing, meal plans, and other campus life. Moreover, many of these students are place-bound due to family and/or job obligations. Finally, many students tend to prefer small campuses due to the personal interactions; lower student-teacher ratios guarantee professional relationships with mentors, personalized attention from advisors, and a collective focus on degree completion. In other words, rural campuses are appealing and vital. Not surprisingly, non-rural students may also find rural campuses even more attractive now, since "crowded city settings or colleges that are on the other side of the country may be less appealing during the pandemic" (Currie, 2020, p. 34). The benefits of a rural college education are evident.

Unfortunately, students in such places lag in college enrollment and matriculation (Rosenboom & Blagg, 2018, as cited in Koricich et al., 2020, p. 61). And perhaps due to their unique challenges, rural students' graduation rates are around forty-two percent (McNamee et al., 2020, p. 12). Of course, such students also face additional issues of accessibility: The pandemic has enforced the need for high-speed internet, a utility often taken for granted at metropolitan, residential, or large main campuses. In other words, these students face unique challenges of equity and inclusivity. The good news is that synchronous (real-time) online learning can actually enhance accessibility, since many rural students live long distances from campuses. And furthermore, synchronous learning can provide students with a friendly, supportive, rigorous academic environment—a welcome bit of silver lining given the challenges these students face.

This chapter addresses these issues by exploring several questions. First, how does geography impact college students' experience, and what particular educational difficulties do rural students traditionally face? How have the pandemic and online learning affected those challenges? What personal struggles do students encounter while learning from home? How is education affected when students must balance their academic and personal lives? Moreover, how can professors rise to meet these challenges? How can the virtual classroom serve to ensure rural students are not further marginalized? In what ways can professors embrace an inclusive pedagogy? Finally, how can higher education, in a broader sense, use this opportunity to evolve and better meet these students' needs? This chapter uses these questions in order to make some proposals for rural higher education in a time of crisis. Specifically, the audience here is faculty who teach synchronous courses in rural higher education, but the proposals have wider significance for educators, students, and policy makers. Even post-pandemic, the future will likely hold more online classes—both synchronous and asynchronous—so these proposals will continue to hold some salience. Ultimately, this chapter argues that simplicity and transparency, along with pedagogical creativity, professional flexibility, and a healthy dose of empathy, can counteract the often impersonal landscape of online learning and the unique challenges faced by rural students. A creative, flexible, and empathetic virtual landscape benefits rural students immensely, increasing the likelihood of their well-being, their retention, and their academic success.

GEOGRAPHY AND INTERNET ACCESS

The most recent U.S. Census data (2010) shows that some sixty million Americans live in rural areas (Koricich et al., 2020, p. 59). Millions of these residents live long distances from a college campus (Weissman, 2020, p. 8), areas which have become known as "education deserts" due to the lack of higher education resources. Additionally, many rural campuses are nonresidential. Therefore, in a typical semester, rural undergraduate students commute some distance to campus in order to meet their professors face to face. At small campuses, such personal connection is vital to student success and retention. However, once students were forced to move fully online in 2020, many nonresidential students faced infrastructure challenges of broadband access and connection instability. A 2018 Urban Institute report found that "[a]pproximately 3 million Americans live more than 25 miles from a public university *and* lack access to high-speed internet" (Weissman, 2020, p. 8, emphasis added). Likewise, the FCC estimates that a quarter of rural citizens do not have broadband access (McNamee et al., 2020, p. 12), and, according to Pew Research Center (2019), "two-thirds of Americans in rural areas have broadband internet access at home, but that's 12 percentage points below Americans overall" (Weissman, 2020, p. 9). Because many students have siblings and parents who also learn or work from home, the competition for broadband can cause difficulty in streaming. Furthermore, the various tiers of broadband lead to further economic stratification. Thus, the shift to online learning can mean that even the strongest of students struggle due to logistics or economics. A power or internet outage or a lapse in connection leads to difficulty communicating with professors, challenges accessing materials, and hurdles to meeting deadlines. In other words, students who might thrive in face-to-face learning are driven into remote classrooms, where a variety of accessibility issues make learning, academic success, and retention more difficult.

CRISIS MANAGEMENT AND HOME LEARNING

In addition to logistical concerns of internet access, rural students face personal crises that are only intensified by the pandemic. Most of these students live with parents, guardians, or grandparents, and now that the students are taking classes from home, the boundary between home and academic life has diminished. Day-to-day personal struggles by necessity encroach on the classroom, and students are expected to transition seamlessly from their kitchen, dining room, bedroom, or basement into an academic mindset. At rural nonresidential campuses, students do not have the luxury of retreating to a quiet dorm room or apartment. When learning from home, they may not even be able to find solitude at the campus library or café. There is no downtime during a commute, no walk to the classroom, no mental or physical space between academia and home. Instead, when learning from home, such students must simultaneously balance matters of the private sphere with their academic endeavors.

Actual scenarios, in just one year of virtual learning, have included the following anecdotes. First, a number of students contracted COVID-19. Some needed to take time off, while some still opted to join class sessions or even participate and give presentations while ill. Several students joined class from a cell phone camera while driving. Many students joined class meetings while caring for children (siblings or their own children), and some took classes while regularly holding an infant. One young father said he was trying to balance learning from home with caring for two young children who were intermittently in and out of school due to pandemic hybrid scheduling. One young mother emailed to say that she was about to deliver her baby, but not to worry—she would take her laptop into the hospital room and join class from there. One student disclosed that two grandparents—both of whom had serious health concerns—had moved in with her and that she was trying to balance coursework with caring for them. Some students reported competing with siblings for home computers and broadband. Several students joined from the workplace (some complete with workplace uniform). Students taking dual enrollment (high school and college) courses discussed the difficulties of high schools meeting in person, or not meeting in person, and the students then having to find a place from which to stream their college courses. Some dual enrollment students talked about challenges related to quarantined high school teachers. Many students reported difficulty with their asynchronous classes—challenges ranging from confusion about assignments to trouble contacting the instructor. (Berry, 2018, discusses the challenges unique to asynchronous classes and offers helpful advice for students, professors, and institutions.) Funerals, sick pets, family members heading to rehab, COVID tests (and later, thankfully, COVID vaccines), doctor's visits; the list goes on. These events would typically keep a student out of the classroom for at least one day, and they may impact the student's success. However, the boundary would be clear, and a decision would have to be made: If the student must go to work, she/he/they simply could not attend class. Now, however, that line is less clear, and it may be possible to sneak in some class time from the restaurant break room. It may be possible to drive a car and listen to a lecture. It may even be possible to get those participation points while laid up in a hospital bed.

These personal crises are exacerbated by the fact that rural students disproportionately face intersectional difficulties related to socioeconomics and mental health. With "central Appalachia's poverty rate stubbornly remain[ing] nearly twice that of the U.S. more broadly" (Snell-Rood & Carpenter-Song, 2018, p. 79), it is no surprise to find higher rates of mental illness and substance abuse:

In this economic context, the high burden of mental health challenges for rural residents is hardly surprising. Epidemiological data on mental health in the rural U.S. shows, in comparison to urban areas, higher rates of suicide (Ivey-Stephenson et al., 2017), depression (Simmons et al., 2008), and opioid use-related deaths (Paulozzi et al., 2012). Yet most rural Americans receive less mental health treatment—in terms of amount and effectiveness—than urban Americans (Adams et al., 2006; Hauenstein et al., 2007). With mental health providers scarce in the rural U.S., most rural mental healthcare consists of pharmacotherapy delivered by primary care providers (Mohatt et al., 2005). (Snell-Rood & Carpenter-Song, 2018, p. 79)

Because mental health and geography so evidently intersect, and because mental healthcare is less accessible in rural communities, major challenges in personal lives are often tantamount to veritable crises.

Of course, some silver lining here is that demographics such as new mothers, who otherwise may be unable to enroll in college courses in a given semester, can now take classes from home. This accessibility is certainly a plus, but the takeaway here is about the pressure to balance personal and academic pursuits. A crisis can always affect the student's academic performance, but prior to the pandemic and virtual learning, the line between personal and academic was clear. Therefore, a demographic that already faces undue burdens ultimately finds that virtual learning may not, in fact, make life easier, despite the time saved by not commuting from remote locations to campus. Therefore, it is incumbent on faculty to find ways to welcome and accommodate students even—or perhaps especially—during times of difficulty.

PROPOSALS FOR EDUCATORS: CREATIVITY, FLEXIBILITY, AND EMPATHY

Considering these accessibility and personal/professional challenges, instructors of rural populations should approach virtual education with an eye toward easing the transition. There may be a perception that such approaches allow for the diminishing of academic rigor, but research suggests otherwise: As Cameron-Standerford et al. (2020) observe, "The perception of lack of quality in online learning persists even after countless studies substantiated findings that there is no significant difference in the learning outcomes between online and face to face courses" (p. 2). In fact, many students find online courses to be more demanding because of the need to balance working from home: "Importantly, online is not perceived as being easier than traditional face-to-face learning by [fully online] students. Most students note that online courses demand more student time per course, even though this class- and work-time scheduling is more controlled by the students in terms of when they do their work" (Berry, 2018, p. 44). Therefore, without fear of compromising their rigor, professors can safely revisit their teaching philosophies and practices to consider how best to approach a student body in crisis. Instructors can make the path smoother for both the students and themselves by exploring a greater degree of creativity, flexibility, and empathy.

First, educators should review the syllabus to determine whether each facet best suits the needs of the students and meets the course objectives. Then, as instructors make these decisions, they should turn their attention to other ways of simplifying. Course materials should be centralized in one location, preferably the institution's learning management system (LMS). Of course, lectures and meetings might take place on another platform, such as Zoom or Microsoft Teams, but even so, links to lectures, meetings, and materials should all be located in the LMS. Such centralization allows students quick and easy access to everything they need for the semester. The more places that a student is expected to navigate—this platform for lecture; that modality for handouts; this link for quizzes; that software for exams; and so forth—the more likely they are to become frustrated and confused. A streamlined LMS homepage, with clear links to course materials and lectures, evident syllabi and posted deadlines, and minimal clutter, can undoubtedly aid students in locating what they need to stay on track and succeed. Next, professors might reconsider assignments to "triage what is important to keep and what can be altered or eliminated" (Cameron-Standerford et al., 2020, p. 3). This is a wonderful opportunity for instructors to reassess course objectives and determine whether all the assignments effectively meet those objectives.

Providing a simplified structure with easy access to materials should not suggest that students need not attend class regularly. In fact, attendance and participation must be integral to the students' grades. Virtual environments can encourage spotty attendance; therefore, to continue student attendance and engagement, instructors can use similar strategies they employ in the traditional classroom. For example, daily roll call is vital. Likewise, students should be aware that, just as in a standard classroom, they should be prepared to answer questions and offer comments during class. A weekly "check-in" might be able to address general concerns that students have: How are classes going? What challenges are students facing? What important dates are coming up on the syllabus or the academic calendar? These general questions can generate surprising feedback. Furthermore, simplifying includes clear and selective communication. Even though many faculty believe that a great deal of correspondence is helpful, "There seems to be a threshold at which an instructor's heavy-handed or overwhelming amount of communication inhibits or discourages learner communication and participation" (Dennen et al., 2007, p. 69). Brief emails with bullet points, highlighted reminders, or bolded keywords might be more conducive than long-winded sets of instructions. Strategies such as roll call, participation grades, check-ins, and simplified correspondence not only encourage participation and attendance, but they also enable a sense of belonging. Students should feel that their presence and contributions are valued. This sense of belonging is positively correlated with retention and mental health (Gopalan & Brady, 2019, p. 136), particularly with respect to first-generation students (Gillen-O'Neel, 2021). By and large, rural students respond that these kinds of strategies set some of their online classes apart, making the students feel that their instructors care about their success.

As instructors work to streamline and simplify in a virtual environment, they should consider ways to approach traditional assignments creatively. In a synchronous environment, students are expected to perform many of the assignments typically given in the face-to-face classroom: discussion posts, class discussions, group work, presentations, and so forth. Instructors can modify these assignments creatively such that the assignments continue meet the course objectives. Rigor remains intact as students work to accomplish similar goals in a new environment. Moreover, students continue to experience many of the traditional classroom benefits.

Following is a very generalized way of creatively (re)conceptualizing some of these assignments:

- **Discussion posts**: Many traditional face-to-face courses have been successfully using LMS discussion posts for years. A fairly typical approach is to have students complete a reading independently, and then to respond to a question or prompt in an online forum outside of class. Whether in a small or large group, students complete the response, and they also review other student posts. Thus, they can reify their understandings or consider new interpretations. Students are typically required to respond to at least one other student, to ensure that the forum is interactive. The value of such an assignment can hardly be overstated. Students tend to keep on track with assignments when they know that a graded post is due. They must also critically engage with said assignment or reading in order to post a concrete and substantive response. Finally, they are able to see other posters' ideas (unlike in a journal-type assignment), which inevitably offer some diversity in opinion. Previous studies have shown the pedagogical value of such forums, and Saleh (2020) summarizes some of these studies and ultimately describes the online discussion forum as "one of the richest learning communities in our present day" (p. 215). Students are encouraged to a higher order of thinking by "reflecting on others' thoughts" and incurring "the risk of being challenged or challenging others' point of view" (Saleh, 2020, p. 215). Hence, this practical and beneficial tool, useful in the traditional classroom, can continue to function as an invaluable way for students to stay engaged and focused, not to mention to have actual personal interactions with other students—a facet of learning that may otherwise be lacking in online environments.
- **Class discussions**: Many college courses rely heavily on class discussion, for it is in the classroom that students encounter new viewpoints, new concepts, and often, new demographics. Rural students in particular tend to come from small communities with more homogenous populations and fewer intellectual, creative, and multicultural opportunities. Therefore, the classroom provides a productive space for these students to encounter and explore new views. In an online environment, this hugely vital experience should not be eliminated. Synchronous teaching allows professors to recreate classroom discussions in creative and equally-effective ways. Creative

modifications might mean allowing students to join in the chat window rather than through their microphones. Or they might mean expanding the in-class conversation to LMS discussion boards or emails after class. Creative modifications that enhance a sense of belonging might include welcoming and encouraging emoji reactions, hand-raising or applause icons, and GIFs. A creative approach plus a healthy sense of humor can make an online class both appealing and beneficial to students. Ultimately, instructors can pose exactly the same kinds of questions that they would in the face-to-face classroom, and they can easily keep a tally of student participation in order to encourage active discussions.

- **Group work**: Some professors rely on group work for collaborative experiences. As a general rule, group work should be reserved for projects in which the whole of the product is greater than the sum of its parts. Lotan (2003), for example, explains best practices with respect to purposeful group work projects. Her "five design features" are the following:
    - They are open-ended and require complex problem solving.
    - They provide students with multiple entry points to the task and multiple opportunities to show intellectual competence.
    - They deal with discipline-based, intellectually important content.
    - They require positive interdependence as well as individual accountability.
    - They include clear criteria for the evaluation of the group's product. (p. 72)

These five features can help determine whether a group work assignment deserves a place in the online classroom. If the content or project merits the transition to online learning, the instructor can find creative ways to adapt. For example, a little proactive prep work allows an instructor to create break-out groups or channels into which students can find their way. Instructions and objectives should be clear verbally and textually. Once students are dismissed to their small groups, they should be able to get right to work with a clear understanding of their goal. In every case, students should know how long they will be in groups and how to return to the larger class when they have completed the assignment. A standing group can be particularly useful, since students can get to know one another and develop professional relationships or even friendships, as often happens in classroom contexts. Messersmith (2015), for example, provides an example of a semester-long standing-group project with clear objectives and concrete products. Such working relationships are invaluable to students' sense of belonging and ultimately to their success and retention.

- **Presentations**: As with group work, instructors should revisit the rationale for presentations. If the instructor believes the objectives for student presentations—such as practicing oral communication skills, or demonstrating mastery of content by "teaching" it to the class—remain sound in the online environment, then some ways of creatively revising the assignment can help students succeed. Offering a presentation can, for many students, be far more nerve-wracking and challenging online than in person. The discomfort arises from unfamiliarity with the online platform and the fear of technical difficulties. What if internet connection is unstable? What if the power goes out? How does one toggle between the class roster and the chat window? Therefore, instructors can provide low-stress ways for students to navigate presentations. A choice of formats is one creative possibility. Is it necessary that the students appear on camera, or could they, instead, share a slideshow, graphic, document, or video? (Some institutions, by the way, have policies against "requiring" students' cameras to be turned on, so instructors should be aware of any such policies.) Similarly, must the students rely on their microphone, or is it acceptable for them to use the chat? Is there a backup plan if they lose power or internet? Could students reschedule if such issues arise, or is there perhaps an alternate assignment they could complete? These are some questions instructors should consider, particularly, again, because of the potentially unstable internet access of rural areas. That said, some strategic and creative planning early on can provide students with a low-stress version of an otherwise unnerving assignment. For instance, Reck (2019) offers some creative and reduced-stress alternatives for online presentations, specifically in larger-enrolled courses. Her suggestions include the possibility of recorded and asynchronous presentations, use of LMS discussions for question-and-answer sessions, and other useful ideas. The real-life applications of adapting such an assignment to the online world are evident. Even before the pandemic changed the face of higher education, many fields had already employed cost-saving measures by instituting virtual interviews and meetings. Therefore, students will see the greater applicability of these creative adaptations well beyond the classroom.

These are just a few suggestions and questions for instructors to consider, and certainly there are more facets to weigh. In any case, any assignment that moves to the online format should be modified such that course objectives are met and rigor remains. Instead of abandoning a requisite assignment, for instance, an instructor can modify the assignment such that students can reasonably complete it and, while gaining the experience, also earn the sense of accomplishment that comes with doing so in a wholly new environment. Instructors can emphasize the practical importance of such experiences alongside the value of meeting the course objectives. If an educator is willing to approach the virtual environment with some extra creativity, the online classroom truly opens up and can, in fact, replicate a similar level of engagement and success.

Because the pandemic has affected students' life so drastically, a greater degree of instructor flexibility is necessary. Professors should modify the virtual environment to allow for the unexpected. In case of internet issues or the crisis scenarios described above, instructors must be prepared to move ahead with class. A few simple proposals here may help make class time smooth no matter what happens. First, policies on attendance, tardiness, and deadlines, for example, deserve revisiting. Many universities have already implemented more flexible pandemic attendance policies, and professors should be willing and able to modify for further flexibility as needed. Next, for assignment submissions, professors might consider a window for submission, say three to seven days, rather than a single-day deadline. For presentations, students might be asked to submit the presentation in some format sometime prior to the presentation date, perhaps twenty-four or forty-eight hours ahead of time. Students could even upload the presentation materials—a handout, a slide show, a group activity, for example—to the LMS before class so that the other students have access to the presentation. In case of an emergency or outage, both instructor and students can begin reviewing the materials. The student presenter can join when internet becomes stable again, or, failing such luck, the instructor can reschedule. A day or so of "to be announced" or "TBA" on the syllabus will accommodate any necessary changes to the schedule.

Instructors must also ensure accessibility for students who traditionally rely on accommodations. Students with disabilities may encounter further difficulties online, so professors must flexibly "provide an accessible education for students with disabilities who, until [the shift to online learning], had relied on the traditional model of live classroom instruction accompanied by accommodations, academic adjustments, and auxiliary aids carefully tailored to their individual circumstances" (Masinter, 2020, p. 3). Student service offices, of course, handle the managerial end as always, but instructors must be prepared to modify logistics. Luckily, the existing technology is already poised to handle some of these accommodations: recorded lectures, live captioning, adaptive release, size and visibility options, and electronically accessible content, to name a few. Innovative accommodations can ensure that students with disabilities succeed. These suggestions for flexibility are not exhaustive, of course, but are meant to spark some thinking on how faculty can evolve their approaches to help all students.

Just as instructors must be creative and flexible, students must be as well. They will need to adapt to the online environment, accept accountability for their work, and expect academic rigor. They will be more willing to make the adjustment if they feel that their instructors care about their success, and so an empathetic approach can help students thrive. A number of researchers base their pedagogies on Carol Gilligan's ethics of care (1982). For example, Deacon (2012) believes that "creating a context of care in a classroom creates a robust environment for student learning; it facilitates better dialogue between students and teachers and allows teachers to draw out individual students and help them achieve their potential" (p. 6). By maintaining high expectations and embracing creativity and flexibility, professors show a level of confidence in and respect for their students. The resulting classroom environment ultimately becomes a place of understanding and mutual respect, and, yes, compassion and empathy. Students working and learning from home are facing a barrage of problems already, and a disinterested instructor should not be one of them. Instructors and students are happier and more engaged when they simply give each other the benefit of the doubt. Instructors should, for instance, honor the truthfulness of absentee excuses. If a student says that she is joining class from the car, as she must take her mother to the hospital, there is no reason not to believe her. Additionally, professors should be willing and able to meet one-on-one. They should allow for student questions during class and should reply quickly to student emails. And perhaps most importantly, they should be willing to make clear the fact that they care about the quality of instruction and about their individual students. As Dennen et al. (2007) argue, "online learners desire an instructor who addresses them as individuals and offers supportive communications" (p. 68).

Students should still be held to the typical quality and quantity of coursework. And just as in the face-to-face classroom, instructors should reach out to students to communicate any concern: "Dear Student, I am concerned about whether you are meeting the course objectives, and I would like to help. I will ask you to do X, Y, or Z from here on out to demonstrate your efforts." A gentle nudge often produces surprising results, particularly in virtual environments that already feel somewhat impersonal. In their survey of students and faculty, VandenAvond et al. (2019) find that "While faculty demonstrate care within their courses, those same actions also contribute to the rigor of the course....[C]are and rigor are not mutually exclusive" (slide 15). The vision here is a classroom environment that facilitates learning by extending care, respect, and understanding. A little compassion goes a long way during an event as genuinely stressful as a global pandemic. And as the world prepares to move beyond the pandemic, creativity, flexibility, and empathy should undoubtedly continue to be the norm.

CONCLUDING SUGGESTIONS AND DISCUSSION

Because of the newness of this environment, it is vital that decision-makers—deans, faculty chairs, university presidents, course schedulers—not move too quickly on any significant changes. A final and very modest proposal offered by this chapter is that all involved in making decisions use caution. A drastic change in course delivery, course cap numbers, curricula, or major and general educational requirements can be to the detriment of students and faculty, and such changes can ultimately affect student success and retention. Therefore, it is vital that those elements that have historically drawn place-bound students to rural, nonresidential campuses—such as student-faculty ratios—stay in place. Now is not the time for increases in class size, revisions in curricula, or drastic changes in staffing. Doing so risks compromising the students' expectations and the quality of their educational experience. All institutions are concerned with retention and with quality education, and Koricich et al. (2020) find that "increasing institutional spending on instruction was the most-impactful factor in increasing a college's first-to-second-year retention rate, even more so than increases in endowment assets or state appropriations alone" (p. 65). In other words, students benefit when classrooms are seen as investments. Better instruction equals better retention, particularly at smaller, teaching-focused campuses. In a time when much of the globe faces uncertainty and change, students should be able to depend on a predictable, high-quality, educational environment.

Ultimately, rural students embody a combination of unique features that can make higher education less accessible: geographical situation within education deserts, absence of stable high-speed internet, the likelihood of being first-generation or employed, economic and mental-health disparities, and so on. Luckily, from an educational standpoint, these challenges are not insurmountable, and decision-makers and educators alike are poised to help these students succeed and even thrive. A reassessment of policies and pedagogies is necessary, with a gentle nod toward simplicity and transparency, along with pedagogical creativity, professional flexibility, and empathy. Many rural students enter the online higher education environment with reduced expectations. They know that online learning can often feel impersonal, so they may be unenthusiastic about it. Fortunately, decision-makers and professors can counteract these expectations with carefully measured modifications to the virtual landscape. Approaching standard course activities such as online discussion posts, class discussions, group work, and presentations using creativity, flexibility, and empathy can have wonderful results for rural populations, whose experiences and unique challenges are often overlooked by academia at large. Maintaining rigor and an ethics of care, instructors can help ensure success for their rural students.

Ultimately, the suggestions contained in this chapter may carry a broader significance to professors, teachers, schools, universities, and educational decision-makers in general. However, the unique situation of rural students generates particular challenges and opportunities that can be met using some of the modest ideas above. As vaccination rates rise and the nation looks towards resuming "normal" life, some of the lessons learned during the pandemic may remain part of daily life. Online learning is here to stay, so the suggestions in this chapter do not fade with the pandemic. As educators weigh a future that entails continued synchronous online education, they might also envision a more inclusive, equitable, personable, and enjoyable online environment. In the meantime, these basic proposals of creativity, flexibility, and empathy hopefully allow for more students to complete their coursework successfully.

REFERENCES

Berry, G. R. (2018). Learning from the learners: Student perception of the online classroom. *Quarterly Review of Distance Education*, *19*(3), 39–56.

Cameron-Standerford, A., Menard, K., Edge, C., Bergh, B., Shayter, A., Smith, K., & VandenAvond, L. (2020). The phenomenon of moving to online/distance delivery as a result of COVID-19: Exploring initial perceptions of higher education faculty at a rural midwestern university. *Frontiers in Education*, *5*, 1–11. doi:10.3389/feduc.2020.583881

Currie, J. (2020). Pandemic raises profile of rural colleges. *Business NH Magazine*, *37*(8), 34–36.

Deacon, A. (2012). Creating a context of care in the online classroom. *Journal of Faculty Development*, *26*, 5–12.

Dennen, V., Darabi, A., & Smith, L. (2007). Instructor-learner interaction in online courses: The relative perceived importance of particular instructor actions on performance and satisfaction. *Distance Education*, *28*(1), 65–79. doi:10.1080/01587910701305319

Gillen-O'Neel, C. (2021). Sense of belonging and student engagement: A daily study of first- and continuing-generation college students. *Research in Higher Education*, *62*(1), 45–71. doi:10.100711162-019-09570-y

Gilligan, C. (2003). In a different voice: Psychological theory and women's development. Harvard University Press.

Gopalan, M., & Brady, S. T. (2019). College students' sense of belonging: A national perspective. *Educational Researcher*, *49*(2), 134–137. doi:10.3102/0013189X19897622

Koricich, A., Tandberg, D., Bishop, B., & Weeden, D. (2020). Doing the same (or more) with less: The challenges regional public universities face in serving rural populations. *New Directions for Higher Education*, *2020*(190), 59–70. doi:10.1002/he.20367

Lotan, R. A. (2003). Group-worthy tasks. *Educational Leadership*, *60*(6), 72.

Masinter, M. R. (2020). Undue burdens in the time of coronavirus. *Disability Compliance for Higher Education*, *25*(11), 3. doi:10.1002/dhe.30853

McNamee, T. Y., Willis, J., Ganss, K. M., Ardoin, S., & Sansone, V. A. (2020). Don't forget about rural higher education students: Addressing digital inequities during COVID-19. *Diverse Issues in Higher Education*, *37*(7), 12–13.

Messersmith, A. S. (2015). Preparing students for 21st century teamwork: Effective collaboration in the online group communication course. *Communication Teacher*, *29*(4), 219–226. doi:10.1080/17404622.2015.1046188

Reck, R. M. (2019). Designing a student video presentation assignment: Considerations for minimizing stress. *2019 IEEE Frontiers in Education Conference (FIE), Frontiers in Education Conference (FIE), 2019 IEEE*, 1–5.

Saleh, A. (2020). Impact of online course discussion on education leadership students' reflections. *Journal of Education and Learning*, *14*(2), 212–216. doi:10.11591/edulearn.v14i2.14607

Snell-Rood, C., & Carpenter-Song, E. (2018). Depression in a depressed area: Deservingness, mental illness, and treatment in the contemporary rural U.S. *Social Science & Medicine*, *219*, 78–86. doi:10.1016/j.socscimed.2018.10.012 PMID:30391873

VandenAvond, L., Menard, K., Johnson, K., Cameron-Standerford, A., Bergh, B., & Edge, C. (2020). *Care and rigor in online courses: An analysis of faculty and student perspectives.* International Council for Distance Education.

Weissman, S. (2020). How will the coronavirus impact enrollment, access for rural students? *Diverse Issues in Higher Education*, *37*(6), 8–9.

Compilation of References

Aaronson, A. L., Backes, K., Agarwal, G., Goldstein, J. L., & Anzia, J. (2018). Mental Health During Residency Training: Assessing the Barriers to Seeking Care. *Academic Psychiatry, 42*(4), 469–472. doi:10.1007/s40596-017-0881-3

Abramo, J. M., & Reynolds, A. (2015). Pedagogical creativity as a framework for music teacher education. *Journal of Music Teacher Education, 25*(1), 37–51. doi:10.1177/1057083714543744

Acharya, B., Maru, D., Schwarz, R., Citrin, D., Tenpa, J., Hirachan, S., & Ekstrand, M. (2017). Partnerships in mental healthcare service delivery in low-resource settings: Developing an innovative network in rural Nepal. *Globalization and Health, 13*(2), 2. Advance online publication. doi:10.118612992-016-0226-0 PMID:28086925

Adcock, T. (2014). Technology Integration in American Indian Education: An Overview. *Journal of American Indian Education, 53*(2), 104–121. Retrieved April 28, 2021, from https://www.jstor.org/stable/43610478es

Aguirre, J., & Carrion, V. (2013). Integrated behavioral health services. *Clinical Pediatrics, 52*(12), 1178–1180. doi:10.1177/0009922812470744 PMID:23299835

Aisbett, D., Boyd, C., Francis, K., & Newnham, K. (2007). Understanding barriers to mental health service utilization for adolescents in rural Australia. *Rural and Remote Health, 7*(1), 624. PMID:17302509

Al-Balushi, S. M., & Al-Abdali, N. S. (2014). Using a moodle-based professional development program to train science teachers to teach for creativity and its effectiveness on their teaching practices. *Journal of Science Education and Technology, 24*(4), 461–475. doi:10.100710956-014-9530-8

Alegría, M., Alvarez, K., Ishikawa, R. Z., DiMarzio, K., & McPeck, S. (2016). Removing Obstacles To Eliminating Racial And Ethnic Disparities In Behavioral Health Care. *Health Affairs (Project Hope), 35*(6), 991–999. doi:10.1377/hlthaff.2016.0029 PMID:27269014

Alismail, H. A., & McGuire, P. (2015). 21st Century standards and curriculum: Current research and practice. *Journal of Education and Practice, 6*(6), 150–154. http://tarjomefa.com/wp-content/uploads/2017/06/6879-English-TarjomeFa.pdf

Almeda, N., García-Alonso, C., & Salvador-Carulla, L. (2021). Mental health planning at a very early stage of the COVID-19 crisis: A systematic review of online international strategies and recommendations. *BMC Psychiatry*, *21*(1), 1–15. doi:10.118612888-020-03015-y PMID:33451305

Amabile, T. M. (1996). *Creativity in context: Update to the social psychology of creativity.* Westview Press.

American Academy of Family Physicians. (2011). *Mental health care services by family physicians* [Position paper]. Retrieved from https://www.aafp.org/online/en/home/policy/policies/m/mentalhealthcareservices.html

American Association of Marriage and Family Therapists. (2015). *Code of Ethics.* https://www.aamft.org/Legal_Ethics/Code_of_Ethics.aspx

American Counseling Association. (2014). *Code of Ethics.* https://www.counseling.org/resources/aca-code-of-ethics.pdf

American Federation of Teachers. (2014). *When poverty comes to school.* Retrieved from https://www.aft.org/periodical/psrp-reporter/winter-2014/when-poverty-comes-school#:~:text=According%20to%20the%20National%20Center,them%20to%20school%20every%20day

American Indian Policy Institute. (n.d.). *Digital Divide.* Retrieved April 28, 2021, from https://aipi.asu.edu/content/digital-divide

American Psychological Association. (2017). *Clinical practice guideline for the treatment of post-traumatic stress disorder: What is psychotherapy.* https://www.apa.org/ptsd-guideline/patients-and-families/psychotherapy

Anagnostopoulos, D., Levine, T., Roselle, R., & Lombardi, A. (2017). Learning to redesign teacher education: A conceptual framework to support program change. *Teaching Education*, *29*(1), 61–80. doi:10.1080/10476210.2017.1349744

Anderson, L. M., & Stillman, J. A. (2013). Student teaching's contribution to preservice teacher development: A review of research focused on the preparation of teachers for urban and high-needs contexts. *Review of Educational Research*, *83*(1), 241–259. doi:10.3102/0034654312468619

Anderson, T. (2008). Towards a theory of online learning. In T. Anderson & F. Elloumi (Eds.), *The theory and practice of online learning* (pp. 45–74). Athabasca University Press.

Andrade, L. H., Alonso, J., Mneimneh, Z., Wells, J. E., Al-Hamzawi, A., Borges, G., Bromet, E., Bruffaerts, R., de Girolamo, G., de Graaf, R., Florescu, S., Gureje, O., Hinkov, H. R., Hu, C., Huang, Y., Hwang, I., Jin, R., Karam, E. G., Kovess-Masfety, V., ... Kessler, R. C. (2014). Barriers to mental health treatment: Results from the WHO World Mental Health surveys. *Psychological Medicine*, *44*(6), 1303–1317. doi:10.1017/S0033291713001943 PMID:23931656

Andrews, T., & Schwarz, G. (2002). Preparing students for the virtual organization: An evaluation of learning with virtual learning technologies. *Journal of Educational Technology & Society*, *5*(3), 54–65. Retrieved June 30, 2021, from http://www.jstor.org/stable/jeductechsoci.5.3.54

Arora, P. G., Godoy, L., & Hodgkinson, S. (2017). Serving the underserved: Cultural considerations in behavioral health integration in pediatric primary care. *Professional Psychology, Research and Practice*, *48*(3), 139–148. doi:10.1037/pro0000131

Ashadi, A., & Rice, S. (2016). High stakes testing and teacher access to professional opportunities: Lessons from Indonesia. *Journal of Education Policy*, *31*(6), 727–741. doi:10.1080/02680939.2016.1193901

Atay, A., & Ashlock, M. Z. (2018). *Millennial culture and communication pedagogies: Narratives from the classroom and higher education*. Lexington Books.

Ateriya, N., Saraf, A., Meshram, V., & Setia, P. (2018). Telemedicine and virtual CONSULTATION: The Indian perspective. *The National Medical Journal of India*, *31*(4), 215. doi:10.4103/0970-258X.258220 PMID:31134926

Atul Gawande. (n.d.). *The checklist manifesto*. http://atulgawande.com/book/the-checklist-manifesto/

Ayre, C., & Scally, A. J. (2014). Critical values for Lawshe's content validity ratio: Revisiting the original methods of calculation. *Measurement & Evaluation in Counseling & Development*, *47*(1), 79–86. doi:10.1177/0748175613513808

Ayudhya, U., & Smithson, J. (2016). Entitled or misunderstood? Towards the repositioning of the sense of entitlement concept in the generational difference debate. *Community Work & Family*, *19*(2), 213–226. doi:10.1080/13668803.2016.1134116

Baker, R. B. (2014). The American Medical Association and race. *AMA Journal of Ethics*, *16*(6), 479–488. doi:10.1001/virtualmentor.2014.16.6.mhst1-1406 PMID:24955603

Barbour, M. K. (2011). *The promise and the reality: Exploring virtual schooling in rural jurisdictions. Keynote paper from the 2010 Annual SPERA Conference*. University of the Sunshine Coast. Retrieved from https://core.ac.uk/download/pdf/231035933.pdf

Barbour, M. K., & Reeves, T. C. (2009). The reality of virtual schools: A review of the literature. *Computers & Education*, *52*(2), 402–416. doi:10.1016/j.compedu.2008.09.009

Barshay, J. (2018). *The debate over students with disabilities, suspensions and race*. The Hechinger Report Retrieved from https://hechingerreport.org/the-debate-over-students-with-learning-disabilities-suspensions-and-race/

Basilaia, G., & Kvavadze, D. (2020). Transition to online education in schools during a SARS-CoV-2 coronavirus (COVID-19) pandemic in Georgia. *Pedagogical Research*, *5*(4), 10. doi:10.29333/pr/7937

Bawden-Davis, J. (2020). How does working from HOME change tax deductions in 2020? In *Business Class: Trends and Insights*. American Express. https://www.americanexpress.com/en-us/business/trends-and-insights/articles/how-does-working-from-home-change-tax-deductions-in-2020/

Benavides-Vaello, S., Strode, A., & Sheeran, B. C. (2013). Using Technology in the Delivery of Mental Health and Substance Abuse Treatment in Rural Communities: *A Review. The Journal of Behavioral Health Services & Research*, *40*(1), 111–120. doi:10.100711414-012-9299-6 PMID:23093443

Bereczki, E. O., & Kárpáti, A. (2018). Teachers' beliefs about creativity and its nurture: A systematic review of the recent research literature. *Educational Research Review*, *23*, 25–56. doi:10.1016/j.edurev.2017.10.003

Berry, G. R. (2018). Learning from the learners: Student perception of the online classroom. *Quarterly Review of Distance Education*, *19*(3), 39–56.

Bhatt, J., & Bathija, P. (2018). Ensuring access to quality health care in vulnerable communities. *Academic Medicine*, *93*(9), 1271–1275. doi:10.1097/ACM.0000000000002254 PMID:29697433

Bischoff, R., Reisbig, J., Springer, A., Schultz, M., Robinson, P., & Olson, R. (2014). Succeeding in rural mental health practice: Being sensitive to culture by fitting in and collaborating. *Contemporary Family Therapy*, *36*(1), 1–16. doi:10.100710591-013-9287-x

Bissell, T. (2004). The Digital Divide Dilemma: Preserving Native American Culture While Increasing Access to Information Technology on the Reservations. *Journal of Law, Technology and Policy, 2004*(1), 129-150. Retrieved from http://illinoisjltp.com/journal/wp-content/uploads/2013/10/bissell.pdf

Blandford, A., Wesson, J., Amalberti, R., AlHazme, R., & Allwihan, R. (2020). Opportunities and challenges for telehealth within, and beyond, a pandemic. *The Lancet. Global Health*, *8*(11), e1364–e1365. Advance online publication. doi:10.1016/S2214-109X(20)30362-4 PMID:32791119

Blat, B., & Kaplan, F. (1974). *Christmas in purgatory: A photographic essay on mental retardation.* Human Policy Press. https://www.canonsociaalwerk.eu/1966_Kerstmis/Xmas-Purgatory.pdf

Blue Cross Blue Shield. (2020) *Telehealth benefits continue in 2021.* https://www.bcbsil.com/provider/education/2020/2020_12_22.html

Bradbury, B. L., Suarez-Sousa, X. P., Coquyt, M., Bockelmann, T. L., & Pahl, A. L. (2020). Teaching Under Crisis: Impact and Implications of the COVID-19 Pandemic on Education in Minnesota. *The Interactive Journal of Global Leadership and Learning*, *1*(2), 2.

Braveman, P., & Gruskin, S. (2003). Defining equity in health. *Journal of Epidemiology and Community Health*, *57*(4), 254–258. doi:10.1136/jech.57.4.254 PMID:12646539

Brenan, M. (2021, August 16). *COVID-19 and remote work: An update.* https://news.gallup.com/poll/321800/covid-remote-work-update.aspx

Breslau, J., Marshall, G. N., Pincus, H. A., & Brown, R. A. (2014). Are mental disorders more common in urban than rural areas of the United States? *Journal of Psychiatric Research*, *56*, 50–55. doi:10.1016/j.jpsychires.2014.05.004 PMID:24857610

Brezenski, T. F. (2014). Interest group formation and mental health care consumers in the United States: The effects of stigma, fiscal policy, and the Affordable Care Act. *Journal of Multidisciplinary Research, 6*(3), 39–52.

Bricker, D. (2020). *Majority (54%) of Canadians Say Physical Distancing has Left Them Feeling Lonely or Isolated.* Ipsos.

Brodersen, M. R., & Melluzzo, D. (2017). *Summary of research on online and blended learning programs that offer differentiated learning options. REL 2017-228.* Regional Educational Laboratory Central.

Brooks, S. K., Webster, R. K., Smith, L. E., Woodland, L., Wessely, S., Greenberg, N., & Rubin, G. J. (2020). The psychological impact of quarantine and how to reduce it: Rapid review of the evidence. *Lancet, 395*(10227), 912–920. doi:10.1016/S0140-6736(20)30460-8 PMID:32112714

Brown v. Board of Education, 347 U. S. Centers for Disease Control and Prevention (2015). *Report to Congress on Traumatic Brain Injury in the United States Epidemiology and Rehabilitation.* National Center for Injury Prevention and Control; Division of Unintentional Injury Prevention.

Brown, B., & Liedholm, C. (2002). Can web courses replace the classroom in principles of microeconomics? *The American Economic Review, 92*(2), 444–448. doi:10.1257/000282802320191778

Bryan, C. (2011). Medical Professionalism Meets Generation X. In Texas Heart Institute Journal (pp. 465–470). Texas Heart Institute.

Bucher, R. D. (2014). *Diversity consciousness: Opening our minds to people, cultures, and opportunities.* Pearson.

Buckner, M. M., & Strawser, M. G. (2016). Millennials and the paralysis of choice: Reigniting the purpose of higher education. *Communication Education, 65*(3), 361–363. doi:10.1080/03634523.2016.1177845

Butler, R. N. (2002). Age, death, and life review. In K. J. Doka (Ed.), *Living with grief: Loss in later life. Hospice Foundation of American.*

Byker, E. J., Putman, S. M., Handler, L., & Polly, D. (2018). Examining elementary education teachers and pre-service teachers' self-efficacy related to Technological Pedagogical and Content Knowledge (TPACK). In C. Hodges (Ed.), *Self-Efficacy in Instructional Technology Contexts* (pp. 119–140). Springer. doi:10.1007/978-3-319-99858-9_8

Camera, L. (n.d.). *Native American students left behind.* Retrieved March 31, 2021, from https://www.usnews.com/news/articles/2015/11/06/native-american-students-left-behind

Cameron-Standerford, A., Menard, K., Edge, C., Bergh, B., Shayter, A., Smith, K., & VandenAvond, L. (2020). The phenomenon of moving to online/distance delivery as a result of COVID-19: Exploring initial perceptions of higher education faculty at a rural midwestern university. *Frontiers in Education, 5*, 1–11. doi:10.3389/feduc.2020.583881

Caniels, M. J., & Rietzschel, E. F. (2015). Organizing creativity: Creativity and innovation under constraints. *Creativity and Innovation Management*, *24*(2), 184–196. doi:10.1111/caim.12123

Capdeferro, N., & Romero, M. (2012). Are online learners frustrated with collaborative learning experiences? *The International Review of Research in Open and Distributed Learning*, *13*(2), 26–44. doi:10.19173/irrodl.v13i2.1127

Cardullo, V., Wang, C. H., Burton, M., & Dong, J. (2021). K-12 teachers' remote teaching self-efficacy during the pandemic. *Journal of Research in Innovative Teaching & Learning*.

Carlisle Indian School Digital Resource Center. (n.d.). Retrieved May 4, 2021 from https://carlisleindian.dickinson.edu/teach/kill-indian-and-save-man-capt-richard-h-pratt-education-native-americans

Cartwright, A. M., & Reeves, E. K. (2018). *Everyday diversity: Developing cultural competency & information awareness*. Cognella Academic Publishing.

Carver, N., Gupta, V., & Hipskind, J. E. (2021 Jan). Medical Error. In *StatPearls*. StatPearls Publishing. Available from https://www.ncbi.nlm.nih.gov/books/NBK430763/

Castagno, A., & Brayboy, B. (2008). Culturally Responsive Schooling for Indigenous Youth: A Review of the Literature. *Review of Educational Research*, *78*(4), 941–993. doi:10.3102/0034654308323036

Castellanos, D., & Leggio, P. (1983). *The best of two worlds: Bilingual-bicultural education in the US*. Academic Press.

Center for Systems Science and Engineering (CSSE) at Johns Hopkins University (JHU) COVID-19 Dashboard. (2021). https://gisanddata.maps.arcgis.com/apps/opsdashboard/index.html#/bda7594740fd40299 423467b48e9ecf6

Center for the Study of Hate and Extremism. (2021). *Fact Sheet: Anti-Asian prejudice March 2021 Center for the Study of Hate & Extremism*. Center for the Study of Hate and Extremism. https://www.csusb.edu/sites/default/files/FACT%20SHEET-%20Anti-Asian%20Hate%20 2020%20rev%203.21.21.pdf

Centers for Disease Control and Prevention. (2019, September 6). *Racial and ethnic disparities continue in pregnancy-related deaths*. Retrieved May 01, 2021, from https://www.cdc.gov/media/releases/2019/p0905-racial-ethnic-disparities-pregnancy-deaths.html

Centers for Disease Control and Prevention. (2020). *Coronavirus (COVID-19)*. https://www.cdc.gov/coronavirus/2019-nCoV/index.html

Centers for Disease Control and Prevention. (n.d.). *Post-covid conditions*. Centers for Disease Control and Prevention. https://www.cdc.gov/coronavirus/2019-ncov/long-term-effects.html

Chandra, S., Chang, A., Day, L., Fazlullah, A., Liu, J., McBride, L., & Weiss, D. (2020). *Closing the K–12 digital divide in the age of distance learning.* Common Sense and Boston Consulting Group. Retrieved on 5/2/2021 from: https://exceptionalchildren.org/sites/default/files/2020-07/AllBriefs_2020.pdf

Chauhan, A., Walton, M., Manias, E., Walpola, R. L., Seale, H., Latanik, M., Leone, D., Mears, S., & Harrison, R. (2020). The safety of health care for ethnic minority patients: A systematic review. *International Journal for Equity in Health, 19*(1), 118. doi:10.118612939-020-01223-2 PMID:32641040

Chinese Americans Civil Rights Coalition, Inc. v. Donald J. Trump. (2021). Cohen, (1972).

Chryssanthou, A., Varlamis, I., & Latsiou, C. (2009). Security and trust in virtual healthcare communities. *Proceedings of the 2nd International Conference on PErvsive Technologies Related to Assistive Environments - PETRA '09*, 1-8. 10.1145/1579114.1579186

Cidade, E. C., Junior, J. F. M., & Ximenes, V. M. (2012). Implicações psicológicas da pobreza na vida do povo latinoamericano. [Psychological implications of poverty in the life of Latin-American people]. *Psicologia Argumento, 30*(68), 87–98. doi:10.7213/psicol.argum.5886

Cigna. (2021, February 4). *Cigna reports fourth quarter and full year 2020 Results, expects continued revenue and earnings growth in 2021.* https://www.prnewswire.com/news-releases/cigna-reports-fourth-quarter-and-full-year-2020-results-expects-continued-revenue-and-earnings-growth-in-2021-301221754.html

Clark, S. K., & Byrnes, D. (2015). What Millennial preservice teachers want to learn in their training. *Journal of Early Childhood Teacher Education, 36*(4), 379–395. doi:10.1080/10901027.2015.1100148

Cocker, F., & Joss, N. (2016). Compassion Fatigue among Healthcare, Emergency and Community Service Workers: A Systematic Review. *International Journal of Environmental Research and Public Health, 13*(6), 618. doi:10.3390/ijerph13060618 PMID:27338436

Cohen, J. (1960). A coefficient of agreement for nominal scales. *Educational and Psychological Measurement, 20*(1), 37–46. doi:10.1177/001316446002000104

Cohen, J. (1988). *Statistical power analysis for the behavioral sciences* (2nd ed.). Erlbaum.

Colao, A., Piscitelli, P., Pulimeno, M., Colazzo, S., Miani, A., & Giannini, S. (2020). Rethinking the role of the school after COVID-19. *The Lancet. Public Health, 5*(7), e370. doi:10.1016/S2468-2667(20)30124-9 PMID:32464100

Cole, B. M. (2020, September 15). 8 Reasons Why Diversity and Inclusion Are Essential to Business Success. *Forbes.* https://www.forbes.com/sites/biancamillercole/2020/09/15/8-reasons-why-diversity-and-inclusion-are-essential-to-business-success/?sh=2280c1ae1824

Colon-Gonzalez, M. C., McCall-Hosenfeld, J. S., Weisman, C. S., Hillemeier, M. M., Perry, A. N., & Chuang, C. H. (2013). "Someone's got to do it" – Primary care providers (PCPs) describe caring for rural women with mental health problems. *Mental Health in Family Medicine, 10*(4), 191–202. PMID:25632302

Communications Act of 1934, 48 Stat. 1064 (1934) (current version at 47 U.S.C. § 151 (2000)

Connolly, S. L., Miller, C. J., Lindsay, J. A., & Bauer, M. S. (2020). A systematic review of providers' attitudes toward telemental health via videoconferencing. *Clinical Psychology: Science and Practice, 27*(2), e12311. doi:10.1111/cpsp.12311

Cook, J. A., Ruggiero, K., Shore, S., Daggett, P., & Butler, S. B. (2007). Public-academic collaboration in the application of evidence-based practice in Texas mental health system redesign. *International Journal of Mental Health, 36*(2), 36–49. doi:10.2753/IMH0020-7411360204

Cooper, S. (2016). Global mental health and its critics: Moving beyond the impasse. *Critical Public Health, 26*(4), 355–358. doi:10.1080/09581596.2016.1161730

Copp, D. T. (2018). Teaching to the test: A mixed methods study of instructional change from large-scale testing in Canadian schools. *Assessment in Education: Principles, Policy & Practice, 25*(5), 468–487. doi:10.1080/0969594X.2016.1244042

Corgnet, B., Espin, A. M., & Hernan-Gonzalez, R. (2016). Creativity and cognitive skills among Millennials: Thinking too much and creating too little. *Frontiers in Psychology, 7*(1626), 1–9. doi:10.3389/fpsyg.2016.01626 PMID:27826268

Correll, J. A., Cantrell, P., & Dalton, W. M. (2011). Integration of behavioral health services in a primary care clinic serving rural Appalachia: Reflections on a clinical experience. *Families, Systems & Health, 29*(4), 291–302. doi:10.1037/a0026303 PMID:22214296

Corrigan, P. W., & Penn, D. L. (2015). Lessons from social psychology on discrediting psychiatric stigma. *Stigma and Health, 1*(S), 2–17. doi:10.1037/2376-6972.1.S.2

Corrigan, P. W., & Shapiro, J. R. (2010). Measuring the impact of programs that challenge the public stigma of mental illness. *Clinical Psychology Review, 30*(8), 907–922. doi:.2010.06.004 doi:10.1016/j.cpr

Corrigan, P. W., Druss, B. G., & Perlick, D. A. (2014). The impact of mental illness stigma on seeking and participating in mental health care. *Psychological Science in the Public Interest, 15*(2), 37–70. doi:10.1177/1529100614531398 PMID:26171956

Costa, R. R. G., Dorneles, J. R., Veloso, J. H., Gonçalves, C. W., & Neto, F. R. (2021). Synchronous and asynchronous tele-exercise during the coronavirus disease 2019 pandemic: Comparisons of implementation and training load in individuals with spinal cord injury. *Journal of Telemedicine and Telecare*, 1357633X2098273. Advance online publication. doi:10.1177/1357633X20982732 PMID:33461399

Council of Chief State School Officers. (2013). *Interstate teacher assessment and support consortium InTASC Model Core Teaching Standards and Learning Progressions for Teachers 1.0: A resource for ongoing teacher development.* Author.

Court Listener. (2021). *Chinese Americans Civil Rights Coalition, Inc. v. Donald J. Trump.* Court Listener. https://www.courtlistener.com/?type=r&q=1%3A21-cv-04548&type=r&order_by=score%20desc

Courtney, E. (2021, April 29). *Remote work Statistics: Navigating the new Normal: FlexJobs.* FlexJobs Job Search Tips and Blog. https://www.flexjobs.com/blog/post/remote-work-statistics/#:~:text=%E2%80%93%20Remote%20Work%20Is%20Here%20to,be%20working%20remotely%20by%202025

Creswell, J. W., & Creswell, J. D. (2018). *Research design: qualitative, quantitative, and mixed methods approaches.* SAGE Publications, Inc.

Creswell, J. W., & Guetterman, T. C. (2015). *Educational research: Planning, conducting, and evaluating quantitative and qualitative research.* Prentiss Hall.

Crunk, A. E., Burke, L. A., & Robinson, E. H. M. III. (2017). Complicated grief: An evolving theoretical landscape. *Journal of Counseling and Development, 95*(2), 226–233. doi:10.1002/jcad.12134

Csikszentmihalyi, M., Montijo, M. N., & Mouton, A. R. (2018). Flow theory: Optimizing elite performance in the creative realm. In S. I. Pfeiffer, E. Shaunessy-Dedrick, & M. Foley-Nicpon (Eds.), *APA Handbooks in Psychology. APA Handbook of Giftedness and Talent* (pp. 215–229). American Psychological Association., doi:10.1037/0000038-014

Currie, J. (2020). Pandemic raises profile of rural colleges. *Business NH Magazine, 37*(8), 34–36.

Darder, A. (1991). *Culture and power in the classroom: A critical foundation for bicultural education.* Greenwood Publishing Group.

De Paepe, L., Zhu, C., & Depryck, K. (2018). Online language teaching: Teacher perceptions of effective communication tools, required skills and challenges of online teaching. *Journal of Interactive Learning Research, 29*(1), 129–142.

Deacon, A. (2012). Creating a context of care in the online classroom. *Journal of Faculty Development, 26,* 5–12.

DeLeon, J. (2018). *Leading the Millennial Generation: An investigation of leadership practices that encourage and support education's youngest teachers* (PhD diss.). Dallas Baptist University.

Delpit, L. (2012). Multiplication is for white people. In *Raising expectations for other people's children.* The New Press.

Dennen, V., Darabi, A., & Smith, L. (2007). Instructor-learner interaction in online courses: The relative perceived importance of particular instructor actions on performance and satisfaction. *Distance Education, 28*(1), 65–79. doi:10.1080/01587910701305319

DeVoe, J. F., & Darling-Churchill, K. E. (2008). *Status and Trends in the Education of American Indians and Alaska Natives: 2008 (NCES 2008-084)*. National Center for Education Statistics, Institute of Education Sciences, U.S. Department of Education.

DiAngelo, R. (2018). *White fragility: Why it's so hard for White people to talk about racism.* Penguin Books Ltd.

Diederich, J., & Song, I. (2014). Mental health informatics: current approaches. Mental Health Informatics, 1-16.

Dinesen, B., Nonnecke, B., Lindeman, D., Toft, E., Kidholm, K., Jethwani, K., Young, H. M., Spindler, H., Oestergaard, C. U., Southard, J. A., Gutierrez, M., Anderson, N., Albert, N. M., Han, J. J., & Nesbitt, T. (2016). Personalized Telehealth in the Future: A Global Research Agenda. *Journal of Medical Internet Research, 18*(3), e53. doi:10.2196/jmir.5257 PMID:26932229

Dinno, A. (2015). Nonparametric pairwise multiple comparisons in independent groups using Dunn's test. *The Stata Journal, 15*(1), 292–300. doi:10.1177/1536867X1501500117

DiRienzo, C., & Lilly, G. (2014). Online versus face-to-face: Does delivery method matter for undergraduate business school learning? *Business Education & Accreditation, 6*(1), 1–11.

Doll, E., Sands, D. J., Wehmeyer, M. L., & Palmer, S. (1996). Promoting the development and acquisition of self-determined behavior. In D. J. Sands & M. L. Wehmeyer (Eds.), *Self-determination across the lifespan: Independence and choice for people with disabilities* (pp. 63–88). Brookes.

Donalek, J. G. (2004). Phenomenology as a qualitative research method. *Urologic Nursing, 24*(6), 516–517. PMID:15658739

Donnellan, E. S. (2017). No Connection: The Issue of Internet on the Reservation. *American Indian Law Journal, 5*(2).

Doucet, A., Netolicky, D., Timmers, K., & Tuscano, F. J. (2020). *Thinking about pedagogy in an unfolding pandemic* (An Independent Report on Approaches to Distance Learning during COVID-19 School Closure). Work of Education International and UNESCO. https://issuu.com/educationinternational/docs/2020_research_covid-19_eng

Douglas, J. (2021, February 4). *Bridging the digital divide in Indian country*. Retrieved April 29, 2021, from https://www.hcn.org/issues/53.4/indigenous-affairs-technology-bridging-the-digital-divide-in-indian-country

Dunn, O. J. (1964). Multiple comparisons using rank sums. *Technometrics, 6*(3), 241–252. doi:10.1080/00401706.1964.10490181

Dyer, J. E. T., & Hagedorn, W. B. (2013). Navigating bereavement with spirituality-based interventions: Implications for non-faith-based counselors. *Counseling and Values, 58*(1), 69–84. doi:10.1002/j.2161-007X.2013.00026.x

Eckert, C. M., Stacey, M., Wyatt, D., & Garthwaite, P. (2012). Change as little as possible: Creativity in design by modification. *Journal of Engineering Design*, *23*(4), 337–360. doi:10.1080/09544828.2011.639299

Egede, L. E. (2006). Race, ethnicity, culture, and disparities in health care. *Journal of General Internal Medicine*, *21*(6), 667–669. doi:10.1111/j.1525-1497.2006.0512.x PMID:16808759

Ellis, V., & Childs, A. (2019). Innovation in teacher education: Collective creativity in the development of a teacher education internship. *Teaching and Teacher Education*, *77*, 277–286. doi:10.1016/j.tate.2018.10.020

Erenli, K. (2016). Generation I (mmersion)–how to meet learner expectations of tomorrow. *International Journal of Advanced Corporate Learning*, *9*(1), 19–25. doi:10.3991/ijac.v9i1.5241

Erlam, G., Smythe, L., & Clair, V. W. (2018). Action research and millennials: Improving pedagogical approaches to encourage critical thinking. *Nurse Education Today*, *61*, 140–145. doi:10.1016/j.nedt.2017.11.023 PMID:29197689

Erskine, J. L. (2014). It changes how teachers teach: How testing is corrupting our classrooms and student learning. *Multicultural Education*, *21*(2), 38–40. https://eric.ed.gov/?id=EJ1045812

Evans, M. K., Rosenbaum, L., Malina, D., Morrissey, S., & Rubin, E. J. (2020). Diagnosing and treating systemic racism. *The New England Journal of Medicine*, *383*(3), 274–276. doi:10.1056/NEJMe2021693 PMID:32521155

Federal Communications Commission. (2003). *Promoting Deployment/Subscribership in Underserved Areas, Including Tribal and Insular Areas*. Retrieved May 3, 2021 from https://www.fcc.gov/Bureaus/Wireless/Factsheets/tribal_lands_fact_sheet.txt

Federal Communications Commission. (2015). *Broadband Progress Report*. Author.

Federal Communications Commission. (2019). *Report on Broadband Deployment in Indian Country, Pursuant to the Repack Airwaves Yielding Better Access for Users of Modern Services Act of 2018*. Retrieved May 03, 2021, from https://aipi.clas.asu.edu/sites/default/files/05011019fccreport_on_broadband_deployment_in_indian_country_pursuant_to_the_repack_airwaves_yielding_better_access_for_users_of_modern_services_act_of_2018.pdf

Federal Communications Commission. (2021). *2.5 GHz Rural Tribal Window*. Retrieved May 03, 2021, from https://www.fcc.gov/25-ghz-rural-tribal-window

Field, A. P. (2018). *Discovering statistics using IBM SPSS Statistics*. Sage.

Fish, J., & Syed, M. (2018). Native Americans in Higher Education: An Ecological Systems Perspective. *Journal of College Student Development*, *59*(4), 387–403. doi:10.1353/csd.2018.0038

Fleiss, J. L., Levin, B., & Paik, M. C. (2013). *Statistical methods for rates and proportions*. John Wiley and Sons.

Fletcher, G. (2016). *Modeling with mathematics through three-act tasks*. Math Tasks to Talk About. Retrieved from: https://www.nctm.org/Publications/TeachingChildren-Mathematics/Blog/Modeling-withMathematics-through-Three-Act-Tasks/

Flexjobs. (2019, December 6). *How remote work supports the fight against climate change: Flexjobs*. FlexJobs Job Search Tips and Blog. https://www.flexjobs.com/blog/post/how-remote-work-supports-un-sustainable-development-goal-no-13-of-combating-climate-change/

Flexjobs. (2021, June 10). *The environmental impacts of remote work: Stats and benefits: Flexjobs*. FlexJobs Job Search Tips and Blog. https://www.flexjobs.com/blog/post/telecommuting-sustainability-how-telecommuting-is-a-green-job/

Fox, T. (2019). *Integrated behavioral healthcare: The future of rural mental health.* Retrieved from https://ttu-ir.tdl.org/handle/2346/85520

Frawley, E. F. (2014). No time for the airy fairy: Teacher perspectives on creative writing in high stakes environments. *English in Australia, 49*(1), 17-26. https://search.informit.com.au/documentSummary;dn=381501270146252;res=IELHSS

Freire, P. (2000). Pedagogy of the oppressed: 30th anniversary edition. *Continuum.*

Fusarelli, L. D., & Fusarelli, B. C. (2018). Instructional supervision in an era of high-stakes accountability. In S. J. Zepeda & J. A. Ponticell (Eds.), *The Wiley Handbook of Educational Supervision* (pp. 131–155). Wiley-Blackwell. doi:10.1002/9781119128304.ch6

Galea, S., Ahern, J., Nandi, A., Tracy, M., Beard, J., & Vlahov, D. (2007). Urban neighborhood poverty and the incidence of depression in a Population-Based cohort study. *Annals of Epidemiology*, *17*(3), 171–179. doi:10.1016/j.annepidem.2006.07.008 PMID:17320784

Garcia, E., & Weiss, E. (2020). *COVID-19 and student performance, equity, and U.S. education policy.* Economic Policy Institute. Retrieved from: https://www.epi.org/publication/the-consequences-of-the-covid-19-pandemic-for-education-performance-and-equity-in-the-united-states-what-can-we-learn-from-pre-pandemic-research-to-inform-relief-recovery-and-rebuilding/

García, E., & Weiss, E. (2020). *COVID-19 and Student Performance, Equity, and US Education Policy: Lessons from Pre-Pandemic Research to Inform Relief, Recovery, and Rebuilding.* Economic Policy Institute.

Garrison, D. R., Anderson, T., & Archer, W. (1999). Critical inquiry in a text-based environment: Computer conferencing in higher education. *The Internet and Higher Education*, *2*(2-3), 87–105. doi:10.1016/S1096-7516(00)00016-6

Gaskin, D. J., Dinwiddie, G. Y., Chan, K. S., & McCleary, R. (2012). Residential segregation and disparities in health care services utilization. *Medical Care Research and Review: MCRR*, *69*(2), 158–175. doi:10.1177/1077558711420263 PMID:21976416

Gee, G. C., & Ford, C. L. (2011). Structural Racism and Health Inequities: Old Issues, New Directions. *Du Bois Review*, *8*(1), 115–132. doi:10.1017/S1742058X11000130 PMID:25632292

Getty, L. A., & Summy, S. E. (2004). The course of due process. *Teaching Exceptional Children*, *36*(3), 40–43. doi:10.1177/004005990403600305

Gillen-O'Neel, C. (2021). Sense of belonging and student engagement: A daily study of first- and continuing-generation college students. *Research in Higher Education*, *62*(1), 45–71. doi:10.100711162-019-09570-y

Gilligan, C. (2003). In a different voice: Psychological theory and women's development. Harvard University Press.

Glauser, W. (2020). Virtual care is here to stay, but major challenges remain. *Canadian Medical Association Journal*, *192*(30), E868–E869. Advance online publication. doi:10.1503/cmaj.1095884 PMID:32719027

Gogia, S. B., Maeder, A., Mars, M., Hartvigsen, G., Basu, A., & Abbott, P. (2016). Unintended Consequences of TeleHealth and their Possible Solutions. Contribution of the IMIA Working Group on Telehealth. *Yearbook of Medical Informatics*, *25*(1), 41–46. doi:10.15265/IY-2016-012 PMID:27830229

Gopalan, M., & Brady, S. T. (2019). College students' sense of belonging: A national perspective. *Educational Researcher*, *49*(2), 134–137. doi:10.3102/0013189X19897622

Gourevitch, D. (1999). The history of medical teaching. *Lancet*, *354*(Suppl), SIV33. https://doi.org/10.1016/s0140-6736(99)90376-8

Grossman, L. V., Creber, R. M. M., Benda, N. C., Wright, D., Vawdrey, D. K., & Ancker, J. S. (2019). Interventions to increase patient portal use in vulnerable populations: A systematic review. *Journal of the American Medical Informatics Association: JAMIA*, *26*(8-9), 855–870. doi:10.1093/jamia/ocz023 PMID:30958532

Gudmundsdottir, G. B., & Hathaway, D. M. (2020). "We always make it work": Teachers' agency in the time of crisis. *Journal of Technology and Teacher Education*, *28*(2), 239–250.

Guerin, E. (2014). Rural Americans have inferior internet access: Does it matter that broadband quality varies so greatly? *High Country News*. Retrieved from https://www.hcn.org/issues/46.2/rural-americans-have-inferior-internet-access

Gunnulfsen, A. E., & Roe, A. (2018). Investigating teachers' and school principals' enactments of national testing policies: A Norwegian study. *Journal of Educational Administration*, *56*(3), 332–349. doi:10.1108/JEA-04-2017-0035

Gutiérrez, R. (2009). Framing equity: Helping students "play the game" and "change the game.". *Teaching for Excellence and Equity in Mathematics*, *1*(1), 4–8.

Hallman, H. L. (2017). Millennial teachers and multiculturalism: Considerations for teaching in uncertain times. *Journal for Multicultural Education*, *11*(3), 194–205. doi:10.1108/JME-10-2016-0055

Hall, S. A., & Gjesfjeld, C. D. (2013). Clergy: A partner in rural mental health? *Rural Mental Health, 37*(1), 50–57. doi:10.1037/rmh0000006

Hall, W. J., Chapman, M. V., Lee, K. M., Merino, Y. M., Thomas, T. W., Payne, B. K., Eng, E., Day, S. H., & Coyne-Beasley, T. (2015). Implicit Racial/Ethnic Bias Among Health Care Professionals and Its Influence on Health Care Outcomes: A Systematic Review. *American Journal of Public Health, 105*(12), e60–e76. doi:10.2105/AJPH.2015.302903 PMID:26469668

Handley, T., Kay-Lambkin, F., Inder, K., Attia, J., Lewin, T., & Kelly, B. (2014a). Feasibility of internet-delivered mental health treatments for rural populations. *Social Psychiatry and Psychiatric Epidemiology, 49*(2), 275–282. doi:10.100700127-013-0708-9 PMID:23689825

Handley, T., Kay-Lambkin, F., Inder, K., Lewin, T., Attia, J., Fuller, J., & Kelly, B. (2014b). Self-reported contacts for mental health problems by rural residents: Predicted service needs, facilitators and barriers. *BMC Psychiatry, 14*(1), 1–23. doi:10.118612888-014-0249-0 PMID:25193400

Harandi, T. F., Taghinasab, M. M., & Nayeri, T. D. (2017). The correlation of social support with mental health: A meta-analysis. *Electronic Physician, 9*, 5212–5222. https://dx.doi.org/10.19082/5212

Hardcastle, L., & Ogbogu, U. (2020). Virtual care: Enhancing access or harming care? *Healthcare Management Forum, 33*(6), 288–292. doi:10.1177/0840470420938818 PMID:32686506

Harris, A., & Ammermann, M. (2016). The changing face of creativity in Australian education. *Teaching Education, 27*(1), 103–113. doi:10.1080/10476210.2015.1077379

Harris, S. (1994). "Soft" and "hard" domain theory for bicultural education in Indigenous groups. *Peabody Journal of Education, 69*(2), 140–153. doi:10.1080/01619569409538769

Hatchuel, A., & Chen, M. K. (2017). Creativity under strong constraints: The hidden influence of design models. *European Review (Chichester, England), 25*(2), 194–207. doi:10.1017/S1062798716000557

Hauck, G., Gelles, K., Bravo, V., & Thorson, M. (2020). Five months in: A timeline of how COVID-19 has unfolded in the U.S. *USA Today.* https://www.usatoday.com/indepth/news/nation/2020/04/21/coronavirus-updates-how-covid-19-unfolded-u-stimeline/2990956001/

Haught, C. (2015). The role of constraints in creative sentence production. *Creativity Research Journal, 27*(2), 160–166. doi:10.1080/10400419.2015.1030308

Haught-Tromp, C. (2017). The green eggs and ham hypothesis: How constraints facilitate creativity. *Psychology of Aesthetics, Creativity, and the Arts, 11*(1), 10–17. doi:10.1037/aca0000061

Haun, J. N., Patel, N. R., French, D. D., Campbell, R. R., Bradham, D. D., & Lapcevic, W. A. (2015). Association between health literacy and medical care costs in an integrated healthcare system: A regional population based study. *BMC Health Services Research, 15*(1), 249. Advance online publication. doi:10.118612913-015-0887-z PMID:26113118

Hayes, S. (2003). *Digital Divide in Indian Country: Obstacles and Opportunities for Native American Communities*. Retrieved April 28, 2021, from http://students.washington.edu/aliss/silverfish/archive/april2003/hayes.pdf

he United States Department of Justice. (2017, May 31). *Electronic health RECORDS vendor to pay $155 million to Settle false Claims Act Allegations*. https://www.justice.gov/opa/pr/electronic-health-records-vendor-pay-155-million-settle-false-claims-act-allegations

Health Resources and Services Administration. (2016). *Health Professional Shortage Areas (HPSAs)*. Retrieved from https://bhw.hrsa.gov/shortage-designation/hpsas

Henriksen, D., Mishra, P., & Fisser, P. (2016). Infusing creativity and technology in 21st Century education: A systemic view for change. *Journal of Educational Technology & Society*, *19*(3), 27–37. https://www.jstor.org/stable/jeductechsoci.19.3.27

Henriksen, D., Mishra, P., & Mehta, R. (2015). Novel, effective, whole: Toward a NEW framework for evaluations of creative products. *Journal of Technology and Teacher Education*, *23*(3), 455–478. https://www.learntechlib.org/p/151574/

Herzer, K. R., & Pronovost, P. J. (2021). Ensuring quality in the era of virtual care. *Journal of the American Medical Association*, *325*(5), 429. doi:10.1001/jama.2020.24955 PMID:33528544

Heward, W. L. (2017). *Exceptional children: An introduction to special education* (11th ed.). Pearson.

Hickling-Hudson, A., & Ahlquist, R. (2003). Contesting the Curriculum in the Schooling of Indigenous Children in Australia and the United States: From Eurocentrism to Culturally Powerful Pedagogies. *Comparative Education Review*, *47*(1), 64–89. doi:10.1086/345837

Hilty, D. M., Gentry, M. T., McKean, A. J., Cowan, K. E., Lim, R. F., & Lu, F. G. (2020). Telehealth for rural diverse populations: Telebehavioral and cultural competencies, clinical outcomes and administrative approaches. *mHealth*, *20*(6). https://doi.org/10.21037/mhealth.2019.10.04

Hilty, D., Feliberti, J., Evangelatos, G., Lu, F., & Lim, R. (2019). Competent cultural telebehavioral healthcare to rural diverse populations: Administration, evaluation, and financing. *Journal of Technology in Behavioral Science.*, *4*. Advance online publication. doi:10.100741347-018-0076-0

Hinchman, T. (2019). Expanding the Curriculum With Creativity. In *Critical Literacy Initiatives for Civic Engagement* (pp. 48–68). IGI Global. doi:10.4018/978-1-5225-8082-9.ch003

Hodges, G. C. (2005). Creativity in education. *English in Education*, *39*(3), 47–61. doi:10.1111/j.1754-8845.2005.tb00624.x

Hoffman, D. A. (2020). Increasing access to care: Telehealth during covid-19. *Journal of Law and the Biosciences*, *7*(1), lsaa043. Advance online publication. doi:10.1093/jlb/lsaa043 PMID:32843985

Hoffman, K. M., Trawalter, S., Axt, J. R., & Oliver, M. N. (2016). Racial bias in pain assessment and treatment recommendations, and false beliefs about biological differences between blacks and whites. *Proceedings of the National Academy of Sciences of the United States of America, 113*(16), 4296–4301. doi:10.1073/pnas.1516047113 PMID:27044069

Holloway, J., & Brass, J. (2017). Making accountable teachers: The terrors and pleasures of performativity. *Journal of Education Policy, 33*(3), 361–382. doi:10.1080/02680939.2017.1372636

Horse, P. G. (2005). *Native American Identity.* Retrieved October 27, 2020, from http://convention.myacpa.org/houston2018/wp-content/uploads/2017/11/Horse_2005.pdf

Hosek, A. M., & Titsworth, S. (2016). Scripting knowledge and experiences for millennial students. *Communication Education, 65*(3), 357–359. doi:10.1080/03634523.2016.1177844

Howard, T. C. (2019). *Why race and culture matter in schools.: Closing the achievement gap in America's classrooms* (2nd ed.). Teachers College Press.

Hussar, B., Zhang, J., Hein, S., Wang, K., Roberts, A., Cui, J., Smith, M., Bullock Mann, F., Barmer, A., & Dilig, R. (2020). *The Condition of Education 2020* (NCES 2020-144). U.S. Department of Education. Washington, DC: National Center for Education Statistics. https://nces.ed.gov/pubsearch/pubsinfo. asp?pubid=2020144

IBM. (n.d.). *Healthcare provider solutions: Watson Health.* https://www.ibm.com/watson-health/solutions/healthcare-provider

IBM. (n.d.). *Healthcare provider SOLUTIONS: Watson Health.* https://www.ibm.com/watson-health/solutions/healthcare-provider

Inder, K. J., Berry, H., & Kelly, B. J. (2011). Using cohort studies to investigate rural and remote mental health. *The Australian Journal of Rural Health, 19*(4), 171–178. doi:10.1111/j.1440-1584.2011.01208.x PMID:21771157

Ioannou, L. (2021, February 6). *1 in 4 Americans will be working remotely in 2021, Upwork survey reveals.* CNBC. https://www.cnbc.com/2020/12/15/one-in-four-americans-will-be-working-remotely-in-2021-survey.html

Jackson, C. S., & Gracia, J. N. (2014). Addressing health and health-care disparities: the role of a diverse workforce and the social determinants of health. *Public Health Reports, 129*(Suppl 2), 57–61. doi:10.1177/00333549141291S211

Jackson, H., Judd, F., Komiti, A., Fraser, C., Murray, G., Robins, G., & Wearing, A. (2007). Mental health problems in rural contexts: What are the barriers to seeking help from professional providers? *Australian Psychologist, 42*(2), 147–160. doi:10.1080/00050060701299532

Jain, N., & Jain, T. K. (2019). Sustainable development through student engagement: Innovations and initiatives in higher education. SSRN *Electronic Journal*, 1-8. doi:10.2139/ssrn.3314509

Jakee, K., & Keller, E. (2017). The price of high-stakes educational testing: Estimating the aggregate costs of Florida's FCAT exam. *Journal of Education Finance*, *43*(2), 123–151. https://muse.jhu.edu/article/688009/summary

Jameson, J. P., & Blank, M. B. (2007). The role of clinical psychology in rural mental health services: Defining problems and developing solutions. *Clinical Psychology: Science and Practice*, *14*(3), 283–298. doi:10.1111/j.1468-2850.2007.00089.x

Jamieson, T., Wallace, R., Armstrong, K., Agarwal, P., Griffin, B., Wong, I., & Bahtia, S. (2015). Virtual Care: A Framework for a Patient-Centric System. *Women's College Hospital Institute for Health Systems Solutions and Virtual Care*. https://www.womenscollegehospital.ca/assets/pdf/wihv/WIHV_VirtualHealthSymposium.pdf

Johns Hopkins Coronavirus Resource Center. (n.d.). *COVID-19 map*. https://coronavirus.jhu.edu/map.html

Johnson, D. D., Johnson, B., Farenga, S. J., & Ness, D. (2008). *Stop high-stakes testing: An appeal to America's conscience*. Rowman and Littlefield.

Kajzer-Mitchell, I., & Walinga, J. (2017). The creative imperative: The role of creativity, creative problem-solving and insight as key drivers for sustainability. *Journal of Cleaner Production*, *140*, 1872–1884. doi:10.1016/j.jclepro.2016.09.162

Kaliya-Perumal, A. K., Omar, U. F., & Kharlukhi, J. (2020). Healthcare virtualization amid COVID-19 pandemic: An emerging new normal. *Medical Education Online*, *25*(1), 1780058. doi:10.1080/10872981.2020.1780058 PMID:32529948

Katsiyannis, A., Yell, M. L., & Bradley, R. (2001). Reflections on the 25th anniversary of the Individuals with Disabilities Education Act. *Remedial and Special Education*, *22*, 324–334.

Kaulback, B. (1984). Styles of learning among native children: A review of the research. *Canadian Journal of Native Education*, *11*(3), 27–37.

Kelly, R. L. (2015). *2015 Kelly Report: Health Disparities in America*. Congressional Black Caucus Health Braintrust.

Kennedy, M. M. (1999). The role of preservice teacher education. In L. Darling-Hammond & G. Sykes (Eds.), *Teaching as the Learning Profession: Handbook of Teaching and Policy* (pp. 54–86). Jossey Bass.

Kettler, T., Lamb, K., Willerson, A., & Mullett, D. (2018). Teachers' perceptions of creativity in the classroom. *Creativity Research Journal*, *30*(2), 164–171. doi:10.1080/10400419.2018.1446503

Kewal Ramani, A., Zhang, J., Wang, X., Rathbun, A., Corcoran, L., Diliberti, M., & Zhang, J. (2018). *Student access to digital learning outside of the classroom. NCES 2017-098*. National Center for Education Statistics.

Kichloo, A., Albosta, M., Dettloff, K., Wani, F., El-Amir, Z., Singh, J., Aljadah, M., Chakinala, R. C., Kanugula, A. K., Solanki, S., & Chugh, S. (2020). Telemedicine, the current COVID-19 pandemic and the future: A narrative review and perspectives moving forward in the USA. *Family Medicine and Community Health, 8*(3), e000530. doi:10.1136/fmch-2020-000530 PMID:32816942

Kim, K. H. (2011). The creativity crisis: The decrease in creative thinking scores on the Torrance Tests of Creative Thinking. *Creativity Research Journal, 23*(4), 285–295. doi:10.1080/10400419.2011.627805

Kim, K. H. (2016). *The creativity challenge: How we can recapture American innovation.* Prometheus Books.

King, M. L. (1967, November 13). *Speech on receipt of Honorary Doctorate in Civil Law* [Address]. University of Newcastle upon Tyne, UK. https://www.ncl.ac.uk/media/wwwnclacuk/congregations/files/Transcript%20of%20Dr%20Martin%20Luther%20King%20Jr%20speech%2013th%20November%201967.pdf

Kleinfeld, J., & Nelson, P. (1991). Adapting instruction to Native Americans' learning styles: An iconoclastic view. *Journal of Cross-Cultural Psychology, 22*(2), 273–282. doi:10.1177/0022022191222007

Klug, B. & Whitfield, P. (2012). *Widening the Circle.* doi:10.4324/9780203616703

Kola, L. (2020). Global mental health and COVID-19. *The Lancet. Psychiatry, 7*(8), 655–657. doi:10.1016/S2215-0366(20)30235-2 PMID:32502468

Koonin, L. M., Hoots, B., Tsang, C. A., Leroy, Z., Farris, K., Jolly, B., Antall, P., McCabe, B., Zelis, C. B. R., Tong, I., & Harris, A. M. (2020). Trends in the use of telehealth during the emergence of the Covid-19 Pandemic— United States, January–March 2020. *MMWR. Morbidity and Mortality Weekly Report, 69*(43), 1595–1599. doi:10.15585/mmwr.mm6943a3 PMID:33119561

Koplan, J. P., Bond, T. C., Merson, M. H., Reddy, K. S., Rodriguez, M. H., Sewankambo, N. K., & Wasserheit, J. N. (2009). Towards a common definition of global health. *Lancet, 373*(9679), 1993–1995. doi:10.1016/S0140-6736(09)60332-9 PMID:19493564

Koricich, A., Tandberg, D., Bishop, B., & Weeden, D. (2020). Doing the same (or more) with less: The challenges regional public universities face in serving rural populations. *New Directions for Higher Education, 2020*(190), 59–70. doi:10.1002/he.20367

kpmg.com. (2020). *Automotives New Reality.* https://assets.kpmg/content/dam/kpmg/br/pdf/2020/09/automotives-new-reality.pdf.

Krahl, D. (2018). Resisting the influence of neoliberalism on Millennial faculty through critical communication pedagogy and critical mentoring. In A. Sutherland & M. G. Strawser (Eds.), *Millennial Culture and Communication Pedagogies: Narratives from the Classroom and Higher Education* (pp. 3–19). Lexington Books.

Kramer, G. M., Kinn, J. T., & Mishkind, M. C. (2015). Legal, regulatory, and risk management issues in the use of technology to deliver mental health care. *Cognitive and Behavioral Practice*, *22*(3), 258–268. doi:10.1016/j.cbpra.2014.04.008

Krippendorff, K. (2004). *Content analysis: An introduction to its methodology*. Sage.

Krishnan, P. (2018). A review of the non-equivalent control group post-test-only design. *Nurse Researcher*, *27*(1), 114–124. https://www.ncbi.nlm.nih.gov/pubmed/30226337 PMID:30226337

Kruse, C. S., Krowski, N., Rodriguez, B., Tran, L., Vela, J., & Brooks, M. (2017). Telehealth and patient satisfaction: A systematic review and narrative analysis. *BMJ Open*, *7*(8), e016242. doi:10.1136/bmjopen-2017-016242 PMID:28775188

Kruskal, W. H., & Wallis, W. A. (1952). Use of ranks in one-criterion variance analysis. *Journal of the American Statistical Association*, *47*(260), 583–621. doi:10.1080/01621459.1952.10483441

Kuehn, S., & Lingwall, A. (2015). In their own words: A thematic analysis of students' comments about their writing skills in mass communication programs. *Journal of Public Relations Education*, *1*(1), 1–12. https://aejmc.us/jpre/wp-content/uploads/sites/25/2015/08/JPRE-11-All-Three-Articles.pdf#page=21

Ladson-Billings, G. (1995). Toward a theory of culturally relevant pedagogy. *American Educational Research Journal*, *32*(3), 465–491. doi:10.3102/00028312032003465

Langarizadeh, M., Tabatabaei, M. S., Tavakol, K., Naghipour, M., Rostami, A., & Moghbeli, F. (2017). Telemental Health Care, an Effective Alternative to Conventional Mental Care: A Systematic Review. *Acta Informatica Medica*, *25*(4), 240–246. doi:10.5455/aim.2017.25.240-246 PMID:29284913

Law, T. (2020). *What Makes COVID-19 Different From the Flu? Time.com.*

Lee, B. X., & Young, J. L. (2018). Clinicians' need for an ecological approach to violence reduction. *AMA Journal of Ethics*, *20*(1), 91–98. doi:10.1001/journalofethics.2018.20.1.msoc1-1801 PMID:29360032

Leedy, P. D., Ormrod, J. E., & Johnson, L. R. (2019). *Practical research: Planning and design*. Pearson Education.

Lee, J. H., Lee, H., Kim, J. E., Moon, S. J., & Nam, E. W. (2021). Analysis of personal and national factors that influence depression in individuals during the COVID-19 pandemic: A web-based cross-sectional survey. *Globalization and Health*, *17*(1), 1–12. https://doi-org.ezp.twu.edu/10.1186/s12992-020-00650-8

Leite, H., Hodgkinson, I. R., & Gruber, T. (2020). New development: 'Healing at a distance'—Telemedicine and COVID-19. *Public Money & Management*, *40*(6), 483–485. Advance online publication. doi:10.1080/09540962.2020.1748855

Lim, R. F., & Lu, F. (2008). Culture and psychiatric education. *Academic Psychiatry*, (32), 269–271. doi:10.1176/appi.ap.32.4.269

Li, S., Wang, Y., Xue, J., Zhao, N., & Zhu, T. (2020). The impact of COVID-19 epidemic declaration on psychological consequences: A study on active Weibo users. *International Journal of Environmental Research and Public Health*, 17.

Liu, N., Zhang, F., Wei, C., Jia, Y., Shang, Z., & Sun, L. (2020). Prevalence and predictors of PTSD during COVID-19 outbreak in China hardest-hit areas: Gender differences matter. *Psychiatry Research*, *287*, 112921.

Li, W., Yang, Y., Liu, Z. H., Zhao, Y. J., Zhang, Q., Zhang, L., Cheung, T., & Xiang, Y. T. (2020). Progression of Mental Health Services during the COVID-19 Outbreak in China. *International Journal of Biological Sciences*, *16*(10), 1732–1738. doi:10.7150/ijbs.45120 PMID:32226291

Lopez Bernal & Shimabukuro. (2020, June 18). Retraction: Cardiovascular disease, drug therapy, and mortality In Covid-19. *New England Journal of Medicine*. https://www.nejm.org/doi/full/10.1056/nejmc2021225

Lorde, A. (1984). *Sister outsider: Essays and speeches*. Ten Speed Press.

Lotan, R. A. (2003). Group-worthy tasks. *Educational Leadership*, *60*(6), 72.

Lustig, T. A. (2012). The role of telehealth in an evolving health care environment: Workshop summary. In *The Role of Telehealth in an Evolving Health Care Environment: Workshop Summary* (pp. 11-16). Washington, DC: National Academies Press

Lynch, M. A.-M., & Franklin, G. V. (2019). Health Literacy: An Intervention to Improve Health Outcomes. *Strategies to Reduce Hospital Mortality in Lower and Middle Income Countries (LMICs) and Resource-Limited Settings*. doi:10.5772/intechopen.86269

Lynch, L., Long, M., & Moorhead, A. (2018). Young Men, Help-Seeking, and Mental Health Services: Exploring Barriers and Solutions. *American Journal of Men's Health*, *12*(1), 138–149. doi:10.1177/1557988315619469 PMID:27365212

Lyne, J., Roche, E., Kamali, M., & Feeney, L. (2020). COVID-19 from the perspective of urban and rural general adult mental health services. *Irish Journal of Psychological Medicine*, *37*(3), 181–186. doi:10.1017/ipm.2020.62 PMID:32434599

Mace, S., Boccanelli, A., & Dormond, M. (2018). *The Use of Telehealth Within Behavioral Health Settings: Utilization, Opportunities, and Challenges*. Behavioral Health Workforce Research Center. Retrieved from https://behavioralhealthworkforce.org/wp-content/uploads/2018/05/Telehealth-Full-Paper_5.17.18-clean.pdf

Mackie, P. F. (2015). Technology in rural behavioral health care practice: Policy concerns and solution suggestions. *Rural Mental Health*, *39*(1), 5–12. doi:10.1037/rmh0000027

MacKinnon, A. A. (1972). Eskimo and Caucasian: A discordant note on cognitive-perceptual abilities. In *Proceedings of the annual convention of the American Psychological Association*. American Psychological Association.

Major, T., & Mulvihill, T. M. (2018). Problem-based learning pedagogies in teacher education: The case of Botswana. *The Interdisciplinary Journal of Problem-Based Learning*, *12*(1), 1–11. doi:10.7771/1541-5015.1543

Maley, A., & Kiss, T. (2017). Pre-conditions for classroom creativity. *Creativity and English Language Teaching,* 205-217. doi:10.1057/978-1-137-46729-4_10

Marsella, A. J. (1998). Urbanization, mental health and social deviancy – a review of issues and research. *The American Psychologist*, *53*(6), 624–634. doi:10.1037/0003-066X.53.6.624 PMID:9633263

Martin, F., Polly, D., & Ritzhaupt, A. (2020). Bichronous online learning: Blending asynchronous and synchronous online learning. *Educause Review*. Retrieved from: https://er.educause.edu/articles/2020/9/bichronous-online-learning-blending-asynchronous-and-synchronous-online-learning

Martin, C., Harbour, K., & Polly, D. (2021). Transitioning the elementary mathematics classroom to virtual learning: Exploring the perspectives and experiences of teachers. In A. Slapac, P. Balcerzak, & K. O'Brien (Eds.), *Handbook of Research on the Global Empowerment of Educators and Student Learning through Action Research* (pp. 343–365). IGI Global. doi:10.4018/978-1-7998-6922-1.ch015

Martin, F., Bacak, J., Dymes, L., & Polly, D. (in press). A systematic review of research on K12 online teaching and learning: A comparison of research from two decades 2000 to 2019. *Journal of Research on Technology in Education.*

Martin, F., Budhrani, K., & Wang, C. (2019). Examining faculty perception of their readiness to teach online. *Online Learning*, *23*(3), 97–119. doi:10.24059/olj.v23i3.1555

Martin, F., & Parker, M. A. (2014). Use of Synchronous Virtual Classrooms: Why, Who, and How? *Journal of Online Learning and Teaching*, *10*(2). Retrieved June 30, 2021, from https://www.researchgate.net/profile/Florence-Martin-7/publication/265335172

Martin, P. C. (2016). Test-based education for students with disabilities and English language learners: The impact of assessment pressures on educational planning. *Teachers College Record*, *118*(14), 1–24. https://eric.ed.gov/?id=EJ1110765

Martz, B., Hughes, J., & Braun, F. (2017). Creativity and problem-solving: Closing the skills gap. *Journal of Computer Information Systems*, *57*(1), 39–48. doi:10.1080/08874417.2016.1181492

Masinter, M. R. (2020). Undue burdens in the time of coronavirus. *Disability Compliance for Higher Education*, *25*(11), 3. doi:10.1002/dhe.30853

McAllum, K. (2016). Managing imposter syndrome among the trophy kids: Creating teaching practices that develop independence in Millennial students. *Communication Education*, *65*(3), 363–365. doi:10.1080/03634523.2016.1177848

McCartin, R., & Schill, W. J. (1977, October). Three modes of instruction. *Journal of American Indian Education*, 14–20.

McCarty, T., & Lee, T. (2014, April). Critical Culturally Sustaining/Revitalizing Pedagogy and Indigenous Education Sovereignty. *Harvard Educational Review*, *84*(1), 101–124. doi:10.17763/haer.84.1.q83746nl5pj34216

McGrail, K. M., Ahuja, M. A., & Leaver, C. A. (2017). Virtual Visits and Patient-Centered Care: Results of a Patient Survey and Observational Study. *Journal of Medical Internet Research*, *19*(5), e177. doi:10.2196/jmir.7374 PMID:28550006

McKibben. (2020). The economic impact of COVID-19. In R. Baldwin & B. Weder di Mauro (Eds.), *Economics in the Time of COVID-19* (pp. 45–53). CEPR Press.

McLesky, J., Barringer, M. D., Billingsley, B., Brownell, M., Jackson, D., Kennedy, M., Lewis, T., Maheady, L., Rodriguez, J., Scheeler, M. C., Winn, J., & Ziegler, D. (2017). *High leverage practices in special education.* Council for Exceptional Children & CEEDAR Center. https://ceedar.education.ufl.edu/wp-content/uploads/2017/07/CEC-HLP-Web.pdf

McNamee, T. Y., Willis, J., Ganss, K. M., Ardoin, S., & Sansone, V. A. (2020). Don't forget about rural higher education students: Addressing digital inequities during COVID-19. *Diverse Issues in Higher Education*, *37*(7), 12–13.

Mechanic, O. J., Persaud, Y., & Kimball, A. B. (2020). Telehealth Systems. In *StatPearls*. StatPearls Publishing.

Medeiros, K. E., Partlow, P. J., & Mumford, M. D. (2014). Not too much, not too little: The influence of constraints on creative problem-solving. *Psychology of Aesthetics, Creativity, and the Arts, 8*(2), 198-210. https://psycnet.apa.org2014-12651-001

Mercier, E., Rattray, J., & Lavery, J. (2015). Twitter in the collaborative classroom: Micro-blogging for in-class collaborative discussions. *International Journal of Social Media and Interactive Learning Environments*, *3*(2), 83–99. doi:10.1504/IJSMILE.2015.070764

Merrett, C. D. (2004). Social justice: What is it? Why teach it? *The Journal of Geography*, *103*(3), 93–101. doi:10.1080/00221340408978584

Messersmith, A. S. (2015). Preparing students for 21st century teamwork: Effective collaboration in the online group communication course. *Communication Teacher*, *29*(4), 219–226. doi:10.1080/17404622.2015.1046188

Meyer, D. (2011). *The three acts of a mathematical story*. Dy/Dan. Retrieved from: http://blog.mrmeyer.com/2011 /the-three-acts-of-a-mathematical-story/

Miller, D. (1999). *Principles of Social Justice*. Harvard University Press.

Mishra, P., & Koehler, M. (2006). Technological pedagogical content knowledge: A framework for teacher knowledge. *Teachers College Record*, *108*(6), 1017–1054. doi:10.1111/j.1467-9620.2006.00684.x

Mnookin, S. (2016). *Out of the shadows: Making mental health a global development priority*. World Bank Group. Retrieved from https://documents.worldbank.org/curated/en/2016/04/26281016/out-shadowsmaking-mental-health-global-development-priority

Monteiro, N. M., Ndiaye, Y., Blanas, D., & Ba, I. (2014). Policy perspectives and attitudes towards mental health treatment in rural Senegal. *International Journal of Mental Health Systems*, *8*(1), 1–16. doi:10.1186/1752-4458-8-9 PMID:24646335

Morgan, H. (2016). Relying on high-stakes standardized tests to evaluate schools and teachers: A bad idea. *The Clearing House: A Journal of Educational Strategies, Issues and Ideas*, *89*(2), 67–72. doi:10.1080/00098655.2016.1156628

Morse, S. (n.d.). *457 hospitals settle for $250 million over IMPROPER implantable cardioverter Defibrillator surgeries*. Healthcare Finance News. https://www.healthcarefinancenews.com/news/457-hospitals-settle-250-million-over-improper-implantable-cardioverter-defibrillator-surgeries#:~:text=than%20%24250%20million.-,The%20biggest%20hospital%20chain%20settlements%20are%20HCA%2C%20at%20%2415.8%20million,cardioverter%20defibrillator%20cost%20about%20%2425%2C000.&text=The%20whistleblowers%20have%20received%20more%20than%20%2438%20million%20from%20the%20settlements

Mouza, C. (2009). Does research-based professional development make a difference? A longitudinal investigation of teacher learning in technology integration. *Teachers College Record*, *111*, 1195–1241.

Munter, C., & Wilhelm, A. G. (2021). Mathematics teachers' knowledge, networks, practice, and change of instructional visions. *Journal of Teacher Education*, *72*(3), 342–354. doi:10.1177/0022487120949836

Nair, L., & Adetayo, O. A. (2019). Cultural Competence and Ethnic Diversity in Healthcare. *Plastic and Reconstructive Surgery. Global Open*, *7*(5), e2219. doi:10.1097/GOX.0000000000002219 PMID:31333951

National Association of Social Workers. (2017). *Code of Ethics*. https://www.socialworkers.org/About/Ethics/Code-of-Ethics

National Council of Teachers of Mathematics. (2014). *Principles to actions: Ensuring mathematics success for all*. National Council of Teachers of Mathematics.

National Institute of Mental Health. (2017). *Mental illness*. Retrieved from https://www.nimh.nih.gov/health/statistics/mental-illness.shtml

Neimeyer, R. A., & Burke, L. A. (2012). Complicated grief and the end-of-life: Risk factors and treatment considerations. In J. T. Werth (Ed.), *Counseling clients near the end-of-life: A practical guide for mental health professions* (pp. 205–228). Springer.

Neimeyer, R. A., Burke, L. A., Mackay, M. M., & Stringer, J. G. (2010). Grief therapy and the reconstruction of meaning: From principles to practice. *Journal of Contemporary Psychotherapy*, *40*(2), 73–78. doi:10.100710879-009-9135-3

Nepomuceno, B., Cardoso, A., Ximenes, V., Barros, J., & Leite, J. (2016). Mental health, well-being, and poverty: A study in urban and rural communities in Northeastern Brazil. *Journal of Prevention & Intervention in the Community*, *44*(1), 63–75. doi:10.1080/10852352.2016.1102590 PMID:26697905

Newton, L., & Beverton, S. (2012). Pre-service teachers' conceptions of creativity in elementary school English. *Thinking Skills and Creativity*, *7*(3), 165–176. doi:10.1016/j.tsc.2012.02.002

Niess, M. L. (2005). Preparing teachers to teach science and mathematics with technology: Developing a technology pedagogical content knowledge. *Teaching and Teacher Education*, *21*(5), 509–523. doi:10.1016/j.tate.2005.03.006

Noddings, N. (2013). Standardized curriculum and loss of creativity. *Theory into Practice*, *52*(3), 210–215. doi:10.1080/00405841.2013.804315

Nong, P., Raj, M., Creary, M., Kardia, S. L., & Platt, J. E. (2020). Patient-Reported experiences of discrimination in the US health care system. *JAMA Network Open*, *3*(12), 1–11. doi:10.1001/jamanetworkopen.2020.29650 PMID:33320264

Nuriddin, A., Mooney, G., & White, A. I. (2020). Reckoning with histories of MEDICAL racism and violence in the USA. *Lancet*, *396*(10256), 949–951. doi:10.1016/S0140-6736(20)32032-8 PMID:33010829

O'Reilly, K. B. (2019). *Federal law cripples telehealth in Medicare. New bill changes that.* Retrieved from https://www.ama-assn.org/practice management/digital/federal-law-cripples-telehealth-medicare-new-bill changes

OECD. (2018), *Equity in Education: Breaking Down Barriers to Social Mobility*. PISA, OECD Publishing. doi:10.1787/9789264073234-en

OECD. (2021). *Teachers by age (indicator).* doi:10.1787/93af1f9d-en

Official Report of the Nineteenth Annual Conference of Charities and Correction (1892), 46–59. Retrieved from http://historymatters.gmu.edu/d/4929/

Oliff, H. (2017, May 16). *Graduation rates and American Indian Education* [Web blog post]. Retrieved from http://nativepartnership.org/graduation-rates-american-indian-education/

Olivant, K. F. (2015). I am not a format: Teachers' experiences with fostering creativity in the era of accountability. *Journal of Research in Childhood Education*, *29*(1), 115–129. doi:10.1080/02568543.2014.978920

Olszewski, P. (2016). Teaching Millennials how to study under the 21st Century sky. *Pyrex Journal of Educational Research and Reviews*, *2*(1), 1-9. https://www.pyrexjournals.org/pjerr

Osborne, A. G., & Russo, C. J. (2021). *Special education and the law: A guide for practitioners.* Corwin.

Ossebaard, H. C., & Van Gemert-Pijnen, L. (2016). EHealth and quality in health CARE: Implementation time. *International Journal for Quality in Health Care, 28*(3), 415–419. doi:10.1093/intqhc/mzw032 PMID:27029590

Pajares, F. (1992). Teachers' beliefs and educational research: Clearing up a messy construct. *Review of Educational Research, 62*(2), 307–332. doi:10.3102/00346543062003307

Pang, W. (2015). Promoting creativity in the classroom: A generative view. *Psychology of Aesthetics, Creativity, and the Arts, 9*(2), 122–127. doi:10.1037/aca0000009

Paris, D. (2012). Culturally sustaining pedagogies: A needed change in stance, terminology, and practice. *Educational Researcher, 41*(3), 93–97. doi:10.3102/0013189X12441244

Paris, D., & Alim, H. S. (2017). *Culturally sustaining pedagogies: Teaching and learning for justice in a changing world.* Teachers College Press.

Park, C., & Kim, D. (2020). Perception of instructor presence and its effects on learning experience in online classes. *Journal of Information Technology Education, 19*, 475–488. doi:10.28945/4611

Park, Y. T. (2016). Emerging New Era of Mobile Health Technologies. *Healthcare Informatics Research, 22*(4), 253–254. doi:10.4258/hir.2016.22.4.253 PMID:27895955

Patel, V. (2014). Why mental health matters to global health. *Transcultural Psychiatry, 51*(6), 777–789. doi:10.1177/1363461514524473 PMID:24595266

Patel, V., & Prince, M. (2010). Global mental health: A new global health field comes of age. *Journal of the American Medical Association, 303*(19), 1976–1977. doi:10.1001/jama.2010.616 PMID:20483977

Paul Keckley. (2020, May 18). *COVID-19 unintended Consequences: UNFLATTERING attention to the gap between "haves" and "have nots".* https://www.paulkeckley.com/the-keckley-report/2020/5/18/covid-19-unintended-consequences-unflattering-attention-to-the-gap-between-haves-and-have-nots

Paul Keckley. (2020, May 18). *COVID-19 unintended Consequences: Unflattering attention to the gap between "haves" and "have nots".* Paul Keckley. https://www.paulkeckley.com/the-keckley-report/2020/5/18/covid-19-unintended-consequences-unflattering-attention-to-the-gap-between-haves-and-have-nots

Paulson, L. (2010). *The lived experience of supervisors in rural settings.* Retrieved from https://www.counseling.org/docs/default-source/vistas/the-lived-experience-ofsupervisors-in-rural-settings.pdf?sfvrsn=1

Peek, M. E., Odoms-Young, A., Quinn, M. T., Gorawara-Bhat, R., Wilson, S. C., & Chin, M. H. (2010). Racism in healthcare: Its relationship to shared decision-making and health disparities: a response to Bradby. *Social Science & Medicine, 71*(1), 13–17. doi:10.1016/j.socscimed.2010.03.018

Pelling, N., & Butler, B. (2015). Rural counselors and psychologists in Australia: A descriptive survey of professionals and practice. *International Journal of Mental Health*, *44*(1/2), 94–106. doi:10.1080/00207411.2015.1009782

Penner, L. A., Blair, I. V., Albrecht, T. L., & Dovidio, J. F. (2014). *Reducing racial health Care Disparities: A social psychological analysis*. Retrieved May 01, 2021, from https://journals.sagepub.com/doi/10.1177/2372732214548430

Penner, L. A., Dovidio, J. F., Gonzalez, R., Albrecht, T. L., Chapman, R., Foster, T., Harper, F. W., Hagiwara, N., Hamel, L. M., Shields, A. F., Gadgeel, S., Simon, M. S., Griggs, J. J., & Eggly, S. (2016). The Effects of Oncologist Implicit Racial Bias in Racially Discordant Oncology Interactions. *Journal of Clinical Oncology*, *34*(24), 2874–2880. doi:10.1200/JCO.2015.66.3658 PMID:27325865

Perrin, P. B., Rybarczyk, B. D., Pierce, B. S., Jones, H. A., Shaffer, C., & Islam, L. (2020). Rapid telepsychology deployment during the COVID-19 pandemic: A special issue commentary and lessons from primary care psychology training. *Journal of Clinical Psychology*, *76*, 1173–1185. https://dx.doi.org/10.1002/jclp.22969

Perry, K., Gold, S., & Shearer, E. M. (2020). Identifying and addressing mental health providers' perceived barriers to clinical video telehealth utilization. *Journal of Clinical Psychology*, *76*, 1125–1134. http://dx.doi .org/10.1002/jclp.22770

Pescosolido, B. A., Martin, J. K., Long, J. S., Medina, T. R., Phelan, J. C., & Link, B. G. (2010). "A disease like any other"? A decade of change in public reactions to schizophrenia, depression, and alcohol dependence. *The American Journal of Psychiatry*, *167*(11), 1321–1330. doi:10.1176/appi.ajp.2010.09121743 PMID:20843872

Pewewardy, C. D., & Hammer, P. (2003). *Culturally Responsive Teaching for American Indian Students*. ERIC Digest.

Pfefferbaum, B., & North, C. S. (2020). Mental Health and the Covid-19 Pandemic. *The New England Journal of Medicine*, *383*(6), 510–512. doi:10.1056/NEJMp2008017 PMID:32283003

Pierce, B. S., Perrin, P. B., Tyler, C. M., & McKee, G. B. (2021). The COVID-19 Telepsychology Revolution: A National Study of Pandemic-Based Changes in U.S. Mental Health Care Delivery. *The American Psychologist*, *76*(1), 14–25.

Pinotti, G. (2021). *Equal Education for All? Equity Issues Impacting Physical Education and Physical Activity During COVID-19 Pandemic*. Academic Press.

Pitman, A., Mann, F., & Johnson, S. (2018). Advancing our Understanding of Loneliness and Mental Health Problems in Young People. *The Lancet Psychiatry*. doi:https://doi.org/10.1016/S2215-0366(18)30436-X

Polly, D. (2011). Developing teachers' technological, pedagogical, and content knowledge (TPACK) through mathematics professional development. *The International Journal for Technology in Mathematics Education*, *18*(2), 83–96.

Polly, D. (2016). Examining Elementary School Teachers' Enactment of Mathematical Tasks and Questions. *Research in the Schools, 23*(2), 61–71.

Polly, D., Byker, E. J., & Colonnese, M. W. (2021). Future directions for K-12 technology enhanced learning environments. *TechTrends, 65*(3), 240–242. doi:10.100711528-021-00602-y PMID:33817701

Polly, D., & Hannafin, M. J. (2010). Reexamining technology's role in learner-centered professional development. *Educational Technology Research and Development, 58*(5), 557–571. doi:10.100711423-009-9146-5

Polly, D., & Hannafin, M. J. (2011). Examining how learner-centered professional development influences teachers' espoused and enacted practices. *The Journal of Educational Research, 104*(2), 120–130. doi:10.1080/00220671003636737

Polson, C. (2018). TAKS-ing students? Texas exit exam effects on human capital formation. *Economics of Education Review, 62*, 129–150. doi:10.1016/j.econedurev.2017.09.009

Population Reference Bureau. (2020). *Children, coronavirus, and the digital divide: Native American, black, and Hispanic students at greater Educational risk during pandemic*. Retrieved April 27, 2021, from https://www.prb.org/coronavirus-digital-divide-education/

Pramling Samuelsson, I., Wagner, J. T., & Eriksen Ødegaard, E. (2020). The coronavirus pandemic and lessons learned in preschools in Norway, Sweden and the United States: OMEP Policy Forum. *International Journal of Early Childhood, 52*(2), 129–144. doi:10.100713158-020-00267-3 PMID:32836368

Pratt, R. H. (n.d.). *Series I: General Correspondence and Official Papers, 1867-1924*. https://archives.yale.edu/repositories/11/archival_objects/435399

Pratt, R. H., Utley, R. M., & Adams, D. W. (2003). *Battlefield and classroom: four decades with the American Indian, 1867-1904*. University of Oklahoma Press.

Procter, N., Ferguson, M., Backhouse, J., Cother, I., Jackson, A., Murison, J., & Reilly, J. (2015). Face to face, person to person: Skills and attributes deployed by rural mental health clinicians when engaging with consumers. *The Australian Journal of Rural Health, 23*(6), 352–358. doi:10.1111/ajr.12204 PMID:26683718

Project Hope. (2018). Foundation funding for telehealth. *Health Affairs, 37*(12), 2096–2097. doi:10.1377/hlthaff.2018.05150 PMID:30508391

Ramirez, M., Castaneda, A., & Herold, P. L. (1974). The relationship of acculturation to cognitive style among Mexican Americans. *Journal of Cross-Cultural Psychology, 5*(4), 424–433. doi:10.1177/002202217400500404

Rathod, S., Pinninti, N., Irfan, M., Gorczynski, P., Rathod, P., Gega, L., & Naeem, F. (2017). Mental Health Service Provision in Low- and Middle-Income Countries. *Health Services Insights, 10*. Advance online publication. doi:10.1177/1178632917694350 PMID:28469456

Ratts, M., Singh, A., Nassar-McMillan, S., Butler, S., & McCullough, J. (2016). Multicultural and social justice counseling competencies: Guidelines for the counseling profession. *Journal of Multicultural Counseling and Development*, *44*, 28–48. doi:10.1002/jmcd.12035

Reay, R. E., Looi, J. C., & Keightley, P. (2020). Telehealth mental health services during COVID-19: Summary of evidence and clinical practice. *Australasian Psychiatry*, *28*(5), 514–516. doi:10.1177/1039856220943032 PMID:32722963

Reck, R. M. (2019). Designing a student video presentation assignment: Considerations for minimizing stress. *2019 IEEE Frontiers in Education Conference (FIE), Frontiers in Education Conference (FIE), 2019 IEEE*, 1–5.

Rivenbark, J. G., & Ichou, M. (2020). Discrimination in healthcare as a barrier to Care: Experiences of socially disadvantaged populations in France from a nationally representative survey. *BMC Public Health*, *20*(1), 31. Advance online publication. doi:10.118612889-019-8124-z PMID:31918699

Robinson, H. A., Kilgore, W., & Warren, S. J. (2017). Care, communication, learner support: Designing meaningful online collaborative learning. *Online Learning*, *21*(4), 29–51. doi:10.24059/olj.v21i4.1240

Robinson, W. D., Geske, J., Backer, E., Jarzynka, K., Springer, P. R., Bischoff, R., & Swinton, J. (2012). Rural experiences with mental illness: Through the eyes of patients and their families. *Families, Systems & Health*, *30*(4), 308–321. doi:10.1037/a0030171 PMID:23088821

Rodriguez, K. (2018). *Native American Education: Building Stronger Families, Communities, and Youth through Cultural Education*. Retrieved from https://digital.sandiego.edu/solesmalscap/22

Rogers, C., & Ng, N. E. (2020). *Memorandum: Report of Initial Findings: COVID-19 Impact on Public Schools*. AASA: The School Superintendents Association.

Ro, J. (2018). Learning to teach in the era of test-based accountability: A review of research. *Professional Development in Education*, *45*(1), 87–101. doi:10.1080/19415257.2018.1514525

Roncoroni, J., Tucker, C. M., & Wall, W. (2014). Patient perceived cultural sensitivity of clinic environment and its association with patient satisfaction with care and treatment adherence. *American Journal of Lifestyle Medicine*, (8), 421–429. doi:10.1177/1559827614521760

Ronquillo, Y., Meyers, A., & Korvek, S. J. (2020). Digital Health. In *StatPearls*. StatPearls Publishing. Available from: https://www.ncbi.nlm.nih.gov/books/NBK470260/

Roskes, M. (2015). Constraints that help or hinder creative performance: A motivational approach. *Creativity and Innovation Management*, *24*(2), 197–206. doi:10.1111/caim.12086

Rosso, B. D. (2014). Creativity and constraints: Exploring the role of constraints in the creative processes of research and development teams. *Organization Studies*, *35*(4), 551–585. doi:10.1177/0170840613517600

Saleh, A. (2020). Impact of online course discussion on education leadership students' reflections. *Journal of Education and Learning*, *14*(2), 212–216. doi:10.11591/edulearn.v14i2.14607

Sanchez, J. G., & Patel, L. (2017). Growing up tested: Teachers' lived experiences of testing as students. *Peabody Journal of Education, 92*(4), 440–444. doi:10.1080/0161956X.2017.1349444

Sanders, J. M., Monogue, M. L., Jodlowski, T. Z., & Cutrell, J. B. (2020). Pharmacologic treatments for coronavirus disease 2019 (COVID-19): A Review. *Journal of the American Medical Association, 323*(18), 1824–1836. doi:10.1001/jama.2020.6019

Santelli, B., Stewart, K., & Mandernach, J. (2020). Supporting high quality teaching in online programs. *Journal of Educators Online, 17*(1), 1–15.

Schiffrin, H. H., & Liss, M. (2017). The effects of helicopter parenting on academic motivation. *Journal of Child and Family Studies, 26*(5), 1472–1480. doi:10.100710826-017-0658-z

Schreiber, J. B. (2018). Creativity and pre-service teacher education: What you see is what you get. *Global Education Review, 5*(1), 5–19. https://ger.mercy.edu/index.php/ger/article/view/421

Schulte, F. (2021, May 17). *Humana health Plan Overcharged Medicare by nearly $200 Million, federal audit finds*. Kaiser Health News. https://khn.org/news/article/humana-health-plan-overcharged-medicare-by-nearly-200-million-federal-audit-finds/

Schulte, F. (2021, May 17). *Humana health PLAN OVERCHARGED Medicare by nearly $200 Million, federal audit finds*. Kaiser Health News. https://khn.org/news/article/humana-health-plan-overcharged-medicare-by-nearly-200-million-federal-audit-finds/

Schwamm, L. H., Estrada, J., Erskine, A., & Licurse, A. (2020). Virtual care: New models of caring for our patients and workforce. *The Lancet. Digital Health, 2*(6), 1–4. doi:10.1016/S2589-7500(20)30104-7 PMID:32382724

Shaw, J., Brewer, L. C., & Veinot, T. (2021). Recommendations for Health Equity and Virtual Care Arising From the COVID-19 Pandemic: Narrative Review. *JMIR Formative Research, 5*(4), e23233. doi:10.2196/23233 PMID:33739931

Shigekawa, E., Fix, M., Corbett, G., Roby, D. H., & Coffman, J. (2018). The Current State Of Telehealth Evidence: A Rapid Review. *Health Affairs (Project Hope), 37*(12), 1975–1982. doi:10.1377/hlthaff.2018.05132 PMID:30633674

Sieck, C. J., Hefner, J. L., & McAlearney, A. S. (2018). Improving the patient experience through patient portals: Insights from experienced portal users. *Patient Experience Journal, l*(5), 47–54. doi:10.35680/2372-0247.1269

Silvestru, C. I., Mehedinţu, G., Silvestru, R. C., & Icociu, V. C. (2019). Millennials and the Need for Adapting e-Learning Methods for Business Performance–a Case Study. In *The International Scientific Conference eLearning and Software for Education* (*Vol. 3*, pp. 225-231). "Carol I" National Defence University.

Simple Practice. (2021). *Support clients anywhere, at any time*. https://www.simplepractice.com/telehealth/?utm_source=google&utm_medium=cpc&utm_campaign=rc-search-non-brand-dsa&utm_term=&gclid=Cj0KCQiApY6BBhCsARIsAOI_GjYIrCOtvVuOQDGy86qC-JcaEnKDG7oRiL9cfFzxmNZYwzPbG5of4xQaAh_OEALw_wcB

Sinclair, C. C., Holloway, K., Riley, A. M. G., Auret, K., & Eysenbach, G. (2013). Online mental health resources in rural Australia: Clinician perceptions of acceptability. *Journal of Medical Internet Research, 15*(9), e193. Advance online publication. doi:10.2196/jmir.2772 PMID:24007949

Skinner, B. F. (1990). *The behavior of organisms: An experimental analysis*. B. F. Skinner Foundation.

Smith, A. C., Thomas, E., Snoswell, C. L., Haydon, H., Mehrotra, A., Clemensen, J., & Caffery, L. J. (2020). Telehealth for global emergencies: Implications for coronavirus disease 2019 (COVID-19). *Journal of Telemedicine and Telecare, 26*(5), 309–313. doi:10.1177/1357633X20916567 PMID:32196391

Smith, M. S., & Stein, M. K. (1998). Selecting mathematical tasks: From research to practice. *Mathematics Teaching in the Middle School, 3*, 344–350. doi:10.5951/MTMS.3.5.0344

Snell-Rood, C., & Carpenter-Song, E. (2018). Depression in a depressed area: Deservingness, mental illness, and treatment in the contemporary rural U.S. *Social Science & Medicine, 219*, 78–86. doi:10.1016/j.socscimed.2018.10.012 PMID:30391873

Soltis-Jarrett, V. (2019). The TANDEM3-PC: The Foundation for an Innovative, Integrated Behavioral Health NP-led Model of Practice in Rural Primary Care. *Archives of Psychiatric Nursing, 33*(1), 2–10. doi:10.1016/j.apnu.2018.08.007 PMID:30663620

Sorkin, D. H., Ngo-Metzger, Q., & De Alba, I. (2010). Racial/Ethnic Discrimination in Health Care: Impact on Perceived Quality of Care. *Journal of General Internal Medicine, 25*(5), 390–396. doi:10.100711606-010-1257-5 PMID:20146022

Sottile, C. (2020, October 16). *Native American Teachers, entrepreneurs seek new ways to close digital divide*. Retrieved April 28, 2021, from https://www.nbcnews.com/news/us-news/native-american-teachers-entrepreneurs-seek-new-ways-close-digital-divide-n1243746

Statista Research Department. (2021, January 20). *Size of urban and rural population of the U.S. 2020*. Retrieved May 01, 2021, from https://www.statista.com/statistics/985183/size-urban-rural-population-us/

Stein, M. K., Remillard, J. T., & Smith, M. S. (2007). How curriculum influences student learning. In F. Lester (Ed.), *Second handbook of research on mathematics teaching and learning* (pp. 319–370). Information Age Publishing.

Stepanikova, I., & Oates, G. R. (2017). Perceived Discrimination and Privilege in Health Care: The Role of Socioeconomic Status and Race. *American Journal of Preventive Medicine, 52*(1S1), S86–S94. doi:10.1016/j.amepre.2016.09.024

Stokes, P. D. (2006). *Creativity From constraints: The psychology of breakthrough*. Springer Publishing Company.

Stokes, P. D. (2008). Creativity from constraints: What can we learn from Motherwell? from Modrian? from Klee? *The Journal of Creative Behavior, 42*(4), 223–236. doi:10.1002/j.2162-6057.2008.tb01297.x

Stokes, P. D. (2009). Using constraints to create novelty: A case study. *Psychology of Aesthetics, Creativity, and the Arts*, *3*(3), 174–180. doi:10.1037/a0014970

Stokes, P. D. (2013). The effects of constraints in a mathematics classroom. *Journal of Mathematics Education at Teachers College*, *4*(2), 25–31. https://jmetc.columbia.edu/

Stokes, P. D. (2014). Thinking inside the box: Creativity, constraints, and the colossal portraits of Chuck Close. *The Journal of Creative Behavior*, *48*(4), 276–289. doi:10.1002/jocb.52

Suarez, R. D. A. (2021). *Collaboration for Healthcare Virtual Teams: Supporting Creativity Through Diversity and Knowledge Management* (Doctoral dissertation). Capella University.

Sue, D. W., Arredondo, P., & McDavis, R. J. (1992). Multicultural counseling competencies and standards: A call to the profession. *Journal of Multicultural Counseling and Development*, *20*(2), 64–88. https://doi.org/10.1002/j.2161-1912.1992.tb00563.x

Summers-Gabr, N. M. (2020). Rural–urban mental health disparities in the United States during COVID-19. *Psychological Trauma: Theory, Research, Practice, and Policy*, *12*(S1), S222–S224. doi:10.1037/tra0000871 PMID:32478541

Tallent, R. J., & Barnes, J. J. (2015). Think bubbles and Socrates: Teaching critical thinking to Millennials in public relations classes. *Universal Journal of Educational Research*, *3*(7), 435–441. doi:10.13189/ujer.2015.030702

Tan, O. S. (2015). Flourishing creativity: Education in an age of wonder. *Asia Pacific Education Review*, *16*(2), 161–166. doi:10.100712564-015-9377-6

Taylor, J. (2019). *Racism, Inequality, and Health Care for African Americans*. The Century Foundation.

Telecommunications Act of 1996, Pub. LA. No. 104-104, 110 Stat.

Texmed. (n.d.). *Texas medicine today*. https://www.texmed.org/tmt/

The United States Department of Justice. (2017, May 31). *Electronic health RECORDS vendor to pay $155 million to Settle false Claims Act Allegations*. https://www.justice.gov/opa/pr/electronic-health-records-vendor-pay-155-million-settle-false-claims-act-allegations

Thompson, T., & Brailer, D. (2004). *Health IT Strategic Framework*. US Department of Health and Human Services.

Tinubu Ali, T., & Herrera, M. (2020). *Distance Learning During COVID-19: Seven Equity Considerations for Schools and Districts*. Southern Education Foundation. https://www.southerneducation.org/covid-19-digital-equity/

Torales, J., O'Higgins, M., Castaldelli-Maia, J. M., & Ventriglio, A. (2020). The outbreak of COVID-19 coronavirus and its impact on global mental health. *The International Journal of Social Psychiatry*, *66*(4), 317–320. doi:10.1177/0020764020915212 PMID:32233719

Torrents-Martin, C., Ric, A., & Hristovski, R. (2015). Creativity and emergence of specific dance movements using instructional constraints. *Psychology of Aesthetics, Creativity, and the Arts*, *9*(1), 65–74. doi:10.1037/a0038706

TSHA. (n.d.). *Pratt, Richard Henry (1840–1924)*. https://www.tshaonline.org/handbook/entries/pratt-richard-henry#:~:text=With%20the%20outbreak%20of%20the,Rosecrans%20and%20George%20H

Tuckson, R. V., Edmunds, M., & Hodgkins, M. L. (2017). Telehealth. *The New England Journal of Medicine*, *377*(16), 1585–1592. doi:10.1056/NEJMsr1503323 PMID:29045204

Turner, C., & Klein, R. (2021, June 16). *After months of special education turmoil, families say schools owe them.* NPR. https://www.npr.org/2021/06/16/994587239/after-months-of-special-education-turmoil-families-say-schools-owe-them

U. S. Department of Agriculture. (2020, October 13). *Racial and ethnic minorities made up about 22 percent of the rural population in 2018, compared to 43 percent in urban areas.* Retrieved April 28, 2021, from https://www.ers.usda.gov/data-products/chart-gallery/gallery/chart-detail/?chartId=99538

U. S. Department of Education. (2006). Assistance to states for the education of children with disabilities and preschool grants for children with disabilities; final regulations. *Federal Register,* 34. CFR parts 300 and 301.

U.S. Census Bureau, International Data Base, & USA Trade Online. (2021). *U.S. and World Population Clock.* U.S. Census Bureau. Retrieved from https://www.census.gov/popclock/world

U.S. Department of Education. (2010). Office for Civil Rights, Free Appropriate Public Education for Students With Disabilities: Requirements Under Section 504 of the Rehabilitation Act of 1973. Washington, DC: Author.

U.S. Food and Drug Administration. (2020). *FDA Takes Additional Action in Fight Against COVID-19 By Issuing Emergency Use Authorization for Second COVID- 19 Vaccine.* https://www.fda.gov/news-events/press-announcements/fda-takes-additional- ction-fight-against-covid-19-issuing-emergency-use-authorization-second-covid

Ujifusa, A. (2020, December 4). *1 in 3 American Indian, black, and Latino children fall into digital divide, study says.* Retrieved April 28, 2021, from https://www.edweek.org/education/1-in-3-american-indian-black-and-latino-children-fall-into-digital-divide-study-says/2020/07

United Nations Children's Fund and International Telecommunication Union. (2020). *How many children and young people have internet access at home? Estimating digital connectivity during the COVID-19 pandemic.* UNICEF.

United Nations. (2020). *Policy brief: Education during COVID-19 and beyond.* United Nations. https://www.un.org/development/desa/dspd/wp-content/uploads/sites/22/2020/08/sg_policy_brief_covid-19_and_education_august_2020.pdf

United Nations. (2020, August). *Policy Brief: Education during COVID-19 and beyond.* Author.

United States Centers for Medicare and Medicaid Services. (2019). *Tele medicine*. Retrieved from https://www.medicaid.gov/medicaid/benefits/ telemedicine/index.html

United States Department of Health and Human Services' Office for Civil Rights. (2020a). *Bulletin: HIPAA Privacy and Novel Coronavirus.* Retrieved from https://www.hhs.gov/sites/default/files/february-2020- hipaa-and-novel-coronavirus.pdf

United States Department of Health and Human Services' Office for Civil Rights. (2020b). *OCR eases HIPAA telehealth enforcement for COVID-19 emergency.* Retrieved from https://www.aha.org/news/headline/2020-03-18-ocr-eases-hipaa-telehealth-enforcement-covid-19-emergency

University of Oxford. (n.d.). *Social media manipulation by political actors an industrial scale problem.* https://www.ox.ac.uk/news/2021-01-13-social-media-manipulation-political-actors-industrial-scale-problem-oxford-report

Urbina, A., & Polly, D. (2017). Examining elementary school teachers' integration of technology and enactment of TPACK in mathematics. *International Journal of Information and Learning Technology, 34*(5), 439–451. doi:10.1108/IJILT-06-2017-0054

Valtorta, N. K., Kanaan, M., Gilbody, S., Ronzi, S., & Hanratty, B. (2016). Loneliness and social isolation as risk factors for coronary heart disease and stroke: Systematic review and meta-analysis of longitudinal observational studies. *Heart (British Cardiac Society), 102*, 1009–1016. https://doi.org/10.1136/heartjnl-2015-308790

Van der Vegt, G. S., & Janssen, O. (2003). Joint impact of interdependence and group diversity on innovation. *Journal of Management, 29*(5), 729–751. doi:10.1016/S0149-2063_03_00033-3

Van Leeuwen, M. S. (1978). A cross-cultural examination of psychological differentiation in males and females 1. *International Journal of Psychology, 13*(2), 87–122. doi:10.1080/00207597808246615

Van Overschelde, J. P., Rawson, K. A., & Dunlosky, J. (2004). Category norms: An updated and expanded version of the Battig and Montague (1969) norms. *Journal of Memory and Language, 50*(3), 289–335. doi:10.1016/j.jml.2003.10.003

VandenAvond, L., Menard, K., Johnson, K., Cameron-Standerford, A., Bergh, B., & Edge, C. (2020). *Care and rigor in online courses: An analysis of faculty and student perspectives.* International Council for Distance Education.

Varker, T., Brand, R. M., Ward, J., Terhaag, S., & Phelps, A. (2019). Efficacy of synchronous telepsychology interventions for people with anxiety, depression, posttraumatic stress disorder, and adjustment disorder: A rapid evidence assessment. *Psychological Services, 16*(4), 621–635. doi:10.1037er0000239 PMID:29809025

Vaughn, S. R., Bos, C. S., & Schumm, J. S. (2018). *Teaching Students Who are at Risk in the General Education Classroom.* Pearson.

Vegesna, A., Tran, M., Angelaccio, M., & Arcona, S. (2017). Remote Patient Monitoring via Non-Invasive Digital Technologies: A Systematic Review. *Telemedicine Journal and E-Health, 23*(1), 3–17. doi:10.1089/tmj.2016.0051

Verhoeven, F., Tanja-Dijkstra, K., Nijland, N., Eysenbach, G., & van Gemert-Pijnen, L. (2010). Asynchronous and synchronous teleconsultation for diabetes care: A systematic literature review. *Journal of Diabetes Science and Technology*, *4*(3), 666–684. doi:10.1177/193229681000400323 PMID:20513335

Vidourek, R. A., King, K. A., Nabors, L. A., & Merianos, A. L. (2014). Students' benefits and barriers to mental health help-seeking. *Health Psychology and Behavioral Medicine*, *2*(1), 1009–1022. doi:10.1080/21642850.2014.963586 PMID:25750831

Volterrani, M. & Sposato, B. (2019). Remote monitoring and telemedicine. *European Heart Journal Supplements*, *21*(Supplement M), M54-M56. doi:10.1093/eurheartj/suz266

Wagner, J. (2021, July 26). Biden says long-term effects of Covid-19 can be considered a disability under federal civil rights laws. *The Washington Post*. https://www.washingtonpost.com/politics/biden-ada-long-covid-disability/2021/07/26/972f2a04-ee20-11eb-a452-4da5fe48582d_story.html

Wang, X. H. F., Kim, T. Y., & Lee, D. R. (2016). Cognitive diversity and team creativity: Effects of team intrinsic motivation and transformational leadership. *Journal of Business Research*, *69*(9), 3231–3239. doi:10.1016/j.jbusres.2016.02.026

Warner-Lambert to pay $430 million to resolve criminal & civil health care liability relating to off-label promotion. (n.d.). https://www.justice.gov/archive/opa/pr/2004/May/04_civ_322.htm

Warren, J. C., & Smalley, K. B. (2020). *Using Telehealth to Meet Mental Health Needs During the COVID-19 Crisis*. Commonwealth Fund. doi:10.26099/qb81-6c84

Washington, H. A., Baker, R. B., Olakanmi, O., Savitt, T. L., Jacobs, E. A., Hoover, E., & Wynia, M. K. (2009). Segregation, civil rights, and health disparities: The legacy of African American physicians and organized medicine, 1910-1968. *Journal of the National Medical Association*, *101*(6), 513–527. doi:10.1016/S0027-9684(15)30936-6 PMID:19585919

Watahomigie, L. J., & Yamamoto, A. Y. (2019). Linguistics in action: The Hualapai bilingual/bicultural education program. In *Collaborative research and social change* (pp. 77–98). Routledge. doi:10.4324/9780429040573-6

Weber, M. C. (2008). *Special education law and litigation treatise* (3rd ed.). LRP Publications.

Webster, P. (2020). Virtual health care in the era of COVID-19. *Lancet*, *395*(10231), 1180–1181. doi:10.1016/S0140-6736(20)30818-7 PMID:32278374

Weinbaum, C., Girven, R., & Oberholtzer, J. (2016). *The millennial generation: Implications for the intelligence and policy communities*. RAND, National Defense Research Institute. doi:10.7249/RR1306

Weissman, S. (2020). How will the coronavirus impact enrollment, access for rural students? *Diverse Issues in Higher Education*, *37*(6), 8–9.

Whalen, J. (2020). Should teachers be trained in emergency remote teaching? Lessons learned from the COVID-19 pandemic. *Journal of Technology and Teacher Education*, *28*(2), 189–199.

Wilcox, K. C., & Lawson, H. A. (2017). Teachers' agency, efficacy, engagement, and emotional resilience during policy innovation implementation. *Journal of Educational Change*, *19*(2), 181–204. doi:10.100710833-017-9313-0

Wilhelm, A. G., Munter, C., & Jackson, K. (2017). Examining relations between teachers' explanations of sources of students' difficulty in mathematics and students' opportunities to learn. *The Elementary School Journal*, *117*(3), 345–370. doi:10.1086/690113

Williams, T. K., McIntosh, R. W., & Russell, W. B. (2021). Equity in Distance Education During COVID-19. *Research in Social Sciences and Technology*, *6*(1), 1–24. doi:10.46303/ressat.2021.1

Wilson, A. T. (2018). *Native American Male Perspectives on Successful Higher Educational Attainment: Primary Motives and Practices in Overcoming Hindering Barriers in Degree Attainment*. UCLA. Retrieved from https://escholarship.org/uc/item/3qv9d1sw

Wilson, M. E. (2004). Teaching, learning, and Millennial students. *New Directions for Student Services*, *2004*(106), 59–71. doi:10.1002s.125

Winzer, M. A. (1993). *History of special education from isolation to integration*. Gallaudwt University Press.

Witkin, H. A., & Berry, J. W. (1975). Psychological differentiation in cross-cultural perspective. *ETS Research Bulletin Series*, *1975*(1), i-100. doi:10.1002/j.2333-8504.1975.tb01043.x

Woolf, S. H., & Aron, L. (2013). U.S. health in international perspective, shorter lives, poorer health. In *U.S. health in international perspective shorter lives, poorer health* (Vol. 4). National Academies Press.

World Health Organization. (2010). *Telemedicine: Opportunities and developments in Member States: report on the second global survey on eHealth*. World Health Organization.

World Health Organization. (2011). *Mental health atlas 2011*. World Health Organization.

World Health Organization. (2020). *Mental Health and Psychosocial Considerations During the COVID-19 Outbreak*. World Health Organization.

World Health Organization. (2021). *WHO Coronavirus disease (COVID-19) dashboard*. Retrieved from https://covid19.who.int/

Wouters, O. J., Shadlen, K. C., Salcher-Konrad, M., Pollard, A. J., Larson, H. J., Teerawattananon, Y., & Jit, M. (2021). Challenges in Ensuring global access To COVID-19 vaccines: Production, affordability, allocation, and deployment. *Lancet*, *397*(10278), 1023–1034. doi:10.1016/S0140-6736(21)00306-8 PMID:33587887

Wright, P. (2015). *In the circuit court of Taylor County, West Virginia*. Retrieved on 4/2/2021 from: https://www.wrightslaw.com/law/caselaw/case_Doe_Withers_Complaint.html

Wykes, T. (2019). Racing towards a digital paradise or a digital hell? *Journal of Mental Health (Abingdon, England)*, *28*(1), 1–3. doi:10.1080/09638237.2019.1581360 PMID:30888284

Yell, M. (2019). *The law and special education*. Pearson.

Yen, P. Y., McAlearney, A. S., Sieck, C. J., Hefner, J. L., & Huerta, T. R. (2017). Health Information Technology (HIT) Adaptation: Refocusing on the Journey to Successful HIT Implementation. *JMIR Medical Informatics*, *5*(3), e28. doi:10.2196/medinform.7476 PMID:28882812

Zohar, A., & Agmon, V. A. (2018). Raising test scores vs. teaching higher order thinking (HOT): Senior science teachers' views on how several concurrent policies affect classroom practices. *Research in Science & Technological Education*, *36*(2), 243–260. doi:10.1080/02635143.2017.1395332

Zubatsky, M., Edwards, T. M., Wakabayashi, H., & Ivbijaro, G. (2018). Integrated behavioral health in primary care across the world: Three countries, three perspectives. *Family Practice*, *35*(6), 645–648. doi:10.1093/fampra/cmy034 PMID:29741628

About the Contributors

Emily Kate Reeves is an Associate Professor in the Curriculum and Learning Department of the West College of Education and the Chair of Human Subject and Review Committee (IRB). She has a BSIS in Interdisciplinary studies, an MA in Educational Leadership, and a PhD in Curriculum and Instruction. Her research interests include literacy, culturally responsive development, and both in-service and preservice teacher education.

Christina Janise McIntyre is an Associate Professor in the Curriculum and Learning Department of the West College of Education and is the program coordinator for Curriculum and Instruction. She has a BA in English, an MEd in Counseling, and a PhD in Instructional Leadership and Academic Curriculum from the University of Oklahoma. Her research interests include preservice teacher education and professional development.

***

Julie Bacak is a Ph.D. student in the Curriculum and Instruction Program at the University of North Carolina at Charlotte. She is a former classroom teacher, instructional coach, and mentor teacher with experience in both elementary and middle grades. Her research interests include practice-based teacher education (PBTE), mathematical discourse, and equitable mathematics teaching practices. Her recent coursework allowed her to explore these research interests through a co-teaching experience in an elementary math methods course. As a graduate assistant in the Department of Educational Leadership at UNC Charlotte, Julie is currently developing curricular resources for elementary student digital safety as part of an NSF-funded grant.

Antoniece J. Bailey is a 4th-grade teacher born and raised in Savannah, Georgia. She graduated with her B. S in Early childhood education from Paine College in 2012. In 2013, she received her Master's in Instructional Technology from Cambridge College. Along with being Gifted and Reading endorsed she has spent the last 9 years teaching in Title 1 schools in urban Georgia. It is her belief that every child deserves a good education no matter who they are and where they come from. In 2020, she graduated from Helms College with a degree in Culinary Arts.

Alexis S. K. Bain recently graduated with a Master's in Health Service Administration from Midwestern State University. She has and continues to haphazardly embark on her lifelong dream of working and changing the scope, practices, and regime of healthcare. She graduated from the College of The Bahamas with a BSc in Biology and a minor in Chemistry. Alexis loves the business part of medicine. She is also a published author with 3 articles and many more to come. She is an inspired world changer that will create new initiatives and avenues for medicine through research and practices.

Angela Cartwright is the Principal Lecturer for Academic Development at the University of Sunderland in London. In this role, she created the institution Academic Development Strategic Plan and leads a team of five Senior Lecturers tasked with implementing the Academic Development Strategic Plan. The Academic Development Team works closely with faculty and staff to enhance student experiences and success. Before joining academic management, Dr Cartwright was a tenured Associate Professor of Teaching and Learning in the United States, and prior to entering higher education, she taught secondary social studies, specializing in at-risk student populations, as well as Literacy and Social Studies with at-risk Elementary and Middle Grades students with AmeriCorps. Dr Cartwright has published and presented extensively in DEI and education. She earned her PhD in Multicultural and Equity Studies in Education and her MA in Drama, Literacy, Language Arts, and Reading, both from Ohio State University, and her BA in Integrated Social Studies Education.

Dennis Cavitt received his Ed.D. from Texas Tech University in Special Education with an emphasis in Autism and High Incidence Disabilities. Cavitt also has a Master's in Education Leadership from Tarleton State University and A Master's in Clinical Psychology from Abilene Christian University. Cavitt is an Assistant Professor in Special Education in the West College of Education at Midwestern State University, Texas (MSU) where he teaches Undergraduate and Graduate courses in Special Education. Cavitt has worked in the field of special education for over 35 years. He began his career as a Licensed Specialist in School Psychology (LSSP).

He was responsible for Educational and Psychological Assessments, Functional Behavioral Assessment and Programing, Case Management, IEP Development, Counseling, And Professional Development. Prior to entering teaching in higher education, he was a director of special services (Dyslexia, Section 504, & Special Education). Before coming to MSU, he taught in the Teacher Education Department at Abilene Christian University. Cavitt is very active in CEC at the local, state, and national/international level. He has been CEC president for Texas CEC, He was Chairperson of the International CEC Representative Assembly. He was the President of CEC International in 2021.

Madelyn Colonnese is an assistant professor in the Reading and Elementary Education Department at the University of North Carolina Charlotte. In this role, Madelyn teaches the elementary mathematics methods courses to pre-service teachers. She earned her PhD in Curriculum and Instruction and MA in teaching from the University of Connecticut. Madelyn currently researches practices to support teachers with the implementation of elementary mathematical writing. She is also interested in ways to support pre-service teachers with eliciting student thinking.

Douglas Graham Fenderson recently completed his M.Ed. in Curriculum and Instruction with a minor in History. He graduated in 2021 from Midwestern State University in Wichita Falls, Texas. His research focused on the transition from in-person to online teaching during the pandemic at the secondary level.

Tara Fox is an assistant professor at Midwestern State University. She has been a clinical mental health provider for almost fifteen years. Dr. Fox has specialized in rural populations and completed her dissertation on utilizing integrated behavioral healthcare in rural areas. She earned her B.A. in Psychology, and an M.A. in Counseling from Wayland Baptist University in Plainview, Texas. Dr. Fox earned her Ph.D. in Counselor Education and Supervision from Texas Tech University in Lubbock, Texas. She is a licensed professional counselor supervisor in the state of Texas.

Alistair Fyfe, a serial over-achiever, physician scientist and educator, was raised on a 5000-acre sheep ranch in Tasmania, commenced Medical School age 17 and founded his first business selling computers at age 19. He was invited to train in Vancouver BC after completing his internship in Hobart, Australia. He then moved to Toronto to study cardiology. Dr. Fyfe then completed a Molecular Biology PhD in less than 3 years. As an Assistant Professor he was the first UCLA medical graduate to be given his own Molecular Biology laboratory, creating Adenoviruses to study transplant and atherosclerosis, and studying at the interface between atherosclerosis and immunology which had previously been considered an anachronism. During this time, he delivered hundreds of lectures on prevention, cholesterol and gene therapy both nationally and internationally. Moving to Dallas in 1999 he migrated to private practice cardiology, continued his research and became an early adopter and proponent of calcium scoring as an early marker for coronary events. In 2013 Dr. Fyfe went to Oxford UK to complete an executive MBA with the specific ambition to find entrepreneurial solutions to healthcare in evolving markets. On return he created a low-cost concierge service; inclusive medical care for $1.00 per day to reduce the barrier to healthcare created by the insurance industry with its insurmountable deductibles. An invitation to joint Cigna insurance in 2016 led to evaluation of this payor side of the healthcare equation with projects evaluating value, re-imbursement, coverage policy and fraud. An early concern in the COVID-19 pandemic required an urgent evaluation of the effects of blood pressure medications on infection, morbidity and mortality which translated into writing what became the insurance industry guidelines on return to work after COVID-19.

Timothy Hinchman is an Assistant Professor in the Gordon T. and Ellen West College of Education at MSU-Texas where he has been a faculty member since 2016. Timothy teaches Science and Social Studies Methodology to elementary school pre-service teachers at MSU-Texas. He completed his Ed.D. at the University of West Florida and his undergraduate studies at Florida State University. His research interests lie in the manipulations of constraints to maximize productivity and creativity. Currently, Timothy is applying his research on designed constraints to maximize NCAA Division II athletes' physical productivity. He has collaborated actively with researchers in several other disciplines of education and art, particularly in the STEM/STEAM disciplines. With a name like Hinchman, it leaves little doubt that he is fascinated with Batman and Gotham City's Rogues' Gallery. Being a comic-book enthusiast, he hopes one day he will realize his dream to present his research at a major Comic-Con.

Kaitlyn O. Holshouser is a doctoral candidate in the Elementary Education strand of the Curriculum and Instruction Program at the University of North Carolina at Charlotte. She is a former fourth and fifth grade teacher who is passionate about supporting both in-service and pre-service teachers. Her research agenda focuses on global education, specifically Education for Sustainable Development (ESD), cross-curricular integration, and teacher development. As a Graduate Assistant in the Reading and Elementary Education Department at UNC Charlotte, Kaitlyn has worked to support and instruct teacher candidates in ELED 4220, Integrating Curriculum for Elementary School Learners.

Pamela June is Associate Professor of English at Ohio University's Eastern Campus. She is the author of two books, Solidarity with the Other Beings on the Planet: Alice Walker, Ecofeminism, and Animals in Literature (Northwestern UP, 2020), and The Fragmented Female Body and Identity: The Postmodern, Feminist, and Multiethnic Writings of Toni Morrison, Theresa Hak Kyung Cha, Phyllis Alesia Perry, Gayl Jones, Emma Pérez, Paula Gunn Allen, and Kathy Acker (Peter Lang, 2010), as well as many scholarly articles on American women's literature. She earned her PhD and MA in English from Indiana University of Pennsylvania and her BA from University of Pittsburgh. June teaches women's literature, American literature, African American Literature, and writing.

Bonnie King is an assistant professor of counseling in the department of Human Development, Family Studies and Counseling at Texas Woman's University. She is a licensed professional counselor and certified school counselor. She earned a Ph.D in Counselor Education at the University of New Orleans, and a Master's of Counseling at Texas State University. Her research interests include counseling in green space, best practices for serving transgender and gender non-conforming clients, and international counseling.

Danielle M. Kovach began her teaching career in 1997 and has taught both general and special education across a wide range of settings. She is currently a third-grade special education teacher and adjunct professor at Centenary University in New Jersey. Danielle is the 2022 President of The Council for Exceptional Children (CEC), the largest international professional organization dedicated to improving the educational success of children and youth with disabilities. She is also the 2011 New Jersey State Teacher of the Year, 2012 NEA Foundation Award for Teaching Excellence recipient, and the 2014 CEC Teacher of the Year.

Wendy Lewis is a Ph.D. student in the Urban Elementary Education Curriculum and Instruction program at the University of North Carolina at Charlotte. She is a former classroom teacher, instructional coach and differentiation specialist in grades 3-6 for fifteen years in North Carolina public schools who has a passion and experience with curriculum differentiation. She enjoys professional development, especially in the area of elementary and middle grades mathematics. She has an interest in researching assessment and teaching pedagogies which focus on creativity, problem posing, and cognitively demanding instruction.

Geoff Paul, as the Assistant Director (Academic) for the University of Sunderland in London, is responsible for the development of academic staff. This includes areas such as teaching and learning standards, innovation in practice and quality management. Through the development of research and enterprise initiatives, he ensures students receive the highest quality academic experience available. Geoff also has an interest in the development of collaborative work between academic and administrative staff in increasing the student experience. He also oversees the development of innovative market-facing academic programmes, serving both industry and the commercial sector. He has responsibility for the implementation of campus-wide standards including the Higher Education Sector, Office for Students and University sector governmental policies. Geoff became a Senior Fellow of Advance HE (formally the Higher Education Academy) in August 2020.

Drew Polly is a professor in the Elementary Education program at the University of North Carolina at Charlotte. His work focuses on examining how to best prepare teacher candidates' and practicing teachers' work enacting learner-centered pedagogies in their mathematics classrooms and how to support teachers effective use of technology. He has published over 60 peer-reviewed chapters and journal articles.

Shatara Sims is a 5th-grade educator who believes that every child deserves excellence in teaching. Her belief as a teacher is that learning must be interesting, innovative, and interactive for all students. She also believes acting as a facilitator is imperative because she creates stimulating discussions and dialogue that foster her students' curiosity and excitement about learning, giving the students a well-rounded educational experience. Sims continues to act on her belief that every child deserves excellence in education. She sacrifices her afternoons and Saturdays tutoring students to ensure that all students reach mastery level in every Mathematical Standard. The time she spends not only improves scores but motivates and builds character and self-esteem to enhance their educational experience. Sims is completing her eighth year as an educator after leaving a seventeen-year career in the medical field. She has led her students each year to significant gains on the state assessment. In 2019, Sims was named Richmond County Teacher of Year for 2020. In 2017, Sims was nominated for the Presidential Award for Excellence in Mathematics and Science Teaching (PAEMST). Sims earned a Bachelor of Arts in Early Childhood Education from Paine College and a Master of Education in Curriculum, Instruction and Assessment from Strayer University, and a certification in the Georgia Governor Teacher Leadership Program. Sims has added several endorsements to her certification. In 2016 she completed her Mathematics Endorsement from Augusta University and in 2017 she completed her Reading Endorsement from the Central Savannah River Area Regional Educational Service Agency (CSRA RESA). Sims continues her desire to teach with excellent by adding the Gifted Endorsement which she will complete in March 2020 (CSRA RESA).

Tim Swagerty has been a Social Studies Instructor in the West College of Education at Midwestern State University since 2018. He was a high school classroom social studies teacher from 2006-2018. He teaches Concepts of Social Studies and Methods of Social Studies for secondary teacher candidates as well as serves as University Field Supervisor for elementary and secondary clinical teaching students. He received his B.F.A. in Metalsmithing and M.A in Public History/Museum Studies from MSU-Texas and his M.Ed. in Secondary Education from Wayland Baptist University. He is currently pursuing his Ed.D. in Curriculum and Instruction from University of Louisiana Monroe. His research interest and focus is on United States Indian Policy and the disequitability of European education models on Native and Indigenous students in the United States. He is a member of the Mvskoke (Creek), Chahta (Choctaw) and Tsalagi (Cherokee) tribes. His personal experiences of a lack of culturally relevant curriculum in modern education has framed his research interests and focus.

Allison Wilson is a second-grade teacher and graduate student born and raised in Wichita Fall, Texas. Allison graduated from MSU-Texas with a BSIS of Special Education in 2017 and is currently pursuing her master's degree in Curriculum and Instruction. For the past three years, she has taught at the elementary school that she attended kindergarten through sixth grade, a Title 1 school. While in college, Allison worked at a local private school with PreK students for three years in afterschool care and yearly summer camps. She has a passion for her horses, donkeys, family, and friends. Allison hopes to one day hold a leadership position on a school campus or position within a university.

Index

www.ingramcontent.com/pod-product-compliance
Lightning Source LLC
LaVergne TN
LVHW081339300926
849161LV00008B/2017